Power Aware Computing

SERIES IN COMPUTER SCIENCE

Series Editor: Rami G. Melhem
University of Pittsburgh
Pittsburgh, Pennsylvania

FUNDAMENTALS OF X PROGRAMMING
Graphical User Interfaces and Beyond
Theo Pavlidis

INTRODUCTION TO PARALLEL PROCESSING
Algorithms and Architectures
Behrooz Parhami

OBJECT-ORIENTED DISCRETE-EVENT SIMULATION WITH JAVA
A Practical Introduction
José M. Garrido

PERFORMANCE MODELING OF OPERATING SYSTEMS USING OBJECT-ORIENTED SIMULATION
A Practical Introduction
José M. Garrido

POWER AWARE COMPUTING
Edited by Robert Graybill and Rami Melhem

Power Aware Computing

Edited by

Robert Graybill
The Defense Advanced Research Project Agency
Arlington, Virginia

and

Rami Melhem
University of Pittsburgh
Pittsburgh, Pennsylvania

Kluwer Academic / Plenum Publishers
New York • Boston • Dordrecht • London • Moscow

Library of Congress Cataloging-in-Publication Data

Power aware computing/edited by Robert Graybill and Rami Melhem.
p. cm. — (Series in computer systems)
Includes bibliographical references and index.
ISBN 0-306-46786-0
1. Electronic digital computers—Power supply. 2. Electric power—Conservation. I. Graybill, Robert. II. Melhem, Rami. III. Series in computer systems.

TK7895.P68 P69 2002
621.39'16—dc21

2002022133

ISBN 0-306-46786-0

233 Spring Street, New York, N.Y. 10013

http://www.wkap.com

10 9 8 7 6 5 4 3 2 1

A C.I.P. record for this book is available from the Library of Congress

Printed in the United States of America

Preface

With the advent of portable and autonomous computing systems, power consumption has emerged as a focal point in many research projects, commercial systems and DoD platforms. One current research initiative, which drew much attention to this area, is the Power Aware Computing and Communications (PAC/C) program sponsored by DARPA. Many of the chapters in this book include results from work that have been supported by the PAC/C program.

The performance of computer systems has been tremendously improving while the size and weight of such systems has been constantly shrinking. The capacities of batteries relative to their sizes and weights has been also improving but at a rate which is much slower than the rate of improvement in computer performance and the rate of shrinking in computer sizes. The relation between the power consumption of a computer system and it performance and size is a complex one which is very much dependent on the specific system and the technology used to build that system. We do not need a complex argument, however, to be convinced that energy and power, which is the rate of energy consumption, are becoming critical components in computer systems in general, and portable and autonomous systems, in particular.

Most of the early research on power consumption in computer systems addressed the issue of minimizing power in a given platform, which usually translates into minimizing energy consumption, and thus, longer battery life. More recently, it was recognized that power is a system resource that should be carefully managed and that minimizing power consumption should be cast in the larger context of power management. For instance, in some systems, the available power may follow a certain profile which is a function of time, and the system should regard this profile as a given constraint. In other systems, the energy available for a certain mission may be constrained, and the system should maximize its performance given the available energy. Yet, in other systems, there may be tradeoffs between power consumption and some performance measures such as timeliness, reliability, throughput or the accuracy of the computation.

Although power is a shared resource, experience in managing other shared resources, such as memory, hardware units and shared code, does not apply directly to managing power. The reason is that power is an expendable resource, much as CPU cycles. In that sense power management can borrow from the different techniques that are used for processor scheduling, which is one form of CPU cycle management. Power consumption can also be managed by bypassing or turning off hardware units that are not used at a particular instant. This is similar to managing the cycle length by controlling the propagation delays through hardware reconfiguration and through bypassing hardware units whose functions are not needed at a particular instant.

From the point of view of resource management, however, managing power is different from managing any other system resource, including expendable resources such as time or CPU cycles. The reason is that power is a component of all other resources. There is hardly a resource that does not consume power either directly or indirectly. Moreover, there is a complex interaction between power and time. For example, we can reduce power consumption by slowing down the operating frequency and supply voltage of a processor. We can also increase the accuracy and resiliency of a computation by using more complex codings or algorithms, thus consuming more time and energy.

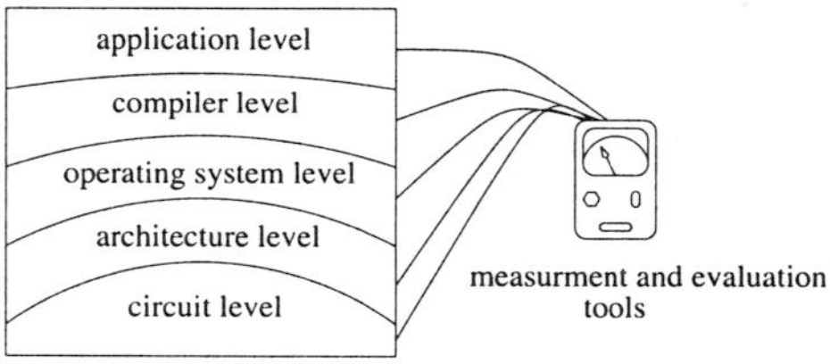

The levels of power management.

The above discussion demonstrates that because power is so intrinsic in the operation of computer systems, a wide variety of techniques can and should be used for power management. Some of these techniques can be at very high and abstract levels, while some other techniques can be at very low circuit levels. In this book, we consider five basic levels of a computer system. Namely, the circuit level, the architecture level, the operating system level, the compiler level and the application level. Power can be managed at each of these levels, and indeed, at many sublevels of these broad levels. The importance of power awareness has developed at different times for each of these levels. For example, low power electronic design has been an established discipline for quite some time, while research on power aware algorithm development is just starting. This book is intended to introduce the reader to some of the most recent techniques for managing power at each of the five levels mentioned above.

The diversity of the different power management techniques is demonstrated in the first fourteen chapters of this book. The fourteen chapters are grouped into five parts, corresponding to the five levels of a computer system. Chapters

1 and 2 in Part I address low power designs of flip-flops and memory cells at the circuit level. The three chapters in Part II address power management at the architecture level. Chapters 3 and 4 propose to reduce CPU energy consumption by carefully designing the instruction issue queues and the control speculation algorithms, respectively, while Chapter 5 discusses Instruction Set Architectures that are specifically designed to facilitate energy management. The three chapters in Part III discuss power aware scheduling algorithms and interfaces and their implementations at the operating system level. Chapters 9, 10 and 11 in Part IV explore compiler optimization techniques that are targeted to power aware systems, Finally, Chapters 12, 13 and 14 in Part V show the need for power management at the application level and introduce specific schemes for such management in three different application domains.

Measurement and evaluation of the different techniques for power management is a very difficult issue. One main reason is that research in power aware systems is a relatively new area of research and no standard measures have been established for evaluating and comparing different power management schemes. The fact that these schemes are distributed across all levels of a computer system makes the standardization of such measures rather difficult since a measure that is suitable for circuit level power management may not be suitable for application level power management. The complexity of the evaluation issue of a power aware system is further complicated by the interaction between the different levels of the system. For instance, the benefits of power management at both the operating systems and the compiler levels may not be linearly combined in a system that manages power at both levels. This is primarily because when two levels are managing the same resource, management is at best complimentary, and can be at worst contradictory. A great deal of coordination is needed to achieve the former and avoid the latter. The last three chapters of this book deal with this issue of measurement and evaluation at the circuit, the architecture and the application levels.

Given the explosive growth of research in power management, we could not cover in this book all the power management techniques that have been recently proposed. For instance, this book did not cover many of the possible techniques for managing power consumption in the storage hierarchy, including caches, main memory and secondary storage. A variety of techniques have been proposed for this purpose at all levels (circuit, architecture, systems, compilers and applications), and only Chapters 2 and 11 cover some of these techniques at the circuit and compiler levels. Power management in multiprocessors is another example of an area that is not covered in this book. Finally, the entire area of power aware communication has been left out of this book. It is a wide and important area which deserves a book of its own.

ROBERT GRAYBILL, RAMI MELHEM

Contents

List of Figures

List of Tables

I

CIRCUIT LEVEL POWER MANAGEMENT

Chapter 1

COMPARATIVE ANALYSIS OF FLIP-FLOPS AND APPLICATION OF DATA-GATING IN DYNAMIC FLIP-FLOPS FOR HIGH SPEED, LOW ACTIVE AND LOW LEAKAGE POWER DISSIPATION

Vamsi Srikantam, Mario Martinez

Electronics Research Lab

Agilent Technologies Laboratories

Palo Alto, CA, 94304

{vamsi_srikantam, mario_martinez}@agilent.com

Abstract Microprocessors, ASICs and DSPs form the core components of digital systems. Power aware computing on such systems necessitates both low power hardware design and software power optimization. Flip-flops are an integral component of digital circuits responsible for data storage. Hence, designing a low power flip-flop is of prime importance. In portable systems, considerable time is spent in idle or sleep mode. In this mode, leakage power is becoming a concern, hence reducing both active and leakage power is a necessity for overall power optimization. Hence, flip-flops which have the following features are desired: 1. Low active power during normal operation, 2. Low inherent leakage power during sleep mode and 3. Not only preventing spurious data to be passed through (data-gating) but also being capable of generating the desired output state for lower leakage power dissipation. Unlike static flops, with dynamic flops clock-gating is not directly possible, as such flops require the clock to be running continuously for proper functioning. This chapter describes a detailed comparison analysis of delay and power including leakage power characteristics of existing flip-flops in literature. Further, the introduction of data-gating in dynamic flops to achieve high speed, low active power and at the same time, setting the output state to reduce leakage power in subsequent blocks is discussed. This shows good potential for active and leakage power optimization in digital CMOS circuits.

Keywords: Flip-flops, active power, leakage power, CMOS circuits, power optimization

1. Introduction

With the advent of portable electronics, power dissipation has become a major concern in designing VLSI circuits. Apart from the active power dissipation, leakage power is a big concern in portable electronics whereby the system is put in standby/sleep mode for a considerable amount of time. In real operating conditions, some blocks of the system are not always active. In these idle periods, these blocks perform unnecessary computation thereby causing spurious power dissipation. This problem has been partially solved by methodologies such as clock gating.

Clock gating is a useful method to reduce active power during idle periods [1, 2]. The inputs to such blocks are not allowed to change by the clock-gated flip-flops and hence unnecessary computation is avoided. Design tools for ASICs are capable of generating the clock gating signals and the sleep signal, which determine the periods of inactivity in a circuit. Essentially, a sleep or enable signal deactivates the clock thereby not allowing the flip-flop to clock new data. A simple form of clock gating is shown in Figure 1.1.

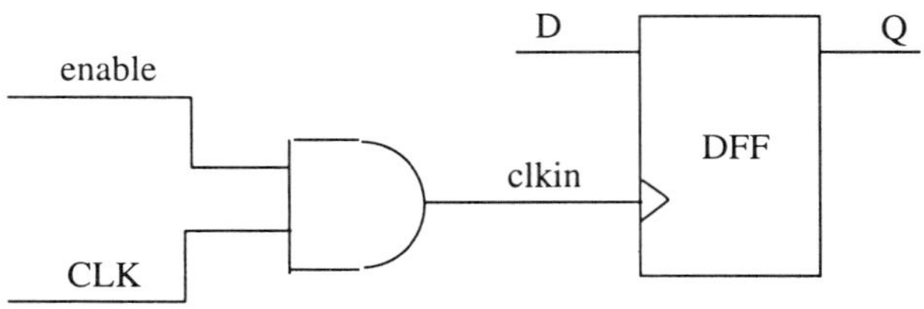

Figure 1.1. Simple Clock-Gating Structure

In Figure 1.1, when enable = 0, clkin = 0 and the flip-flop does not clock any data thereby saving power. However, this form of clock gating is applicable only to static flip-flops since the static master and slave latches retain the data when the clock is disabled. Furthermore, there are issues with testability. In Figure 1.1, the enable signal is not testable as the other input of the AND gate is the CLK itself. Hence, fault coverage is affected. In dynamic flops, the clock has to be running continuously in order for data to be latched. If clock is disabled, the internal nodes could float since there is no feedback thereby causing the output to float.

An alternate design style is to gate the data rather than the clock as shown in Figure 1.2. When enable = 0, the flop retains the output. When enable = 1, new data is allowed to be clocked in. This is called "Data-Gating". The effect is the same as clock gating and also has no testability problems. This form of data-gating is applicable to dynamic flops directly. Dynamic flops are used extensively in high performance designs. This chapter describes specific dynamic flip-flops with data gating structures along with the comparative analysis of active and leakage power with other flip-flops proposed in the literature.

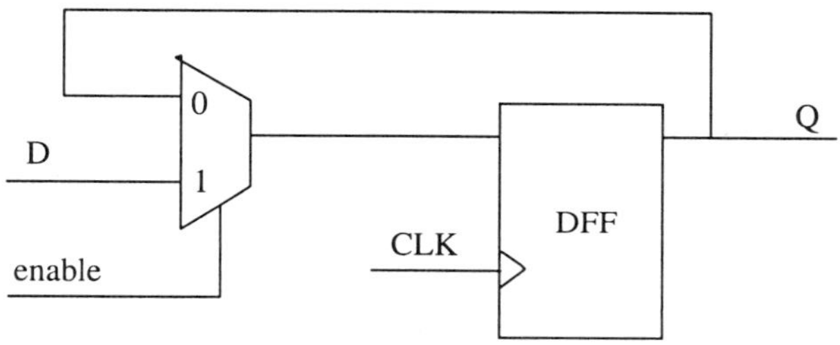

Figure 1.2. Data-Gating Circuit

Leakage power is becoming a concern in deep submicron technologies. Nowhere is this more evident than in portable systems whereby the inactive blocks are put in a standby mode. In such a mode of operation, a suitable input vector applied by the flip-flop can substantially reduce leakage power [3, 4] as shown in Figure 1.3 since leakage power is a function of the input vectors. The works in [3, 4] also present algorithms to determine the lowest leakage vector for a combinational block. Clock-gating or data-gating is useful for subblocks, which can be put to sleep in order to reduce active power. However, they can also be used to reduce the leakage power of the sub-block by generating the required low leakage input vector as shown in Figure 1.4.

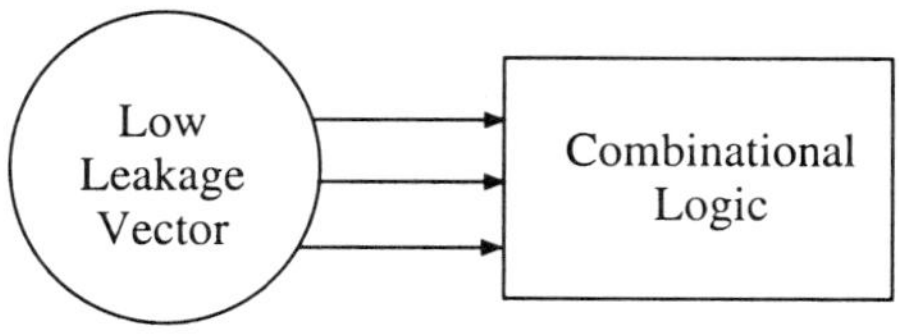

Figure 1.3. Low Leakage Input Vector

The concepts of low leakage vector and clock-gating have been applied on static flops for power reduction. However, clock-gating is not directly applicable to dynamic flops. The work in [12] proposes the design of high speed, low

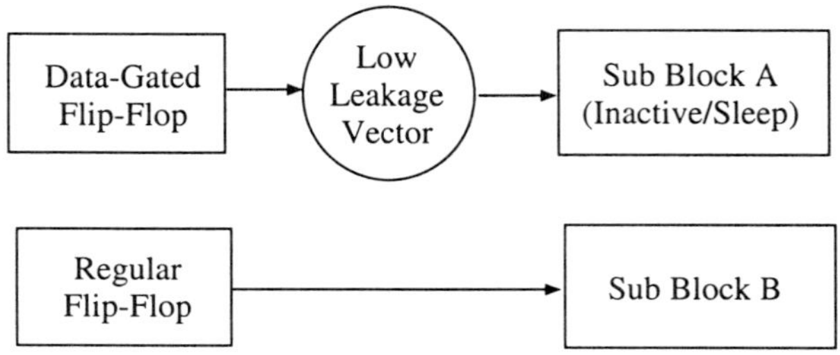

Figure 1.4. Method to Reduce Leakage Power in Sleep Mode in Inactive Sub-Blocks of the Design System

active and low leakage power dynamic D flip-flops with data-gating. Not only do these flip-flops prevent spurious data from being passed, but also they are also capable of generating the desired output state for lower leakage power dissipation. Experimental results with 0.25u CMOS process designs show from 25% to 48% reduction in active power as compared to other flip-flops. An 8x8 pipelined multiplier has been designed using the flip-flops described in [12]. The multiplier dissipates 0.25mW at 1.8V and operating frequency of 100MHz. Also, when operated using the methodology shown in Figure 1.4, the leakage power was reduced by a factor of 3.

2. Various Static and Dynamic Flip-Flops

The work in [5] presents a comparative analysis of various flip-flops. However, there is no comparison of leakage power dissipation. As process technologies shrink, leakage power is becoming a dominant factor. The analysis in this chapter includes the comparison of leakage power apart from active power. Further, others flops not included in the work of [5] are also presented. Although, the focus is on dynamic flip-flops, comparisons to static flip-flops also are presented. The different static and dynamic flip-flop designs that have been proposed previously are described below.

- StrongArm 110 Flip-Flop [5], Sense Amp FF [5], DSTC [5], Hybrid Latch Flip-Flop (HLFF)[5], Semi-dynamic Flip-Flop (SDFF) [5], TSPC (True Single Phase Clocking) Flip-Flop [6], PTTFF (Pulse-Triggered True Phase Flip-Flop) [8], 8TDFF [8], 9TDFF [8], TI's Low Power Flip Flop [9], and Lee and Do's High Speed D Flip-Flop [10] are the dynamic and semi-dynamic structures used extensively in high performance designs. However, these flip-flop designs fail to decouple any transition between the latching structure itself and the output, causing potentially unnecessary and undesired transitions from a power perspective during idle mode.

As described before, data-gating as against clock-gating is suitable for this purpose.

- Fully static Flip-Flop designs such as PowerPC 603 Flip-Flop [5], Modified C^2MOS Flip-Flop [5], Push-Pull Flip-Flop [7], Conventional Flip-Flop [6], [11], and Ko's Low Power Flip-Flop [7] are the basic inverter feedback based structures. These flops are able to alleviate the problem of charging and discharging the intermediate nodes at every clock cycle by means of external clock gating circuitry as shown in Figure 1.1.
- The above mentioned flip-flops have the disadvantage of not allowing control of the output (independently of the input) to be a fixed value. This is especially required in the standby/sleep mode where leakage power is dominant.
- The work in [12] proposes high speed dynamic flip-flop designs with data-gating, namely, LPSDFF (Low Power Sleep mode with output pull Down Flip-Flop) and LPSPFF (Low power Sleep mode and output Pull-up Flip-Flop). These flip-flops are modifications of the 8TDFF [8].

The next few sections describe a comparative analysis of the various static and dynamic flip-flops in terms of power and delay. Further, the results seen from a test case design using LPSPFF and LPSDFF is presented.

3. Dynamic Flip-Flop Design and Operation

The following is a description of LPSPFF and LPSDFF. The circuit structures are shown in Figure 1.5 and Figure 1.7. Owing to the true single-phase-clocking these flip-flops have the advantage of compact and simple clock distribution without any extra timing considerations such as additional clock skew.

Under non-sleep conditions, this design operates as a normal single (rising) edge triggered D-flip-flop would with the sleep signal being high. Following conventional setup and hold time rules, data is allowed into the first stage via Mp1 and Mn1. Assuming the clock is low, this data propagates to the next stage via Mpclk1 (either pulling the second stage high if the data is high or low if the data is low). Mnclk1 controls the data from further propagating to the third stage. Once the clock transitions high, data is allowed to pass to the third stage via Mp3 or Mn3 (depending on the data value) where the output is an inverted form of the input. In order to produce a representation of the data, a further inversion is necessary (which also serves as a benefit in driving further loads).

In sleep mode, the sleep signal is low and two main functions are performed. First, the input data signal is prevented from affecting the previous mode of the flip-flop. This is achieved by means of the complementary pass-gate located at

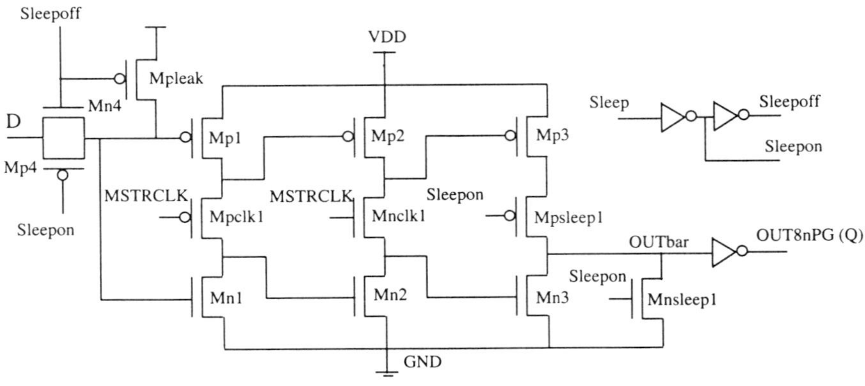

Figure 1.5. Design Structure of LPSPFF

the input, Mp4 and Mn4. This has the effect of not allowing any further transitions within the flip-flop itself, therefore reducing any unnecessary power from these transitions. This is how "data-gating" is achieved. Secondly, the output, node $OUT8nPG(Q)$ is pulled to V_{dd} in the case of LPSPFF and node $OUT8pPG(Q)$ is pulled to V_{ss} in the case of LPSDFF. The circuit operation of these designs is shown in Figures 1.6 and 1.8. Furthermore, if the previous latched data were an opposite value of what the output is driven to, there is no contention between the two signals. This is due to Mpsleep1 and Mnsleep2, which effectively disable the pull-up portion of the LPSPFF and similarly the pull-down portion of the LPSDFF respectively. If the internal data value at the output is the same as the desired sleep-mode output, the data as well as the sleep mode transistor (Mnsleep1 or Mpsleep2) are working together to pull node OUTbar to V_{ss} in the case of LPSPFF or V_{dd} in the case of LPSDFF, so there is an additional benefit to this design. To further assist in this effort, keeper transistors Mpleak and Mnleak are used to prohibit any subsequent leakage. Therefore, these keeper transistors not only assist in setting the output to its predetermined state, but also help in leakage prevention. The ultimate effect this method has is to set the final output $OUT8nPG(Q)$ to V_{dd} (PRESET) for LPSPFF and output $OUT8pPG(Q)$ to V_{ss} (CLEAR) for LPSDFF.

4. Comparative Analysis and Experimental Results

All the flip-flops are implemented using the Agilent/HP 0.25*um* CMOS technology. Optimal transistor sizes to drive the load that allowed correct flip-

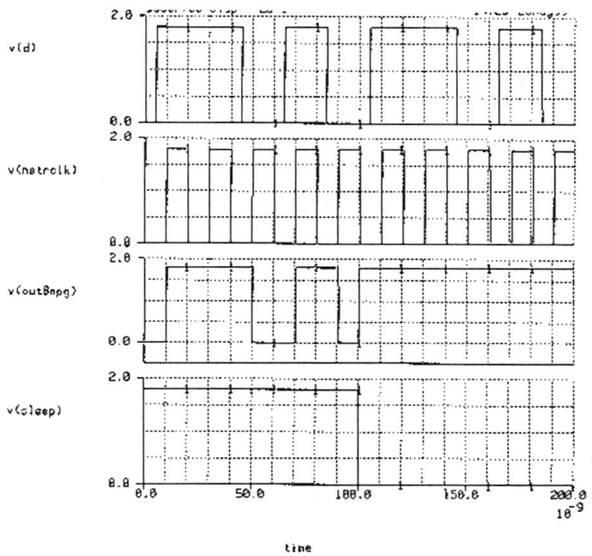

Figure 1.6. LPSPFF Output Waveform: Active Mode 0-100ns, Sleep Mode 100-200ns

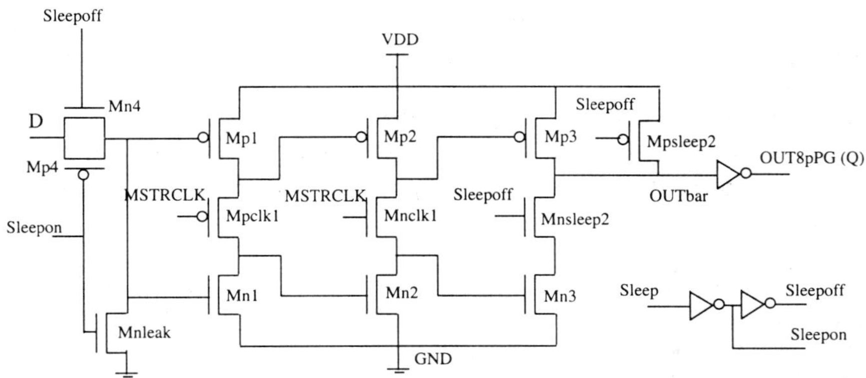

Figure 1.7. Design Structure of LPSDFF

flop operation were chosen for the implementations and results were obtained from HSPICE simulations. The data and clock inputs are loaded and the input

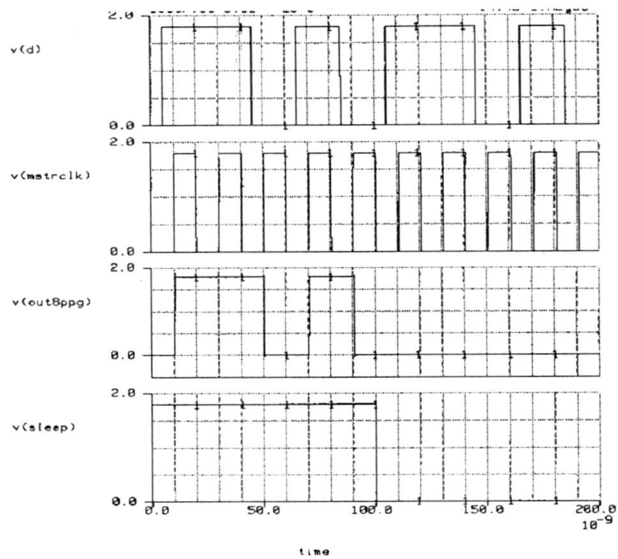

Figure 1.8. LPSDFF Output Waveform: Active Mode 0-100ns, Sleep Mode 100-200ns

vector of data (D) consisted of a combination of continuous ones and zeros and consecutive toggling to simulate realistic conditions. Two buffering inverter load for each input (data and clock) are connected for this purpose. Table 1.1 lists the transistor count, total power (Ptotal), average Clk-Q delay and the power-delay product under no loading at the output. All simulations were performed at a supply voltage of 1.8V and operating frequency of 50MHz. Also, the superscript 1 in the tables denote sleep-mode opeartion for half the simulation cycle.

Table 1.2 lists the transistor count, total power (Ptotal), average Clk-Q delay and the power-delay product with a $0.2pf$ load at the output. Table 1.3 lists the leakage power (Pleak) under no load conditions. As all flip-flops other than LPSDFF and LPSPFF do not have a sleep mode, the leakage power was obtained by holding the data constant (simulations were done with data both high and low). From the tables, it can be seen that LPSDFF and LPSPFF consume the minimum active power as compared to all flip-flops except 8TDFF whereby the active power of 8TDFF is slightly lower. A similar trend is observed in the power delay product of the proposed flip-flops. During sleep mode, the tables show that both LPSDFF and LPSPFF consume the lowest active power as

Table 1.1. Total power dissipation and delay comparison of proposed flip-flop with prior static and dynamic flip-flops, unloaded.

Name	*# of transistors*	*Ptotal (uW)*	*Ave. Delay (ps)*	*PDPtot* ($x10^{-18}$ *J*)
HighSpeedDFF	9	216.7	81.25	17605.79
8TDFF	10	2.783	146.05	406.46
9TDFF	11	2.687	131.70	353.88
DSTC	14	16.67	72.59	1209.99
PTTFF	18	12.53	165.95	2079.35
TILowPowerDFF	18	223.6	93.07	20809.33
ConventionalFF	20	5.221	183.90	960.14
LowPowerDFF	20	6.231	234.75	1462.73
PushPullDFF	20	8.003	165.75	1326.50
HLFF	20	14.03	108.14	1517.13
SenseAmpFF	21	10.13	151.45	1534.19
StrongArm110FF	22	15.08	162.40	2448.99
PPC603	22	6.723	213.10	1432.67
SDFF	23	31.99	138.31	4424.38
Modified CCMOS	24	9.397	184.15	1730.46
TSPCFF	26	7.638	240.65	1838.08
LPSDFF	19	3.057	182.30	557.29
LPSPFF	19	3.297	205.9	678.85
LPSDFF[1]	19	1.687	182.30	307.54
LPSPFF[1]	19	2.205	205.9	454.01

compared to any other flip-flop. Further, in sleep mode, LPSPFF and LPSDFF demonstrate the best power and delay tradeoff as compared to other designs. Also, they have low leakage power, but other flip-flops dissipate lower leakage power.

In low-power high-performance applications, high speed is as important as achieving low power. The power-delay product illustrates this metric in the tables 1.1 and 1.2. Taking an average of the PDPtot values for both LPSPFF and LPSDFF it is found that no other design has a lower PDPtot product, meaning this design is extremely beneficial in low-power high-speed designs as well. Some flops like the HighSpeedDff have high short-circuit power dissipation due to their circuit structure. For some input patterns there is a direct path from V_{dd} to *Gnd* for a considerable time as captured by our simulations. HighSpeedDff has the lowest delay value for the unloaded case. It should be noted that the layouts of LPSDFF and LPSPFF are very compact. This is due to a lesser number of transistors and substantial diffusion sharing permitted by the transistor connectivity. Similar compact layout structures occur in High-

Table 1.2. Total power dissipation and delay comparison of proposed flip-flop with prior static and dynamic flip-flops with 0.2pf capacitive load.

Name	*# of transistors*	*Ptotal (uW)*	*Ave. Delay (ps)*	*PDPtot* $(x10^{-18} J)$
HighSpeedDFF	9	282.2	1407.00	397055.40
8TDFF	10	15.55	420.60	6540.33
9TDFF	11	51.14	1434.50	73360.33
DSTC	14	114.8	2732.5	313691.00
PTTFF	18	65.78	5565.00	366065.70
TILowPowerDFF	18	229.60	4197.05	963642.68
ConventionalFF	20	16.22	974.70	15809.63
LowPowerDFF	20	54.63	4104.50	224228.84
PushPullDFF	20	79.17	1578.50	124969.85
HLFF	20	80.53	1515.00	122002.95
SenseAmpFF	21	78.25	1451.00	113540.75
StrongArm110FF	22	90.51	1459.50	132099.35
PPC603	22	55.09	4107.00	226254.63
SDFF	23	92.75	2097.90	194580.23
Modified CCMOS	24	80.95	3377.50	273408.63
TSPCFF	26	56.01	4109.00	230145.09
LPSDFF	19	15.74	462.15	7274.24
LPSPFF	19	15.84	490.6	7771.1
LPSDFF[1]	19	8.031	462.15	3711.53
LPSPFF[1]	19	11.7	490.6	5740.02

SpeedDff, DSTC flop and the 8T and 9T flops. Other flops have complex structures and connectivity like the StrongArm flop which even uses logic gates in their design. This increases the layout complexity and thereby the area. It should be noted that 8T, 9T, DSTC, LowPowerDFF, PushPullDFF, PPC603 and TSPCFF have very low leakage power dissipation.

In the sleep mode, the output can be set to a desired value. This gives a capability to input a low leakage input vector to the subsequent circuits, a capability that does not exist in any of the other designs. This is demonstrated using a test case of an an 8x8 multiplier, described in the next section. LPSDFF and LPSPFF have a setup time of 0.15ns and a hold time of 0.05ns. The total power values for these flip-flops, when not in sleep mode, are comparable to that of the 8TDFF, being within 2% for almost twice as many transistors. The real benefit comes when the flip-flop is used in its sleep mode for a fraction of the total operation cycle. From Table 1.2, it is seen that these designs consume from 25% to 48% less power when compared to 8TDFF (with a $0.2pF$ load at

Table 1.3. Leakage power dissipation comparison of proposed flip-flop with prior static and dynamic flip-flops.

Name	*# of transistors*	*CLK & D (=Vss) Pleak (nW)*	*CLK & D (=Vdd) Pleak (nW)*
HighSpeedDFF	9	463800	409400
8TDFF	10	0.1837	4.792
9TDFF	11	4.193	3.904
DSTC	14	1.494	1.469
PTTFF	18	18.83	18.57
TILowPowerDFF	18	180800	199500
ConventionalFF	20	0.8063	0.7679
LowPowerDFF	20	1.076	2.654
PushPullDFF	20	2.587	1.161
HLFF	20	2.011	3.764
SenseAmpFF	21	31.19	23.47
StrongArm110FF	22	35.14	27.7
PPC603	22	1.172	2.77
SDFF	23	19.11	21.92
Modified CCMOS	24	9.934	10.48
TSPCFF	26	2.079	1.869
LPSDFF	19	0.091	5.443
LPSPFF	19	0.1057	5.382

the output).

Based on the LPSDFF and LPSPFF, a novel methodology for overall power optimization is shown in Figure 1.9. The designer can choose between various flops depending on other constraints. If power is critical and the designer wants to use the same flip-flop throughout the design, then the regular flops can be substituted by LPSDFF or LPSPFF. However, care should be taken to disable the sleep operation through the sleep signal. If speed is the concern in sub-block B, then HLFF or 8TDFF can be used. But for blocks such as sub block A which will be inactive for some time, a suitable combination of the data-gated LPSDFF and LPSPFF not only gives low active power during normal operation, but also reduces the leakage power of the whole sub-block A by inputting suitable vectors. Hence, overall active and leakage power is reduced.

5. 8x8 Pipelined Multiplier Design

An 8x8-pipelined multiplier was designed using LPSDFF and LPSPFF. The structure of this multiplier is shown in Figure 1.10. This multiplier consists of the basic full adder cell and flip-flops at every pipeline stage [8]. The work in

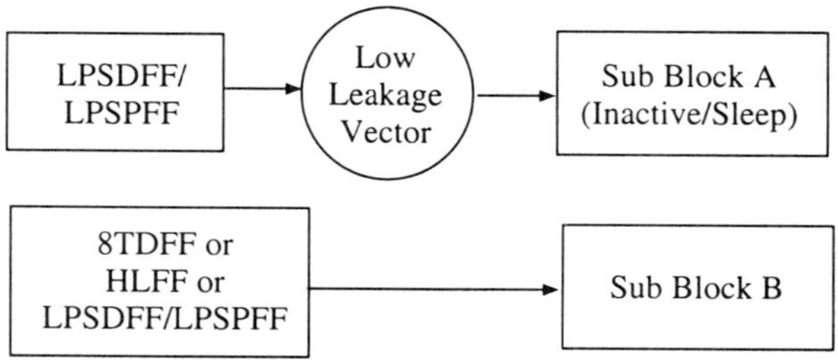

Figure 1.9. Methodology to Reduce Overall Power

[8] designed the multiplier using TSPC flops and as shown in Table 1.2, TSPC flops consume about 4 times more power than LPSDFF and LPSPFF flops.

There are 432 flip-flops in this design. Each input of every gate and the full adder is driven by a flip-flop. The full adder is simulated for all possible input vectors and the lowest leakage input vector is determined as shown in Table 1.4. The lowest leakage vector is $A = 1, B = 0, C = 1$. Then for each full adder in the circuit, LPSPFF is connected to inputs A and C and LPSDFF is connected to input B. This way, in sleep mode, each full adder is put in the lowest leakage mode. Two experiments are conducted for this design. First in the normal mode, the circuit is continuously performing multiplications. The operating frequency is 100MHz at a supply voltage of 1.8V. The average power dissipated in this mode is $0.25mW$. This shows that designing with LPSDFF and LPSPFF results in extremely low power circuits. In the second experiment, the multiplier is operating for half the time and is in sleep mode for the other half. During sleep mode, the low leakage input vector is input to each full-adder cell. It is observed that the leakage power reduces by a factor of 3 or 66%. Hence, LPSDFF and LPSPFF can be used for extremely low active and low leakage power design of digital circuits. This capability is not provided in other dynamic flops.

6. Summary

This chapter presents comparative analysis of flip-flops in terms of their delay and power (including leakage) characteristics. Further, specific design structures of low active power, low leakage power, dynamic D flip-flops, namely LPSDFF and LPSPFF are discussed in detail. LPSDFF and LPSPFF flip-flop designs are capable of reducing the overall system power dissipation substantially as compared to other designs. In low-power high-performance applications, high speed is as important as achieving low power. The power-delay

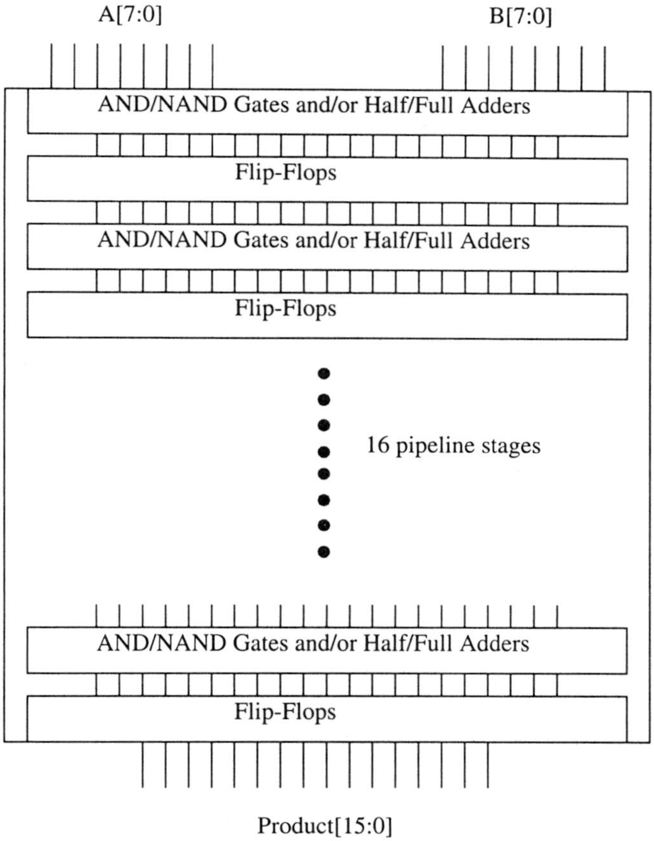

Figure 1.10. Structure of an 8x8 Pipelined Multiplier

product illustrates this metric and both LPSPFF and LPSDFF have significantly low power-delay product, making them extremely beneficial in low-power high-speed designs as well. Experimental results with 0.25*um* CMOS process designs show from 25% to 48% reduction in active power as compared to other flip-flops. An 8x8 pipelined multiplier designed using LPSDFF and LPSPFF dissipates 0.25mW at 1.8V and operating frequency of 100MHz. Also, when operated in sleep mode and by inputting the low leakage vector, the leakage power was reduced by a factor of 3. The comparative analysis enables a designer to judiciously choose the flip-flops that best suit the particular design constraints. Future work in this area would include research methods for ensuring the system to come back to the original state before entering the

Table 1.4. Leakage Power dissipation of the full adder for different input vectors.

A	*B*	*C*	*Leakage Power (nW)*
0	0	0	87.8
0	0	1	109.2
0	1	0	90.04
0	1	1	93.32
1	0	0	50.36
1	0	1	35.01
1	1	0	62.78
1	1	1	85.75

sleep mode if it was different. Further, analysis of current spikes as a result of switching between sleep and normal modes should be investigated.

Acknowledgments

The authors wish to thank Dieter Vook, Bob Taber and Jim Hollenhorst for their encouragement and support. The authors also wish to thank their colleagues in Agilent Technologies Laboratories.

References

[1] Strollo, A.G.M., Napoli, E., and De Caro, D., "New Clock-Gating Techniques for L ow-Power Flip-Flops," *Proceedings of International Symposium on Low Power Electronics and Design,* 2000, pp.114-119.

[2] Benini, L., and G. De Micheli, "Automatic Synthesis of Low-Power Gated-Clock Finite-State-Machines," *IEEE Transactions on Computer-Aided Design of Integrated Circuits and Systems,* 1996, pp. 630-643.

[3] Z. Chen, M. Johnson, L. Wei and K. Roy, "Estimation of standby leakage power in CMOS circuits considering accurate modeling of transistors stacks," *Proceedings of International Symposium on Low Power Electronics and Design,* 1998, pp 239-244.

[4] Halter, J.P., Najm, F.N., "A gate-level leakage power reduction method for ultra-low-power CMOS circuits," *Proceedings of Custom Integrated Circuits Conference,* 1997, pp 475-478.

[5] V. Stojanovic, V. Oklobdzija, "Comparative Analysis of Master-Slave Latches and Flip-Flops for High-Performance and Low-Power Systems," *IEEE J. Solid-State Circuits,* Vol. 34, NO. 4, April 1999, pp. 536-548.

[6] S. Hsu, S. Lu, "A Novel High-Performance Low-Power CMOS Master-Slave Flip-Flop," *Proceedings of IEEE International ASIC Conference,* 1999, pp.340-343.

[7] U. Ko, P. Balsara, "High-Performance Energy-Efficient D-Filp-Flop Circuits," *IEEE Transactions on Very Large Scale Integration (VLSI),* Vol. 8, February 2000, pp. 94-97.

[8] J.-S. Wang, P.-H. Yang, and D. Sheng., "Design of a 3-V 300-MHz Low-Power 8-b x 8-b Pipelined Multiplier Using Pulse-Triggered TSPC Flip-Flops," *IEEE J. Solid State Circuits,* Vol. 35, Apr. 2000, pp 583-592.

[9] Mahant-Shetti, S. Shivaling, R. Landers, "Low Power Flip-Flop," *U.S. Patent, Number 5789956.*

[10] D. Lee, K. Do, "High Speed D Flip-Flop," *U.S. Patent, Number 6060927.*

[11] M. Hamada, et. al., "Flip-flop selection technique for power-delay trade-off," *IEEE International Solid-State Circuits Conference,* 1999, pp 270-271.

[12] M. Martinez and V. Srikantam, "Low operational power, low leakage power D-type flip-flop," *U.S. Patent, Number 6275083.*

Chapter 2

LOW POWER SANDWICH/SPIN TUNNELING MEMORY DEVICE

Jim Daughton, Arthur Pohm, Russell Beech
NVE Corporation
11409 Valley View Road
*Eden Prairie, MN 55344**

Abstract ABSTRACT A new nonvolatile storage element has potential for greatly reducing power consumption when used in memory or as a reconfiguration element. The element uses magnetic material hysteresis to store data and tunneling magnetoresistance to read data. Advantages compared to other magnetic and semiconductor nonvolatile storage cells are cited. The design of the cell and some limited characterization data are presented along with conceptual circuit and memory architecture desings. Smaller, niche memory and reconfiguration circuits are the most probable power aware applications areas. Extension to very high density memories may be possible through using heat and magnetic field for writing data.

1. Introduction

In order to reduce the energy required by a system, it is highly desirable to be able to turn off an electronic function when it isn't being used. A nonvolatile element - one that retains data without being powered - can be used in two ways toward that end. It can store data in an "off" mode until the data is needed, or it can be incorporated in combination with logic to turn off a logic block until it is needed.

Historically, battery-powered CMOS circuitry has provided these functions in systems where some power drain is tolerable and where battery life is not an issue. However, as semiconductor devices shrink to 100 nm dimensions and

*The work was largely carried out under DARPA Contract # F29601-00-C- 0194 entitled "Ultra Low Power Enabling Technologies for Adaptive Reconfigurable Power Aware Computing and Communications as part of the Power Aware Computing and Communications Program.

below, transistor leakage currents due to the body effect and to gate leakage can consume significant power in systems, even in the standby mode. Nonvolatile semiconductor cells such as EEPROM cells and flash memory cells provide the true nonvolatility required, but suffer from two problems - they wear out with about one million write operations, and they write very slowly (in about 10 μs). The slow write cycle leads to a relatively high total write energy. Although these memories are used very successfully in many "read mostly" applications, they are not satisfactory for applications that require many or fast write operations.

Magnetoresistive Random Access Memory (MRAM) [1, 2, 3] is the object of several large development activities. This technology uses the hysteresis of magnetic materials to store data, and the magnetoresistive properties of certain new materials to read data out. Memory cells are deposited on integrated circuit substrates, which provide bit selection and sensing circuit functions. MRAM has the potential to be the "ideal, all purpose" memory because it is relatively fast and dense, and is nonvolatile. The primary MRAM development challenge is obtaining sufficient yield of memory cells, with the primary limiting factor being the requirement for very high cell uniformity. Longer term, the challenge is to obtain stability at high density (lithography at 0.1 μm or less) with reasonably low drive currents [4].

Most current MRAM developments use Magnetic Tunnel Junctions (MTJs), and this kind of memory will be described first. Then, a new kind of MRAM approach - a sandwich/spin tunneling cell that uses an extra transistor in the cell for write select - will be described. It can eliminate a major limitation on cell yields, and provide a path to moderate density MRAM with lower operating energy consumption per write operation than other solid state nonvolatile memory cells.

2. Magnetic Tunnel Junction MRAM Cell

Most MRAM designs use a 2D array of cells with two sets of orthogonal current drive lines as shown in Figure 2.1. A bit is selected for writing by passing currents through both the corresponding horizontal and vertical current drive lines. Unlike Dynamic Random Access Memory (DRAM), a transistor for selection is not needed. However, in order for this technique to select only the desired cell, the "half select" currents encountered by the cells not written must not change the state of these unselected cells. The tight current tolerances required in this mode of operation is the primary source of yield problems faced by MRAM developers.

Spin tunneling has been recently reduced to practice [5]. Two magnetic layers separated by a thin dielectric barrier layer have a resistance which is very sensitive to the alignment of the magnetizations in the two layers. When the magnetizations are aligned, the tunneling current for a given voltage is larger than when they are antiparallel for the Magnetic Tunnel Junction (MTJ) shown Figure 2.2. One of the two magnetic layers is "pinned" (doesn't change with magnetic field) by coupling with an antiferromaget, while the other may be magnetized along either of two directions.

The memory cell recently described by Motorola [2] is shown in Figure 2.3. A "one" or a "zero" is represented by the direction of magnetization of the "free" layer in the MTJ shown. There is one select transistor in each memory cell. The select transistor is turned on in order to read data from a specific bit in a 2D array. If a current is injected through the tunnel junction, the voltage across the MTJ will be higher or lower depending on the magnetic state of the free layer. To write, the select transistor is turned off, and the 2D selection scheme that was described previously is used to write the desired bit, with the current direction through the "digit" line determining the memory state. The 256K MRAM memory that was demonstrated by Motorola had a read access time of about 50 ns.

3. Sandwich/Spin Tunneling Cell

The sandwich/spin tunneling cell is similar to other tunneling cells [2, 3] with two exceptions. First, a sandwich containing two magnetic films and a thin interlayer is used for the write/storage element rather than a single layer. Second, a select transistor is used to write with a single current instead of two coincident currents as in a 2D scheme.

In addition to greatly reducing the required write current, the use of a select transistor for writing makes the manufacturing tolerances for devices very wide because there are no "half select" currents, the major yield obstacle in other MRAM developments. A tunneling barrier overlays the sandwich cell, and a pinned magnetic layer lies on top of that. The top magnetic layer of the sandwich and the pinned layer are then two elements sandwiching a tunnel barrier, and the magnetoresistance properties of the device can be used to read data from the cell through a read select transistor. The cell concept is shown in Figure 2.4.

When a current I passes through a conductor stripe of width W, the magnetic field near the stripe is about I/2W and is directed across the stripe, as indicated by the right hand rule. In Figure 2.4, the write current flows in both the inner layer (copper or aluminum)of the sandwich and through the two permalloy layers (80Ni20Fe or a similar alloy). To the extent that the current flows through the permalloy, the effective magnetic field is reduced. If all of the current flows through the permalloy and none flows through the interlayer metal, then the effective magnetic field acting to switch the magnetic layers is I/4W, or one half the field if all current flows through the interlayer. If the required switching field is 15 Oe (1200 A/m), and if the line width is 2 microns, and if all the current flows through the stripe, then the current required to write the cell should be approximately 4.8 mA. This is consistent with experimental results. Note that as the line widths decrease or as the stiffness of the magnetic materials is lessened, the required current decreases. The switching time is approximately 2 ns, and a 3 ns write current pulse width is sufficient. Thus, with a 3 Volt supply, the switching energy for such a cell is approximately 40 pJ. It is possible to share write currents with other cells, which would reduce the write energy when several cells are written simultaneously.

Tunnel junctions have two properties that determine the design of the read circuits: junction resistance variations and junction voltage limitations. Because tunneling resistance varies as an extremely strong function of oxide thickness, there can be considerable variation in the tunneling resistance in manufacturing, perhaps a factor of two. However, this variation is very small over short distances on a wafer. Also, the voltage applied across the barrier should be less than about 200 mV for reliability considerations, and the magnetoresistance of the tunnel junctions decreases somewhat even above 100 mV. The current source for sensing the state of the cell should ideally supply a current that results in a barrier voltage of between 100 mV and 200 mV. This can be accomplished by using a tunnel junction in a feedback mode in the current source circuit design in order to account for resistance variations. With a nominal 150 mV potential across the junction and a 30% tunneling magnetoresistance, the signal difference between a "1" and a "0" would be about 45 mV. This is approximately the level of signal used in semiconductor memory - the voltage across a capacitor in the case of DRAM and the offset voltage on internal nodes in an SRAM cell.

An improvement to the sandwich/tunneling cell makes use of a strong antiparallel coupling between two magnetic layers. That coupling can be induced by depositing a thin interlayer of a selected material, typically ruthenium [6], between the two magnetic layers in the sandwich. If an interlayer of approximately 9 of ruthenium is used, as shown in Figure 2.5, there is an antiparallel coupling field between the two magnetic layers of approximately 5000 Oe (400,000 A/m). The relatively high resistance of the very thin ruthenium layer forces more of the current into the permalloy layers, which reduces to some extent the field conversion efficiency of the cell. In very small cells, this efficiency loss is more that compensated for by increased stability of the cell and the elimination of demagnetizing and stray magnetic field effects during switching. Cells as small as 0.1 micron in width have been simulated, and uniform switching was predicted at very low currents. Figure 2.6 shows a plot of switching current derived from simulations done at Carnegie Mellon University by Dr. Jimmy Zhu. Note that the switching current projected at 0.1 micron is less than one mA.

Sandwich/spin tunneling cells have been designed, fabricated, and tested. A top view of the mask layouts for a pair of cells is shown in Figure 2.7. A pair of cells was used to facilitate the tests by taking the signal as the difference between the potentials across two junctions. The deposition processes use standard sputtering equipment. The photolithography which was used had 1 micron resolution, which greatly constrained

Density density and current required by the device. Switching speeds of less than 3 ns have been measured at currents through the sandwich of 2.5 mA (the current will be shared by two cells in the memory). This demonstrated write energy of less than 12 pJ per cell is less than 1/1000 of the energy that is require to write an EEPROM cell.

Even more favorable data is projected as higher resolution lithography is used. The elimination of the most serious yield detractor in MRAM development, "half-select" current tolerance needed for the 2D selection scheme, is also a very important advantage. It should be noted that the added write select transistor increases memory cell area, and although the sandwich/tunneling cell should provide high speed and very low power, it is not the highest density MRAM.

4. Memory Circuits/Architecture

A memory cell can be configured from a single sandwich/spin tunneling element, with the signal across the junction compared to a reference in order to determine the memory state, or from two elements which are written into complementary states, with the difference signal detected between the tunnel junctions in the two elements. The first cell configuration gives higher density, and the second cell gives higher speed.

To date, the circuits described here are simulated concepts. With the exception of the demonstration of some critical circuits using MOSIS, the circuits have not been fabricated and tested.

In the 1JC (single SDT junction memory cell) architecture, a read compares the cell voltage, that is developed across a selected SDT junction when a fixed current is passed through the junction, to a reference voltage. The reference voltage is obtained from two parallel SDT junctions, one that is written to the high resistance state and one that is written to the low resistance state. Because of the relatively small difference between the cell voltage and the reference voltage, an amplifier is required to boost the signal prior to being able to latch the data. The write operation simply passes a fixed current through the write terminals of the SDT memory cell - where the polarity of the write current determines the state that the cell is written to. Though it does not provide quite as fast an access time as the 2JC memory architecture, which is discussed later in this section, the 1JC architecture provides higher density in larger arrays. Figure 2.8 illustrates this 1JC memory architecture.

The circuit blocks that are needed for the 1JC memory architecture include: 1) Sense amplifier: this circuit block is the largest and most complex in the architecture. At the input, this circuit block contains a differential amplifier with an auto-zero capability. The amplifier is required to boost the small memory cell signal and the auto-zero is included to null the amplifier offsets, prior to application of the memory cell signal to the amplifier inputs. At the center of this circuit block is a comparator. The comparator is connected to the amplifier output, and further conditions the read signal - giving a logic level output. At the output, a digital latch is used to latch the output of the comparator. Additionally, the sense amplifier contains several delay blocks and associated logic gates that provide the timing signals that control the auto-zero and output latch.

2) Write Driver: this circuit block provides the write current that is used to switch the storage film between its two states. As mentioned earlier, this circuit must be able to drive a fixed current through the SDT memory cell - in both directions, where the data being written determines the direction. This circuit must be capable of switching the write current in a very short time in order to maintain a low cycle time for the memory. Because this current will flow only through the selected memory cell, there are no half-select disturb margin problems. This eases the design constraints on the write current magnitude, and the turn-on/turn-off speed is the only primary design consideration. 3) Sense Current Driver: this circuit block provides the fixed currents that are passed through the selected memory cell and the reference cell when a read operation is performed. Because the reference cell consists of two SDT junctions that are in parallel, yielding a resistance that is nominally 1/2 that of the memory cells, the current that is supplied to the reference cell must be twice that of the current that is supplied to the selected memory cell. In addition, the two currents, the reference cell current, and the memory cell current, must track each other very closely so that transient differences do not appear as a false signal to the sense amplifier. 4) Addressing and control: in addition to the three blocks mentioned above, there will be some additional digital circuitry needed, even for a very small array, to decode the address and condition the control inputs.

The 2JC (two SDT junction memory cell) architecture is the another possible memory architecture. This architecture uses a relatively simple latch cell circuit with one SDT junction in each of the two halves of the latch cell. The two SDT devices are written to opposite states - one high resistance and one low resistance. To read the memory, the latch is momentarily shorted, so that both outputs are at the same potential, and then released. When the short is removed, the different resistance that is presented to the two halves of the latch, by the oppositely written SDT junctions, causes the latch output to rapidly drive to a known logic level based on which SDT junction is in a higher resistance state. This architecture is very fast and, for small arrays, has higher density than the 1JC architecture because of the lower overhead electronics. The reason that this architecture has higher density for small arrays is that the area that is required for overhead circuitry, such as the sense amplifier, is much larger for the 1JC architecture than for the 2JC architecture. The 2JC architecture also has less stringent requirements on SDT junction uniformity since the two SDT junctions within a given cell act as local "references" for each other - there is no need to match closely to a reference cell that may be located a relatively large distance away. Figure 2.9 illustrates the 2JC memory architecture.

There are only two primary circuit blocks in the 2JC architecture, the write driver and the latch. The write driver is the same as that required for the 1JC cell - which was discussed earlier. The latch consists of two cross-coupled inverters and a shorting switch that connects the two latch outputs together. Additionally, the latch may contain a current limiting transistor that is connected between the positive supply rail and the two inverters. The current limiting transistor is needed to provide the proper bias across the SDT junctions. Alternately, the inverter transistors can be sized to limit the current. In either case, the maximum current through the inverters, and therefore hence the SDT junctions, is a critical design parameter.

For both the 1JC and 2JC memory architectures, the standby power is virtually nil. Simulations indicate that the read access time and write times will be about 10 ns to 15 ns for the 2JC memory and 25 ns to 40 ns for the 1JC design. Write energies will depend strongly on the ultimate write current for the cell, but a goal of 10 pJ is reasonable.

5. Potential Higher Density Sandwich/Tunneling Memory

As all magnetic memory cells shrink in size (volume), eventually they become thermally unstable, i.e. the depth of the energy storage well for the cell is not sufficiently above kT to insure the data with enough certainty over a long enough time for practical storage. There is the possibility of modifying the design of the sandwich/tunneling cell for very high densities, and at the same time eliminate the write select transistor described in the previous architectures.

In magnetic materials the strong inter-atomic forces that align internal atomic spins (parallel for ferromagnetism and anti-parallel for anti-ferromagnetism) diminish with increasing temperature and are overcome completely by thermal agitation at the Curie temperature (TC) for ferromagnets and at the Neel temperature (TN) for anti-ferromagnets.

When a ferromagnet and an anti-ferromagnet are deposited in sequence so they are in contact, there can be coupling forces at the interface such that the ferromagnet's magnetization is restored to its rest direction after even very large magnetic fields are applied. The magnetic field can be 5,000 Oe or more, and the ferromagnet's magnetization will still be restored to its initial direction. These coupling forces can be used to create a very deep energy well (and attain very high densities). Raising the temperature of the ferromagnetic material above its Curie point and cooling it while a magnetic field is on should also allow writing of such materials in a memory cell, even with small currents.

The cell shown in Figure 2.4 could use magnetic materials in the sandwich layers that have a Curie point that is lower than the Neel temperature of the pinning layers. If an antiferromagnetic pinning layer is added to the cell so as to pin the bottom layer of the sandwich, a deep energy storage well could be formed. The sandwich layer could be written with a current that causes the structure to heat above the Curie point of the ferromagnetic layers in the sandwich. The low Curie point can be achieved, for example, by adding a small amount of chromium to a nickel iron alloy - on the order of 5% to get the Curie point to about 250 C. The thin ruthenium layer acts to make the two magnetic layers antiparallel with an effective field of several thousands of Oe. Tunneling currents could be used to determine memory state as described previously.

In theory, a 2D selection scheme (no write select transistor would be needed) can be realized by combining a heating element with the sandwich layer and hooking a number of sandwich layer segments (memory cells) in series. A heating element can be placed orthogonal to the series connections of memory cells so that the cells under the heating element have an elevated temperature. If the heat from the combination of current through the cell and the heat from the heating element raise the temperature of the cell above the Curie point, then that cell could be written with a very small magnetic field as the current in the heating element is reduced. Heating from the heating element alone, and heating from the current in the sandwich alone must not raise the temperature of the cell above the Curie temperature of the cell providing the energy well. The concept is illustrated in Figure 2.10. This switching mechanism has been demonstrated on micron size devices having a surprisingly short 3 ns time constant [4], and simulations on submicron size devices indicate a 1ns or faster thermal time constant.

6. Potential Applicatons

Because the technology is still in its infancy, it is too early to project definite applications; however, there are some applications areas with promise.

Unattended ground sensors, such as those used in perimeter protection, are usually battery-operated, and energy management is critical. These systems are off most of the time and operate in bursts. Furthermore, while operating, memory operation consumes a significant amount of power. Projections of higher leakage currents in static semiconductor memories could make the energy loss from convention SRAMs intolerable. The new memory technology could provide the user an option with virtually no standby power.

Most of today's programmable gate arrays are programmed once, and the reprogrammable versions use EEPROMs. EEPROMs write very slowly (in about 10 microseconds) and wear out within about a million write cycles. There may be applications for an infinitely writeable programmable gate array, and this new technology could potentially provide an embedded reconfiguration element for this application. A cell library of reconfiguration elements has been designed and is in process.

Contactless RF identification tags have many potential uses, including identification of crated items in military logistics operations. They rely on RF power from a remote source to activate the tag. Writing information into such tags, such as when the article was processed, or when it passed an inspection point, is very difficult with high energy consuming components, such as EEPROMs. This new technology could make writeable RF ID tags viable.

Digital Signal Processors store weighting constant data, which can frequently change. DSP Architectures has projected that the new technology described here could be used to store this data locally on-chip and save chip area while improving throughput. The high speed (less than 10 ns read/write times) would be critical.

References

[1] J.M. Daughton,*Thin Solid Films*, 216 p 162 (1992).

[2] R.E. Scheuerlein et al.,*2000 IEEE International Solid State Circuits Conference. Digest of Technical Papers (Cat. No.00CH37056)*, p 128 (2000).

[3] P. Naji, M. Durlam, S. Tehrani, J. Cader, and M. DeHerrera,*2001 ISSCC Digest of Technical Papers*, p 122, (2001).

[4] R. Beech, J. Anderson, A. Pohm, and J. Daughton, *J. Appl. Phys.* 87, No. 8 p 1 (2000).

[5] J. Moodera et al, *J.Appl.Phys.* 79, No. 8, p 4724 (1996).

[6] S. Parkin and D. Mauri,*Phys. Rev.* B 44 7131 (1991)

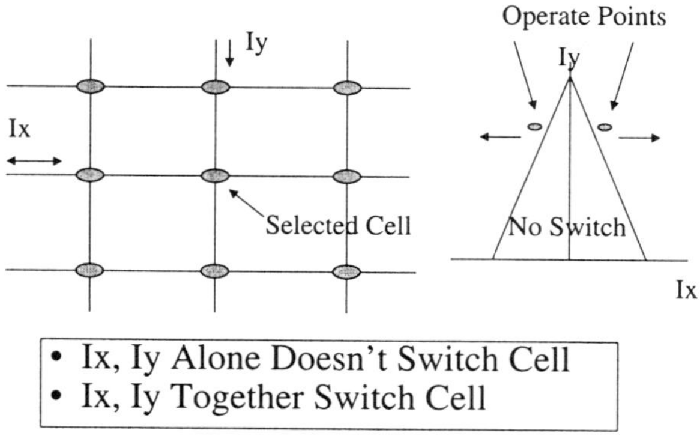

Figure 2.1. 2D Memory Select Used In Most MRAM Designs.

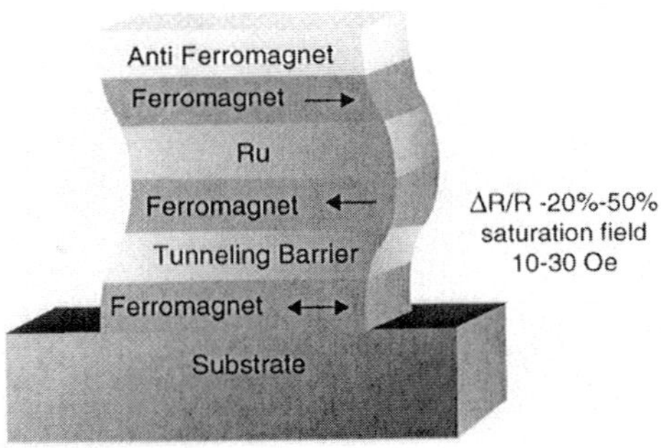

Figure 2.2. Magnetic Tunnel Junction (MTJ) Cross-Section.

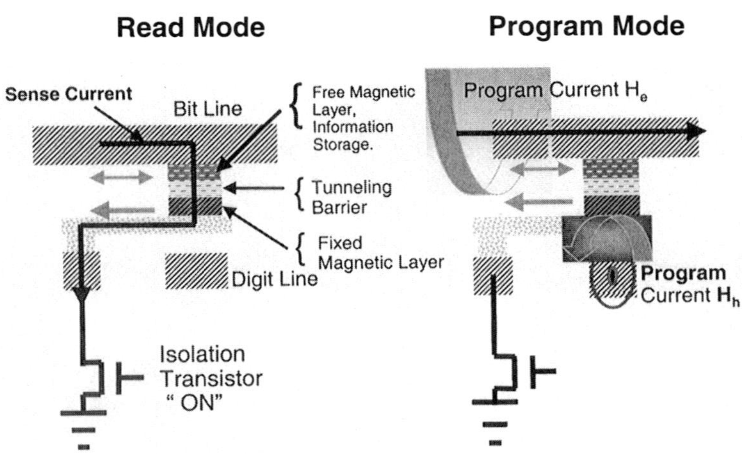

Figure 2.3. MTJ Cell (Motorola)[2]

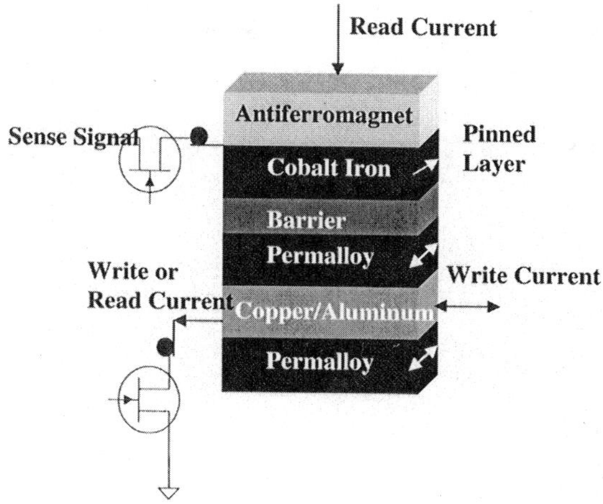

Figure 2.4. Sandwich/Tunneling MRAM Cell

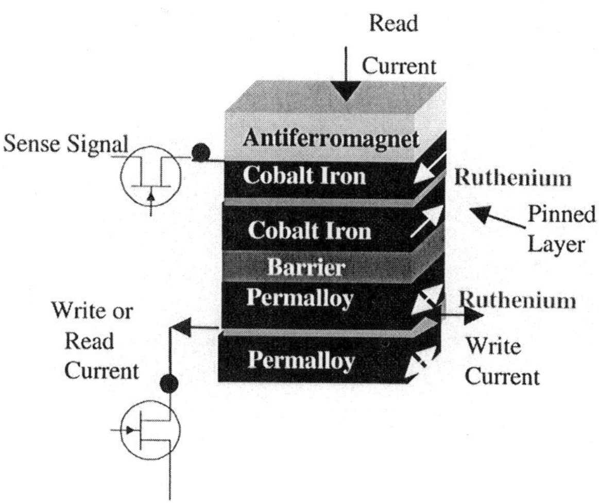

Figure 2.5. Sandwich Tunneling Cell With A Thin Layer of Ruthenium As An Interlayer In The Permalloy Sandwich.

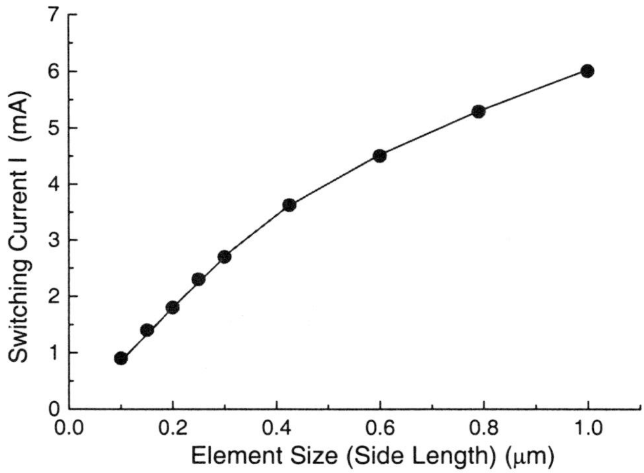

Figure 2.6. Switching Current Required As A Function Of Cell Width For A Tightly Antiparallel Coupled Sandwich.

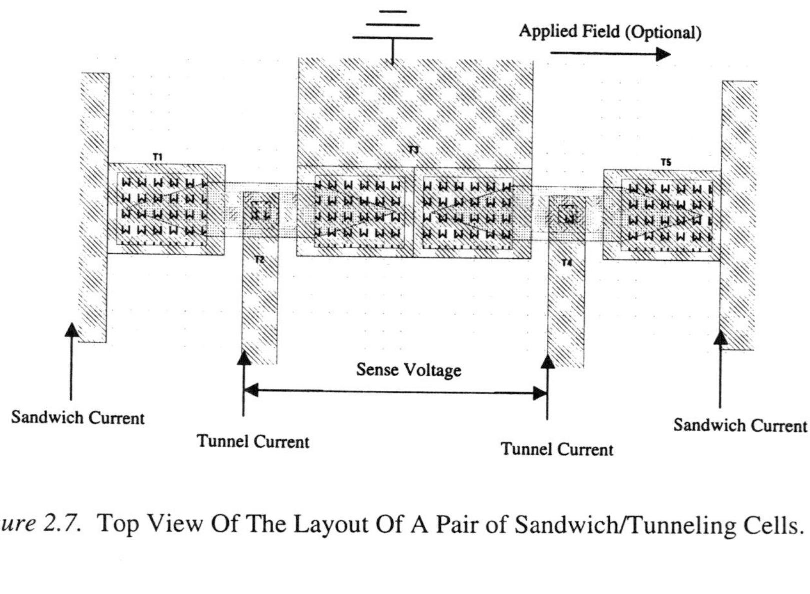

Figure 2.7. Top View Of The Layout Of A Pair of Sandwich/Tunneling Cells.

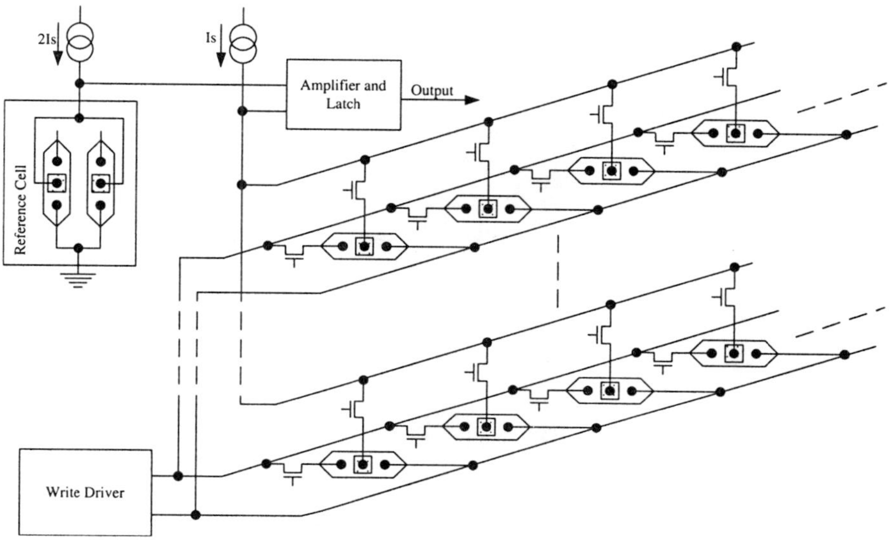

Figure 2.8. Diagram of the 1JC memory architecture

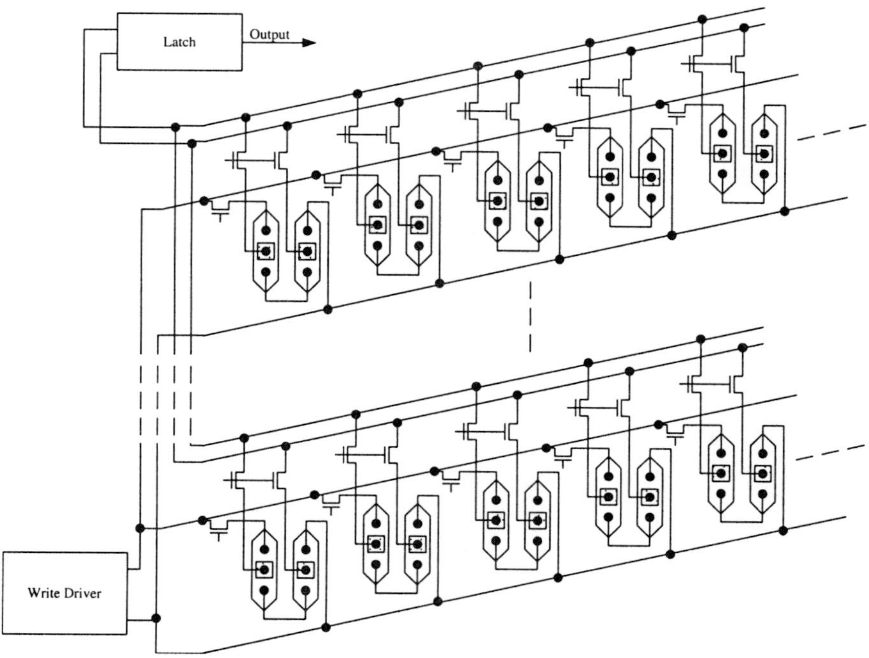

Figure 2.9. Diagram of the 2JC memory architecture.

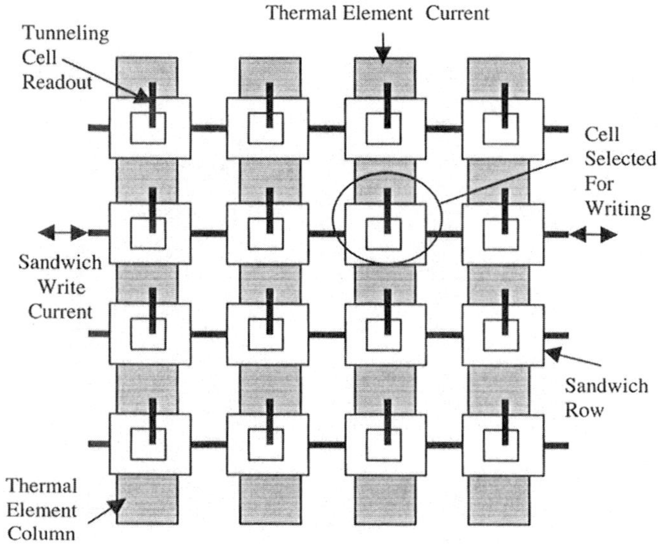

Figure 2.10. 2D Array of Cells Using Thermal Write Lines and Current Through The Sandwich Layers Of A Sandwich/Tunneling Cell.

II

ARCHITECTURE LEVEL POWER MANAGEMENT

Chapter 3

POWER-EFFICIENT ISSUE QUEUE DESIGN

Alper Buyuktosunoglu⋆, David H. Albonesi⋆, Stanley Schuster*, David Brooks*, Pradip Bose*, Peter Cook*

⋆ *Department of Electrical and Computer Engineering*

University of Rochester

∗ *IBM T.J. Watson Research Center*[*]

Abstract Increasing levels of power dissipation threaten to limit the performance gains of future high-end, out-of-order issue microprocessors. Therefore, it is imperative that designers devise techniques that significantly reduce the power dissipation of the key hardware structures on the chip without unduly compromising performance. Such a key structure in out-of-order designs is the issue queue. Although crucial in achieving high performance, the issue queues are often a major contributor to the overall power consumption of the chip, potentially affecting both thermal issues related to hot spots and energy issues related to battery life. In this chapter, we present two techniques that significantly reduce issue queue power while maintaining high performance operation. First, we evaluate the power savings achieved by implementing a CAM/RAM structure for the issue queue as an alternative to the more power-hungry latch-based issue queue used in many designs. We then present the microarchitecture and circuit design of an adaptive issue queue that leverages transmission gate insertion to provide dynamic low-cost configurability of size and speed. We compare two different dynamic adaptation algorithms that use issue queue utilization and parallelism metrics in order to size the issue queue on-the-fly during execution. Together, these two techniques provide over a 70% average reduction in issue queue power dissipation for a collection of the SPEC CPU2000 integer benchmarks, with only a 3% overall performance degradation.

1. Introduction

Power dissipation has become a major microprocessor design constraint, so much so that it threatens to limit the amount of hardware that can be included

*This work was supported in part by DARPA/ITO under AFRL contract F29601-00-K-0182, by NSF grants CCR-9701915 and CCR-9811929, and by an IBM Partnership Award.

year	1999	2002	2005	2008	2011	2014
feature size (nm)	180	130	100	70	50	35
logic trans/cm^2	6.2M	18M	39M	84M	180M	390M
clock (MHz)	1250	2100	3500	6000	10000	16900
chip size (mm^2)	340	430	520	620	750	900
power supply (V)	1.8	1.5	1.2	0.9	0.6	0.5
high-perf power (W)	90	130	160	170	175	183

Table 3.1. SIA Roadmap for Power Dissipation in Current and Future Microprocessors

on future microprocessors how fast it can be clocked [11, 12, 14]. Designers of handheld devices are already quite familiar with the difficulties in meeting ever-increasing performance demands while maintaining reasonable battery life and small product form-factors. However, the increasing packaging, cooling, and power distribution costs, as well as reliability issues, associated with increasing power dissipation in high-end systems also threatens their future viability.

Table 3.1 summarizes the SIA Roadmap [18] for power dissipation in current and next generation high-performance microprocessors. According to these projections, power dissipation will increase from 90 watts in 1999 to 170 watts in 2008. Clearly, rapid increases in both the clock frequency and chip functionality (complexity) are outpacing circuit and microarchitectural design attempts to maintain reasonable power dissipation limits.

One of the major contributors to the overall power consumption in a modern out-of-order superscalar processor, like the Alpha 21264 and Mips R10000 [13, 16], is the issue queue. The issue queue holds decoded and renamed instructions until they issue to appropriate functional units. The choice of an issue queue size requires striking a careful balance between the ability to extract instruction-level parallelism (ILP) from common programs and achieving high frequency operation. The size of the issue queue represents the *window* of instructions that may be eligible at any given cycle for issue. An instruction residing in the issue queue becomes eligible for issue (or *woken up*) when both of its source operands have been produced and an appropriate functional unit has become available. The *selection logic* determines which instructions (up to maximum issue width of the processor) should issue out of those woken up on a given cycle. Many superscalar processors such as the Alpha 21264 [13] and MipsR10000 [16] use multiple issue queues tailored to the type of instruction (*e.g.*, integer, floating point, and memory). Because the issue queue can be a major contributor to overall power dissipation (for instance, the integer queue on the Alpha 21264 is the highest power consumer on the chip [15]), new issue queue approaches must be devised that significantly reduce power dissipation without unduly compromising clock speed or instructions per cycle (IPC) performance.

There have been several prior proposals for reducing issue queue power dissipation. *Dynamic adaptation* of the issue queue size to match application demands is proposed in [1, 2] in order to increase performance and reduce power dissipation. The basic idea is to exploit the fact that the issue queue size needed to obtain most of the achievable IPC performance (within reasonable implementation limits) differs from application to application, and therefore the size of the queue can be adjusted (in *increments* of n entries) to fit the application. However, in [1, 2], it was assumed that the best issue queue size for a given application was known *a priori*, no attempt was made to adapt within an individual application, and the circuit-level design issues associated with an adaptive issue queue were not addressed in detail.

Gonzalez et al. [8, 9, 10] propose several schemes that attempt to reduce the issue queue control logic complexity [8] or its power dissipation [9, 10] without significantly impacting IPC performance. In [8], the authors propose and evaluate two different schemes. In the first approach, the complexity of the issue logic is reduced by having a separate *ready queue* which only holds instructions with operands that are determined to be fully available at decode time. Thus, instructions can be issued *in-order* from the ready queue at reduced complexity, without any associative lookup. A separate *first-use table*, indexed by unavailable operand register specifiers, holds those instructions that are the first-time consumers of these pending operands. Instructions which are deeper in the dependence chain simply stall or are handled separately through a separate issue queue. The dependence link information connecting multiple instances of the same instruction in the first-use table is updated after each instruction execution is completed. At the same time, if a given instruction is deemed to be *ready* it is moved to the in-order ready queue. Since none of the new structures require associative lookups or run-time dependence analysis, and yet, instructions are able to migrate to the ready queue as soon as their operands become available, this scheme significantly reduces the complexity of the issue logic.

The second approach relies on static scheduling. Here, the main issue queue holds instructions with pre-determined availability times of their source operands. Because the queue entries are time-ordered (due to known availabilities), the issue logic can use simple, in-order semantics. Instructions with operands which have unknown availability times are held in a separate *wait queue* and get moved to the main issue queue only when those times become definite. In both approaches described in [8], the emphasis is on reduction of the complexity of the issue control logic. The added (or augmented) support structures in these schemes may actually cause an increase of power, despite the simplicity and elegance of the control logic.

In [9, 10], the main focus is on power reduction. The issue queue is designed to be a circular queue structure, with head and tail pointers, and the effective size is dynamically adapted to fit the ILP content of the workload during different periods of execution. In both [8] and [9, 10], the authors show that the IPC loss is very small with the suggested modifications to the issue queue structure and logic. Also, in [9, 10], the authors use a trace-driven power-performance simulator (based on the model by Cai [7]) to report substantial power savings on dynamic queue sizing. However, a detailed circuit-level design and simulation of the proposed implementations are not reported in [8] or [9, 10]. Without such an analysis, it is difficult to assess whether the cycle time overhead of the extra circuitry required for dynamic adaptation, or its power dissipation overhead, are overridden by the power dissipation benefits of these approaches.

In this chapter, we propose two techniques that achieve significant issue queue power savings while maintaining high speed operation and good IPC performance. First, we propose a CAM/RAM-based issue queue as an alternative to the latch-based design used in many microprocessors. We then propose modifications that divide the queue into four sections, each of which can be individually enabled/disabled, and algorithms that adapt the issue queue size on-the-fly during the running of an individual application. Our scheme is simpler than that reported in [8] in that it does not require any new data storage or access structure (like the first-use table or the wait queue). Rather, we leverage the CAM/RAM structure that we propose as an alternative to a latch-based design. Although our approach of dynamically adapting the issue queue size to match workload demands is conceptually similar to the method proposed in [9, 10], our adaptation is more coarse-grained to limit circuit complexity and our control logic is quite different. Also, unlike [9, 10], we perform a detailed circuit-level evaluation of the power and performance overheads of the adaptive logic in addition to a microarchitecture analysis.

The rest of this chapter is organized as follows. In the next section, we compare the power and performance characteristics of latch-based and CAM/RAM-based issue queues. We then present in Section 3 the adaptive issue queue design in detail, both in terms of circuit-level design and algorithms for adaptive control. We also perform a circuit and microarchitecture-level analysis of the power benefits, and performance costs, of dynamic adaptivity. Finally, we conclude in Section 4.

2. Latch and CAM/RAM-Based Issue Queues

As previously mentioned, some microprocessors, such as the Alpha 21264 [11, 13, 15], implement latch-based out-of-order issue queues. As shown in

Figure 3.1, each entry in a latch-based queue consists of a set of latches that hold the required instruction information, a pair of comparators for detecting source operand availability (source operand register specifiers are held in the left latch in Figure 3.1), and selection logic to select instructions to issue from the ready pool. In addition, after an instruction issues, all entries behind the issued one are shifted forward to fill the slot. We refer to this as queue *compaction*. This is the function of the multiplexer shown in Figure 3.1, which can either hold the current instruction information or pass instruction information forward from the prior entry to fill a "hole" caused by an issuing instruction. Thus, a latch-based issue queue with compaction has the advantage of maintaining the oldest to youngest program order, making it straightforward to give issue priority to the oldest instructions. Furthermore, implementing the issue queue with latches and multiplexers is attractive due to its design simplicity, modularity, and resulting ease of verification.

However, the major disadvantage of latch-based designs is their high power dissipation due to the high power of the latches themselves and the power consumed in compaction. Compaction entails shifting instructions in the queue every cycle and can be a major source of power consumption. Furthermore, studies have shown that overall performance is largely independent of what selection policy is used (oldest first, position based, *etc.* [6]). As such, not only may a compaction strategy be unsuitable for low power operation, it may not be critical to achieving good performance. Note that a non-compacting issue queue can be implemented with a slight modification to the latch-based design shown in Figure 3.1. The connections between entries (from the output of the latch to the right multiplexer input of the next entry) are removed and instead the input bus (from dispatch) is connected to the right multiplexer input of all entries.

An alternative to a latch-based design is a CAM/RAM-based issue queue (shown in Figure 3.1) in which the source operand numbers are placed in the CAM structure and the remaining instruction information is placed in the RAM structure. The number of entries in the CAM/RAM structure corresponds to the size of the issue queue. The CAM/RAM structure is arguably more complex in terms of design and verification time. Also, it does not support compaction, if that feature is deemed necessary for performance. However, because of the lower power density of CAM/RAM logic relative to random logic, the CAM/RAM-based issue queue approach has the potential to eliminate hot spot problems in addition to reducing the average power dissipation of the queue.

While potentially consuming less power than a latch-based solution, a CAM-/RAM-based issue queue brings a new set of power-related issues. CAM and RAM structures require precharging and discharging internal high capacitance lines and nodes for every operation. The CAM needs to perform tag matching operations every cycle. This involves driving and clearing high capacitance

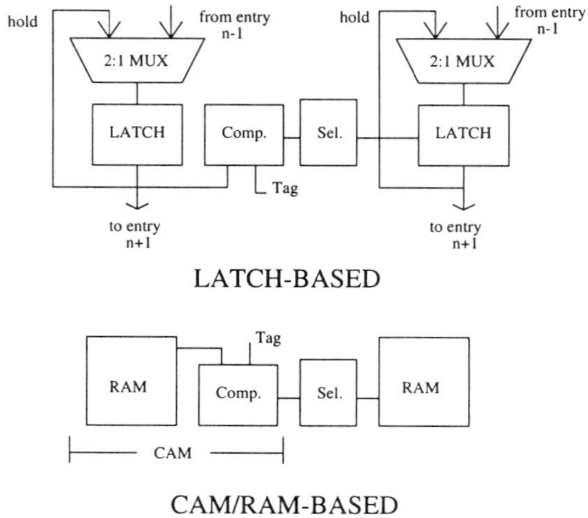

Figure 3.1. Latch-Based (with Compaction) and CAM/RAM-Based Issue Queues

tag-lines, and also precharging and discharging high capacitance matchline nodes every cycle. Similarly, the RAM also needs to charge and discharge its bitlines for every read operation. Thus, considerable additional effort is required to create a power-efficient CAM/RAM design. Thus, a considerable advantage (speed and/or power) must be demonstrated to justify this additional design and verification effort.

The operation of a CAM/RAM-based issue queue is illustrated in Figure 3.2. During the first clock phase of issue, new instructions are placed into the queue

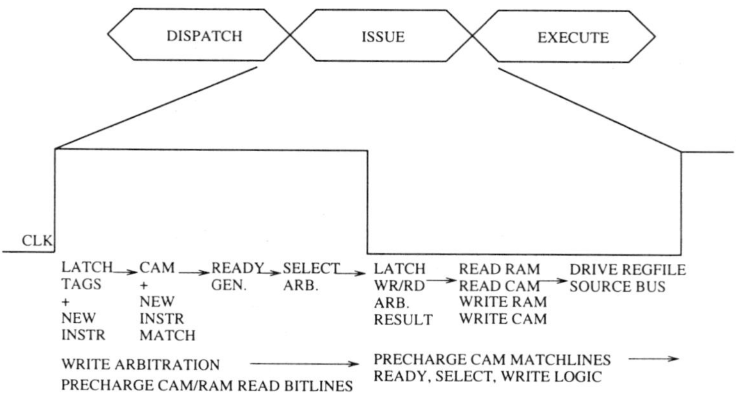

Figure 3.2. Operations Performed by the CAM/RAM-Based Issue Queue

from the dispatch stage, operand specifiers are compared and ready flags are set, and instruction selection occurs. In parallel, the CAM/RAM read bitlines are precharged. During the second clock phase, the instruction information of issued instructions is read from the CAM/RAM and the register file source bus is driven. In parallel, the CAM matchlines and ready, select, and write logic are precharged in preparation for phase one.

In order to understand the power dissipation tradeoffs in latch versus CAM-/RAM-based issue queue design, we performed a detailed comparative power analysis using the IBM AS/X circuit simulation tool [17] and next-generation process parameters. Figure 3.3 presents the relative power dissipation of the latch-based design with compaction, the latch-based design without compaction, and the CAM/RAM-based design. Our first observation is that eliminating

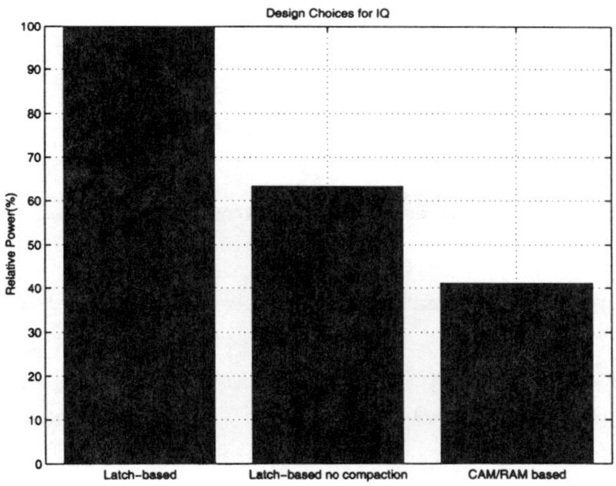

Figure 3.3. Relative Power Dissipation of Issue Queue Alternatives

compaction has a significant impact on power dissipation, a reduction of over 35% according to our analysis. Second, the CAM/RAM-based design is significantly more power efficient than the latch-based design without compaction, dissipating about a third less power. Our conclusion is that the added complexity of implementing a CAM/RAM-based issue queue may be justified, especially for power-constrained designs. For this reason, we selected the CAM/RAM design as the baseline for our second power-saving technique: dynamic adaptation of the issue queue size.

3. Dynamic Adaptation of the Issue Queue

Conventional issue queues have fixed-size resources in an attempt to achieve good overall performance over a range of applications. However, an individual application whose requirements are not well-matched to this particular hardware organization may underutilize issue queue resources. Even a single application may exhibit enough variability that causes uneven use of the issue queue resources during different execution phases. Thus, a fixed issue queue wastes power unnecessarily in the entries that are not in use. For example, Figure 3.4 shows utilization data for one of the queue resources within a high performance

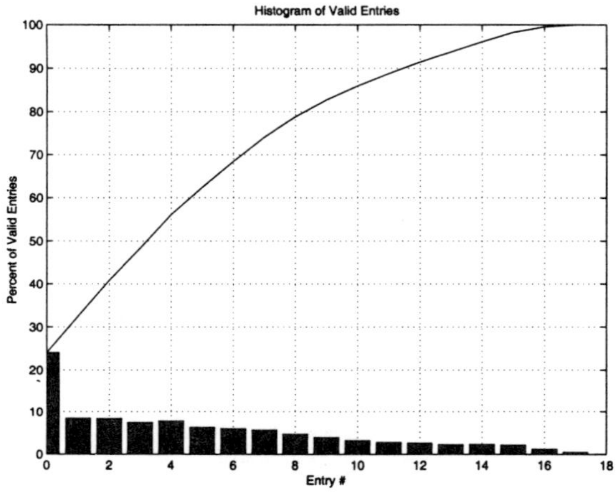

Figure 3.4. Histogram of Valid Entries for an Integer Queue Averaged over SPECint95

processor core when simulating the SPEC95 integer benchmarks. From this figure, we see that the upper 9 entries contribute 80% of the valid entry count for the SPEC95 integer suite. One option to save power is to clock-gate each issue queue entry on a cycle by cycle basis. However, in a CAM/RAM-based design, clock gating does not address some of the largest components of the issue queue power such as the CAM taglines, the CAM/RAM precharge logic, and CAM/RAM bitlines.

Adaptive design ideas (*e.g.*, [1, 2, 3, 4]) exploit workload variability to dynamically adapt the machine resources to match the program characteristics. For the issue queue, shutting down the queue in *chunks* based on application usage addresses those components that are not affected by clock gating in a CAM/RAM design, and thus has the potential to produce significant additional power savings. This idea of dynamic issue queue adaptation forms the basis of the design described in the remainder of this chapter.

3.1 Partitioned CAM/RAM Array Design and Evaluation

The partitioning of the CAM/RAM structure into chunks that can be individually disabled is illustrated in Figure 3.5. The effective sizes of the individual arrays can be changed at run-time by adjusting the enable inputs that control the transmission gates. For our circuit-level implementation and simulation study, a 32-entry issue queue is assumed which is partitioned into four 8-entry chunks. Thus, the queue can operate with 8, 16, 24, or 32 entries enabled. Note that the chunk at the bottom of Figure 3.5 is always enabled, and enabled chunks must be adjacent.

Note also that particular attention must be paid to the taglines to avoid a cycle time impact, and therefore we take a different approach than with the bitlines. As shown in Figure 3.5, a global tag-line is traversed through the CAM array and its local tag-lines are enabled/disabled depending on the control inputs. The sense amplifiers and precharge logic are located at the bottom of both arrays. Another feature of the design is that these CAM and RAM structures are implemented as self-timed blocks. The timing of the structure is performed via an extra dummy bitline, which has the same layout as the real bitlines, within the datapath of the CAM/RAM structures. A logic zero is stored in every dummy cell. Reading the selected cell creates a logical one to zero transition on the dummy bitline that controls the set input of the sense amplifier. (Note that the dummy bitline is precharged each cycle as with the other bitlines.) This work assumes a latching sense amplifier that is able to operate with inputs near Vdd as shown in Figure 3.5. When the set input is high, a small voltage difference from the memory cell passes through the NMOS pass gates of the sense amplifier. When the set signal goes low, the cross-coupled devices amplify this difference to a full rail signal as the pass gates turn off to avoid bitline loading from the cross-coupled structure.

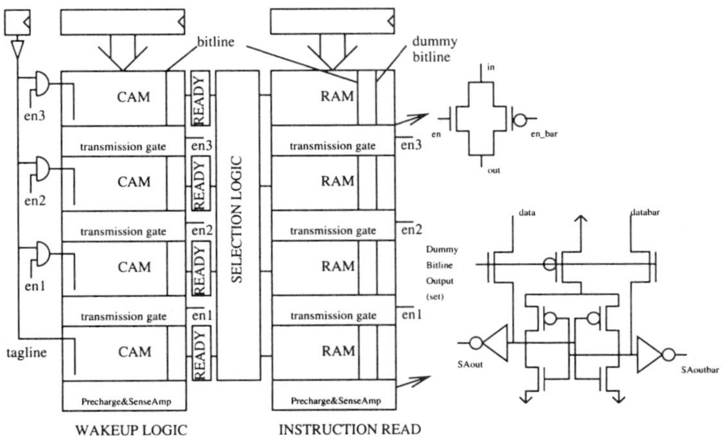

Figure 3.5. Adaptive CAM/RAM Array Partitioning

Figure 3.6 shows data from CAM read AS/X simulations. Here, the issue queue size is successively downsized from 32 entries to 8 entries and the data is correspondingly read. (The third, fourth and fifth signals from the top of Figure 3.6 correspond to the en3, en2 and en1 signals, respectively, in Figure 3.5.) The sixth signal waveform from the top of the figure shows the variation in latencies. When the issue queue size is 8, a faster access time is achieved because of the 24 disabled entries. This occurs because the dummy bitline enables the sense amplifiers at the exact time the data becomes available. AS/X simulations demonstrate a 56% decrease in the cycle time of the CAM array read with only 8 entries enabled. However, in this chapter we do not explore options for exploiting the variable cycle time nature of the design, but focus only on its power-saving features.

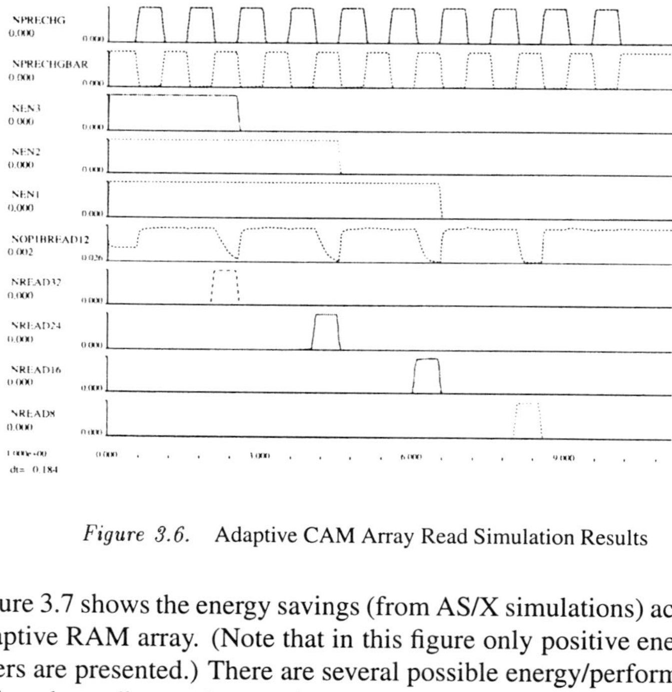

Figure 3.6. Adaptive CAM Array Read Simulation Results

Figure 3.7 shows the energy savings (from AS/X simulations) achieved with an adaptive RAM array. (Note that in this figure only positive energy savings numbers are presented.) There are several possible energy/performance tradeoff points depending on the transistor width of the transmission gates. A larger transistor width results in less cycle time impact, although more energy is dissipated. By reducing the transistor width to 0.39um, one can obtain an energy savings of up to 44%. These numbers are inferred from the energy dissipation corresponding to one read operation of a 32-entry conventional RAM array and that of various alternatives of the adaptive RAM array. An interesting feature of the adaptive design is that it achieves energy savings even with 32 entries enabled. This is because the transmission gates in the adaptive design reduce the signal swing, resulting in less energy dissipation. The adaptive RAM array delay is illustrated in Figure 3.8 for various numbers of enabled entries and transmission gate transistor widths. It is important to note that the delay of the additional circuitry did not affect the overall target frequency of the processor across all widths. This was true also for the CAM structure, and thus our goal of maintaining high frequency performance is achieved.

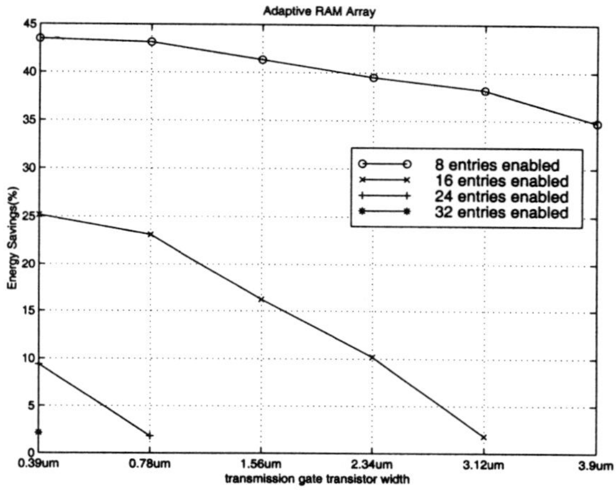

Figure 3.7. Adaptive RAM Array Energy Savings

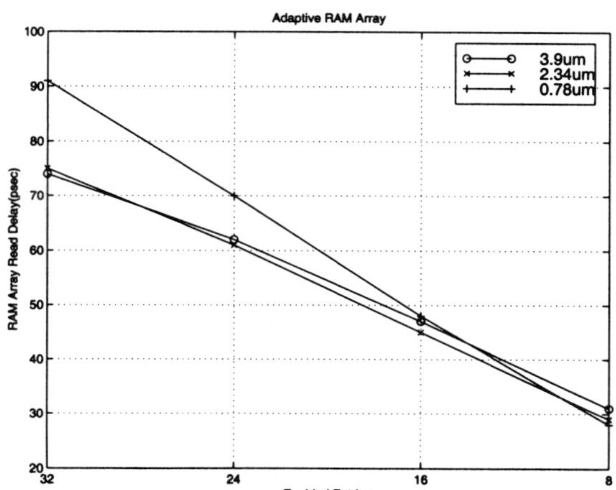

Figure 3.8. Adaptive RAM Array Delay Values

The adaptive CAM array energy and delay values are presented in Figures 3.9 and 3.10, respectively, for various numbers of enabled entries and transmission gate transistor widths. These values account for the additional circuitry that generates the final request signal for each entry (input to the arbiter logic). With this structure, a 75% savings in energy dissipation is achieved by downsizing from 32 entries to 8 entries. Furthermore, as previously demonstrated, the cycle time of the CAM array read is reduced by 56%. It should be noted that a 32 entry conventional CAM structure consumes roughly the same amount of energy as the adaptive CAM array with 32 entries. Because the CAM array dissipates ten times more energy than the RAM array (using a 2.34um transmission gate transistor width), a 75% energy savings in the CAM array corresponds to a 70% overall issue queue energy savings. (The energy savings takes into account the overhead of the shutdown logic described in Section 3.3.)

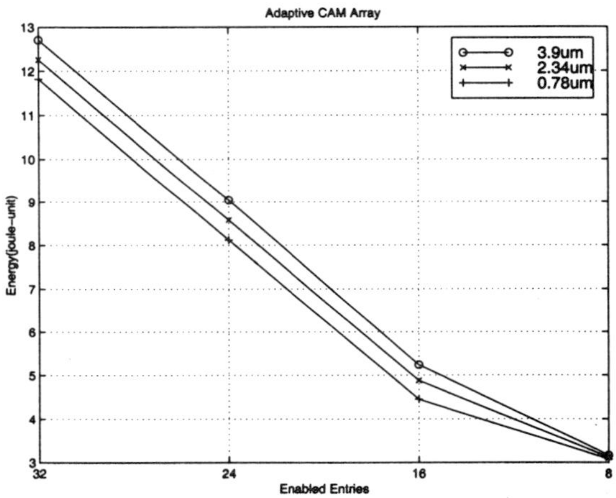

Figure 3.9. Adaptive CAM Array Energy Values

3.2 Dynamic Adaptation Algorithms

Thus far, we have explored the potential power savings via dynamic adaptation of the issue queue size. In other words, we have designed a specific, circuit-level solution that allows the possibility of such adaptation, and we have quantified the energy savings potential when the queue is downsized. In our simulations, we have always factored in the overhead of the extra transistors which result from the run-time resizing hardware. In this section, we explore some of the alternate algorithms (or *decision logic*) one may use in determining what chunks of the queue should be disabled at different points of execution.

The issue unit (in conjunction with the upstream fetch/decode stages) can be thought of as a *producer*. It feeds the subsequent execution unit(s) which act as consumer(s). Assuming, for the moment, a fixed (uninterrupted) fetch/decode bandwidth, the issue queue will tend to fill up when the issue logic is un-

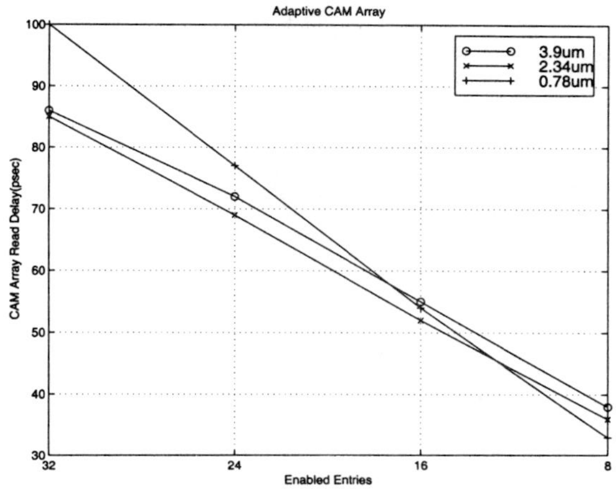

Figure 3.10. Adaptive CAM Array Delay Values

able to sustain a matching issue bandwidth. This could happen because: i) the program dependency characteristics are such that the average number of *ready* instructions detected each cycle is less than the fetch bandwidth seen by the receiving end of the issue queue; or, ii) the execution pipe backend (*the consumer*) experiences frequent stall conditions (unrelated to register data dependencies), causing issue slot *holes*. This latter condition (ii) could happen due to exception conditions (*e.g.*, data normalization factors in floating point execution pipes, or address conflicts of various flavors in load/store processing, *etc.*). On the other hand, the *issue-active* part of the queue will tend to be small (around a value equal to the fetch bandwidth or less) if the consuming issue-execute process is faster than or equal to the producing process. Obviously, this would happen during stretches of execution when the execution pipe stalls are minimal and the issue bandwidth is maximal, as plenty of *ready* instructions are available for issue each cycle. However, one may need a large issue queue window just to ensure that enough *ready* instructions are available to maximize the issue bandwidth. On the other hand, if the stretch of execution involves a long sequence of relatively independent operations, one may not need a large issue queue. So, it should be clear, that even for this trivial case, where we assume an uninterrupted flow of valid instructions into the issue queue, the decision to resize the queue (and in the right direction) can be complicated. This is true even if the consideration is limited only to IPC performance, *i.e.*, if the objective is to always have *just enough* issue queue size to meet the execution needs and dependency characteristics of the variable workload. If the emphasis is more on power reduction, then one can perhaps get by with a naive *utilization-based* algorithm for size adaptation, provided that the average IPC loss across workloads of interest can be kept within acceptable limits.

The pseudocode for one such utilization-based algorithm is listed below:

utilization-based algorithm {
  if ($present\_IPC < factor * last\_IPC$)
    increase_size;
  else if ($counter <= threshold1 \times cw$)
    issue_queue_size=8;
  else if ($(counter > threshold1 \times cw) \&\& (counter <= threshold2 \times cw)$)
    issue_queue_size=16;
  else if ($(counter > threshold2 \times cw) \&\& (counter <= threshold3 \times cw)$)
    issue_queue_size=24;
  else
    issue_queue_size=32;
}

The *counter* is a measure of the number of valid entries in the queue during a fixed period of time (in cycles) called the *cycle window*, and denoted *cw* in the algorithm. At the end of the cycle window, the counter is read and there are four possible actions. The issue queue size is increased to the next highest size if the present IPC is a *factor* lower than the last IPC during the last *cycle window*. This guarding mechanism attempts to limit the performance loss of adaptation. Otherwise, depending on the comparison of counter values with certain threshold values, the decision logic may do the following: i) increase issue queue size by enabling higher order entries, ii) retain the current size, or iii) decrease the size by disabling the highest order entries. Note that a simple NOR of all the active instructions in a chunk ensures that all entries are issued before the chunk is disabled.

An alternative approach to the decision logic, related to that proposed in [10], is to devise a *parallelism-based* metric for issue queue adaptation. Depending on the level of parallelism (distant or close), the issue queue is upsized or downsized. The level of parallelism is estimated by observing where in the Reorder Buffer (ROB) instructions are issued from. Namely, if instructions are frequently issued from deep in the ROB, then the application has distant parallelism (for which a large issue queue may be needed), while the opposite is true if most instructions issued are the oldest (for which a small issue queue may suffice). A counter is associated with each of four sections of a 64-entry portion of the ROB. A particular counter is incremented for every instruction that issues from that ROB section. At the end of the cycle window, the counter values are read, added to form a total count, and acted upon according to the following algorithm:

parallelism-based algorithm {
  if ($present\_IPC < factor * last\_IPC$)
    increase_size;
  else if ($counter\_0\_15 > threshold1 \times total\_count$)
    issue_queue_size=8;
  else if ($counter\_16\_31 > threshold2 \times total\_count$)
    issue_queue_size=16;
  else if ($counter\_32\_47 > threshold3 \times total\_count$)
    issue_queue_size=24;
  else if ($counter\_48\_63 > threshold4 \times total\_count$)
    issue_queue_size=32;
  else retain_current_size;
}

As with the utilization-based algorithm, the parallelism-based approach increases the issue queue size if the IPC of the current interval has degraded by *factor* relative to the IPC of the last interval. If not, the issue queue size is set based on the fraction of instructions that have issued from each section of the 64-entry portion of the ROB. Note that the algorithm favors small-size issue queue configurations over larger ones, in that the *counter* values for the oldest section of the ROB are examined first, and if the condition is met, the size is set irregardless of the *counter* values of other sections. An alternative is to take the opposite approach: walk from the "distant parallelism" section of the ROB towards the "nearby parallelism" section.

Both decision logic schemes require event counts to be gathered within a cycle window. We call the combination of the decision logic and these hardware counters the *shutdown logic*. A primary goal in designing the shutdown logic is to keep the transistor count and power dissipation overheads to within tolerable limits. To evaluate whether this goal can be achieved, we perform a circuit-level design and characterization of the shutdown logic for the utilization-based approach, and estimate the overhead for the parallelism-based scheme as well.

3.3 Shutdown Logic

Figure 3.11 illustrates the high-level operation of the shutdown logic. It consists of bias logic at the first stage followed by the statistics process&storage stage. The activity information is first filtered by the bias logic and then it is fed to the process&storage stage where the information is fed to counters. At the end of the *cycle window*, this data passes through the decision logic to generate the corresponding control inputs.

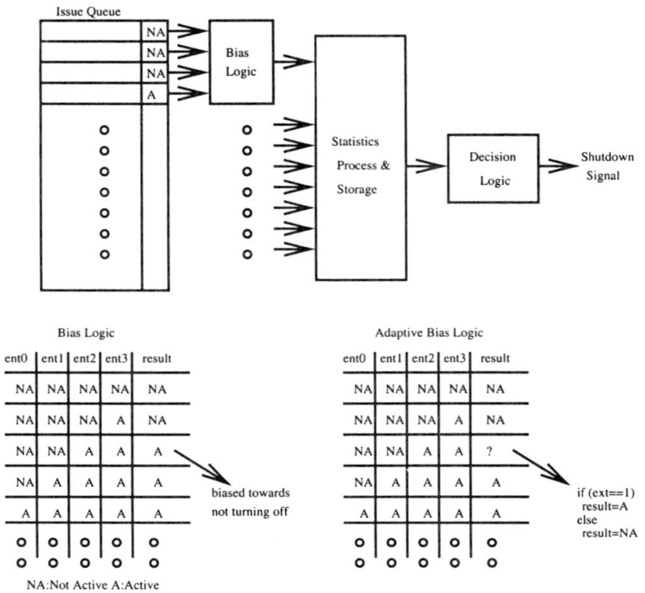

Figure 3.11. High-Level Structure of Shutdown Logic and Logic Table for Bias Logic

The 32-entry issue queue is partitioned into 8-entry chunks that are separately monitored for activity. The bias logic block monitors the activity of the issue queue in 4-entry chunks. This scheme is employed to decrease the fan-in of the bias logic. The bias logic simply gathers the activity information over four entries and averages them over each cycle. The activity state of each instruction may be inferred from the *ready flag* of that particular queue entry. One particular state of interest is when exactly half of the entries in the monitored chunk are active. One alternative (the Bias Logic table in Figure 3.11) is to statically choose either active or not active in this particular case. Another approach (the Adaptive Bias Logic in Figure 3.11) is to dynamically change this choice by making use of an extra logic signal variable. We chose the former approach due to its simplicity.

The statistics process&storage stage, which is shown in Figure 3.12, is comprised of two different parts. The detection logic provides the value that will be added to the final counter. It gathers the number of active chunks from the bias logic outputs and then generates a certain value (*e.g.*, if there are two active 4-entry chunks, the detection logic will generate binary two to add to the final counter). The second part, which is the most power hungry, is the flip-flop and adder pair (forming the counter). Each cycle, this counter is incremented by the number of active clusters (4 entry chunks). In this figure one can also see the function of the detection logic. The zeros in the inputs correspond to the

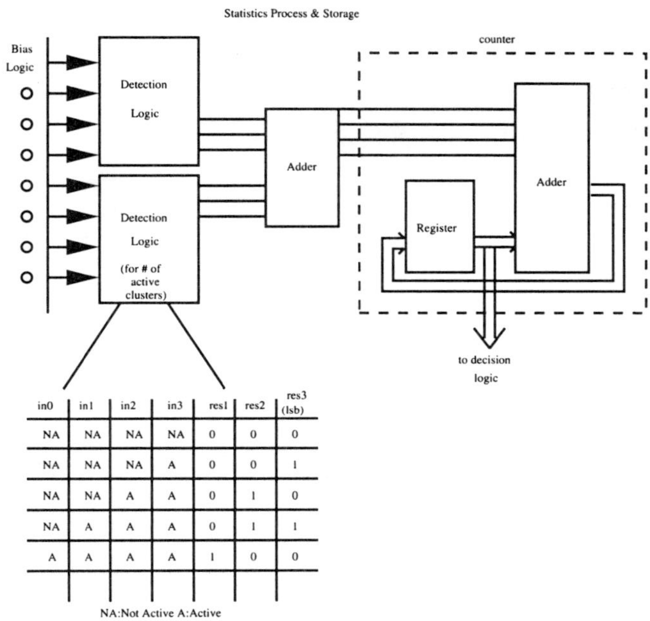

Figure 3.12. Shutdown Logic Statistics Process and Storage Stage

Issue Queue Number of Entries	Transistor Counts Issue Queue	Transistor Counts Shutdown Logic	Complexity of Shutdown Logic
16	28820	802	2.8%
32	57108	1054	1.8%
64	113716	1736	1.5%
128	227092	2530	1.1%

Table 3.2. Complexity of Shutdown Logic in Terms of Transistor Count

non-active clusters and the ones to active clusters. The result section shows which value in binary should be added. For 32 entries, two of these detection circuits and a small three-bit adder are required to produce the counter input. One of the detection logic units covers the upper 16 entries and the other one covers the bottom 16 entries.

Table 3.2 shows the complexity of the shutdown logic in terms of transistor count. From this table it is clear that the extra logic adds only a small amount of complexity to the overall issue queue. AS/X simulations show that this

Branch predictor	comb. of bimodal and 2-level Gag
Fetch and Decode Width	16 instructions
Issue Width	8
Integer ALU/Multiplier	4/4
Floating Point ALU/Multiplier	2/2
Memory Ports	4
L1 Icache, Dcache	64KB 2-way
L2 unified cache	2MB 4-way

Table 3.3. Simplescalar Simulator Parameters

extra circuitry dissipates on average 3% of the power dissipated by the whole CAM/RAM structure.

We have also estimated the transistor count overhead for the parallelism-based algorithm. This approach uses four counters, each of which counts the number of active entries in a 16-entry section of the ROB, and adds them together to form a total count. The transistor count overhead is almost 8%, or four times that of the utilization-based algorithm.

These overheads can be tolerated assuming that significant power savings can be realized through dynamic adaptation without significant performance degradation. In the next section, we present a microarchitectural analysis to evaluate this premise.

3.4 Microarchitecture Simulation-Based Results

We used SimpleScalar-3.0 [5] to simulate an aggressive 8-way superscalar out-of-order processor. The simulator has been modified to model separate integer and floating point queues. We chose a workload of six of the SPEC2000 integer benchmarks (each of which is run for 400 million instructions), and characterize the performance and power dissipation of conventional and adaptive 32-entry integer queues. The simulation parameters are summarized in Table 3.3. We first analyze the utilization-based algorithm with different *factor* and *cycle window* values to find the values that work best over all of the benchmarks. We also experimented with different *threshold* values, eventually settling on values of 7, 15, and 23 for $threshold1$, $threshold2$, and $threshold3$, respectively. Figures 3.13 and 3.14 show the power savings and performance degradation for each benchmark as well as the overall average with different *factor* and *cycle window* values. To estimate the power savings, we assumed a power variation profile which is essentially linear in the number of entries, based on the circuit-level simulation data reported earlier. We also take into account the shutdown logic overhead. The performance degradation and power savings are both relative to a fixed 32-entry integer issue queue.

The different *factor* and *cycle window* values present different power-performance tradeoff points. As expected, a decrease in *factor* increases the power savings but also the performance degradation. The effect of varying the *cycle window* provides less consistent results. However, note the negative power savings results with *mcf* using the larger *cycle windows* of 8K and 16K. This occurs because at this coarse level of dynamic adaptation, the 32-entry configuration is always selected. The use of the smaller *cycle windows* allows the dynamic adaptation algorithm to capture the finer-grain phase-change behavior of *mcf*, resulting in smaller configurations being selected. Note also that over all of these benchmarks, the use of smaller *cycle windows* results in a higher power savings and a lower performance degradation than when larger *cycle windows* are used.

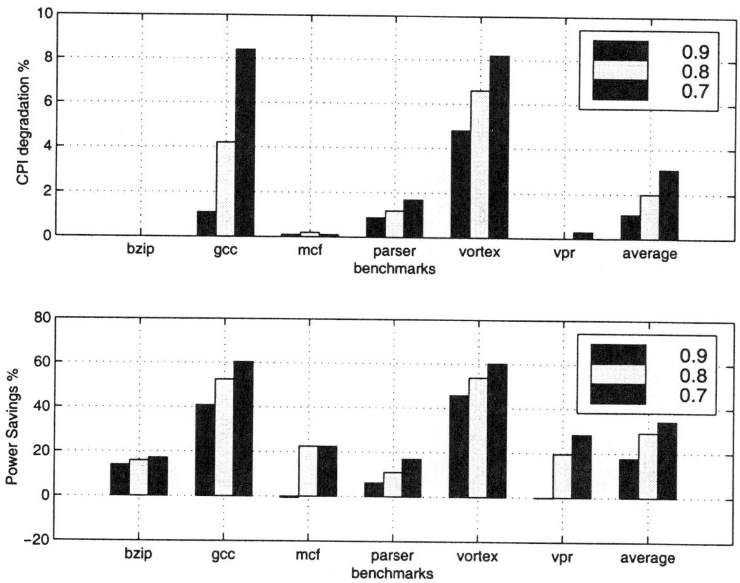

Figure 3.13. CPI Degradation and Power Savings are Plotted over All the Benchmarks. *Cycle Window*=4K. *Factor* Varies as 0.9, 0.8, 0.7

Figure 3.15 shows the percentage of time each queue size is used with the dynamic algorithm. These results demonstrate a broad range of workload variability. For *mcf*, the 24-entry issue queue configuration is used throughout almost its entire execution whereas, for *vortex* and *gcc*, only 8 and 16 entries are largely used. The remaining benchmarks as well display characteristics that differ from these three benchmarks.

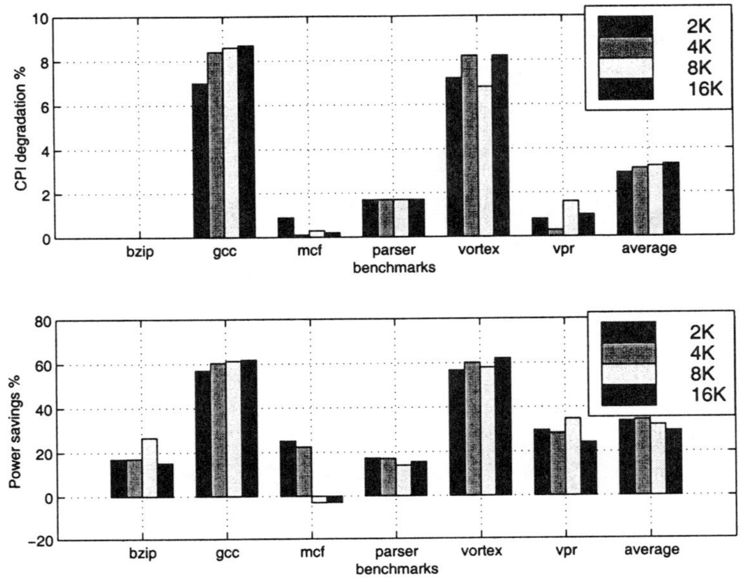

Figure 3.14. CPI Degradation and Power Savings are Plotted over All the Benchmarks. *Factor*=0.7. *Cycle Window* Varies as 2K, 4K, 8K, 16K

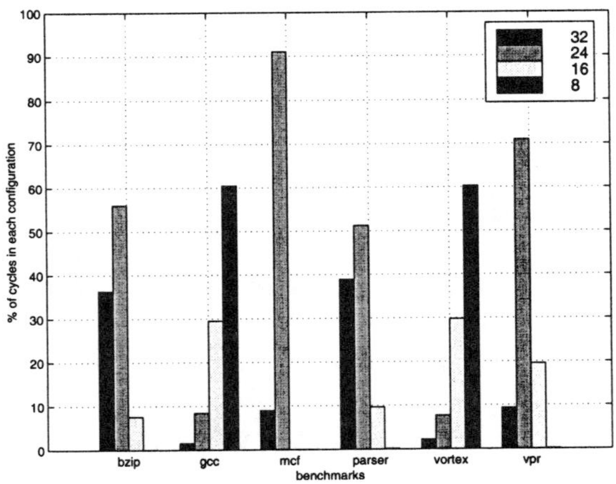

Figure 3.15. Percentage of Utilization for Each Queue Size with the Dynamic Adaptation (*Factor*=0.7, *Cycle Window*=4K)

Figure 3.16 illustrates the variation in queue usage during execution of an Unmanned Airborne Vehicle (UAV) application for a conventional queue, and the queue size chosen by the utilization-based adaptive algorithm. This graph demonstrates how the algorithm is able to identify phase changes and adapts the issue queue size accordingly during the execution of this application.

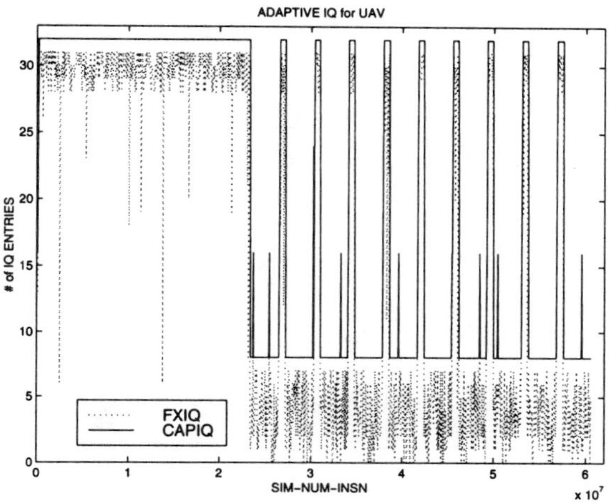

Figure 3.16. Conventional Issue Queue Utilization and Adaptive Queue Size Decisions for the UAV Application

We also experimented with a number of threshold values corresponding to the parallelism-based algorithm, ultimately finding that values of 0.9, 0.3, 0.1, and 0.05 for the oldest to youngest ROB instruction sections worked best over all of the benchmarks. Figure 3.17 shows a comparison of the utilization and parallelism-based algorithms in terms of power savings and performance degradation for each benchmark as well as the overall averages. We make several observations from these results. First, the results vary significantly among benchmarks, with some (like *gcc*) exhibiting greater power savings and performance degradation with the utilization-based algorithm, while others (like *mcf*) experiencing the opposite effect. We expect that fine-tuning the *factor*, *cycle window*, and *threshold* values to each benchmark, potentially via profiling and feedback optimization, would yield significantly more consistent results. Second, the parallelism-based algorithm exhibits slightly better power savings with half of the performance degradation of the utilization-based approach. However, recall that the parallelism-based approach has a four-fold greater transistor count overhead than the utilization-based approach.

Overall, the utilization-based approach achieves a 35% savings in issue queue power with only a 3% average performance degradation as compared with a standard CAM/RAM design. Compared with the latch-based design with compaction, the CAM/RAM design with dynamic adaptation achieves over a 70% reduction in issue queue power dissipation with only a 3% average performance degradation. Such a dramatic power reduction has the potential

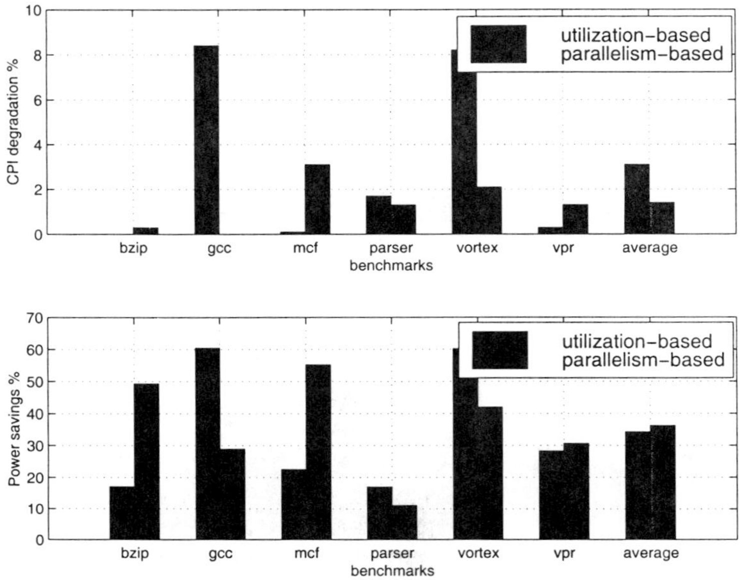

Figure 3.17. Comparison of the Utilization and Parallelism-Based Algorithms (*Factor*=0.7, *Cycle Window*=4K)

to greatly reduce overall chip power in designs where the issue queue power dominates the overall power consumption as well as reduce thermal hot spot problems, leading to significant power-related cost savings.

4. Conclusions

We present two techniques for increasing the power efficiency of the issue queues in out-of-order microprocessors. The first is the replacement of the latch-based issue queue used in many processors with a CAM/RAM-based structure. We find that a CAM/RAM-based design affords a significant power savings over the latch-based design, which may justify its added design and verification complexity in power-sensitive environments.

The second approach is on-the-fly dynamic adaptation of the issue queue size to match application characteristics. Through a detailed circuit-level analysis, we determine that for a 32-entry adaptive issue queue that can be divided into four equal-size chunks, we conclude that the delay associated with the extra circuitry required to partition the queue is small enough so as not to unduly impact processor cycle time. Furthermore, we find that the transistor count and power dissipation overheads of the adaptive control algorithm logic was small compared to the significant power savings that were realized. When combined, these techniques achieve over a 70% reduction in issue queue power dissipation with only a 3% average performance degradation.

Future work includes exploring alternate hardware algorithms for queue-size adaptation, pursuing improvements at the circuit level that provide better configuration flexibility, investigating methods for exploiting the self-timed issue queue capability, and exploring feedback-based optimizations that tailor the adaptive control algorithms to the application.

Acknowledgements

We wish to thank John Wellman, Prabhakar Kudva, Victor Zyuban and Hans Jacobson for many interesting discussions and helpful hints. We also wish to thank Rajeev Balasubramonian for his help in building up the microarchitectural simulation infrastructure.

References

[1] D. H. Albonesi. Dynamic IPC/Clock Rate Optimization. Proc. ISCA-25, pp. 282-292, June/July 1998.

[2] D. H. Albonesi. The Inherent Energy Efficiency of Complexity-Adaptive Processors. Proc. ISCA Workshop on Power-Driven Microarchitecture, June 1998.

[3] R. Balasubramonian, D.H. Albonesi, A. Buyuktosunoglu, and S. Dwarkadas. Memory Hierarchy Reconfiguration for Energy and Performance in General-Purpose Processor Architectures. 33rd International Symposium on Microarchitecture, pp. 245-257, December 2000.

[4] R. Balasubramonian, D.H. Albonesi, A. Buyuktosunoglu, and S. Dwarkadas. Dynamic Memory Hierarchy Performance Optimization. Proc. ISCA Workshop on Solving the Memory Wall Problem, June 2000.

[5] D. Burger and T. Austin. The Simplescalar toolset, version 2.0. Technical Report TR-97-1342, University of Wisconsin-Madison, June 1997.

[6] M. Butler and Y.N Patt. An investigation of the performance of various dynamic scheduling techniques. Proc. ISCA-92, pp. 1-9, May 1992.

[7] G. Cai. Architectural level power/performance optimization and dynamic power estimation. Proc. of the Cool Chips Tutorial, in conjunction with Micro-32, 1999.

[8] R. Canal and A. Gonzalez. A low-complexity issue logic. Proc. ACM Int'l. Conference on Supercomputing (ICS), pp. 327-335, June 2000.

[9] D. Folegnani and A. Gonzalez. Reducing the power consumption of the issue logic. Proc. ISCA Workshop on Complexity-Effective Design, June 2000.

[10] D. Folegnani and A. Gonzalez. Energy-Effective Issue Logic. Proc. ISCA-01, pp. 230-239, June 2001.

[11] M. K. Gowan, L. L. Biro, D. B. Jackson. Power considerations in the design of the Alpha 21264 microprocessor. Design Automation Conference, June 1998.

[12] L. Gwennap. Power issues may limit future CPUs. Microprocessor Report, 10(10), August 1996.

[13] R. Kessler. The Alpha 21264 microprocessor. IEEE Micro, 19(2): 24-36, March/April 1999.

[14] V. Tiwari, D. Singh, S. Rajgopal, G. Mehta, R. Patel, F. Baez. Reducing power in high-performance microprocessors. Design Automation Conference, June 1998.

[15] K. Wilcox and S. Manne. Alpha Processors: A history of power issues and a look to the future. Proc. of the Cool Chips Tutorial, in conjunction with Micro-32, 1999.

[16] K.Yeager. The Mips R10000 superscalar microprocessor. IEEE Micro, 16(2): 28-41, April 1996.

[17] AS/X User's Guide. IBM Corporation, 1996.

[18] The National Technology Roadmap for Semiconductors, Semiconductor Industry Association, 1999.

Chapter 4

MICRO-ARCHITECTURE DESIGN AND CONTROL SPECULATION FOR ENERGY REDUCTION

Dirk Grunwald
Dept. of Computer Science
University of Colorado
Boulder, CO, 80309-0430
grunwald@cs.colorado.edu

Abstract Conventional wisdom states that the best way to design an energy-efficient microprocessor is to design it for high performance, since a high performance processor will complete a task quicker than an energy-conscious design. However, our research group has found ways to reduce energy without impacting performance by controlling the amount of speculation used by the processor in its quest for performance.

1. Introduction

As the issues and concerns facing power consumption grow to include high performance microprocessors as well as the pervasive hand-held and wireless computing market, novel approaches to power conservation at the architectural level are being sought. In the same vein, ubiquitous computing is becoming more prevalent, and thus the performance demands made upon these devices is increasing, leading toward more complex microarchitectures.

Although much work on energy reduction has taken place in the circuit and device technology domains [3], there has been an increasing emphasis on *designing* for power efficiency at the architectural level. This implies that the desired energy and power goals must be targeted early in the design cycle and that the system microarchitecture must work in concert with advances in circuit technology to reduce power demands. In this chapter, we describe some architectural techniques appropriate for reducing power and energy in processor designs; we will focus on designs appropriate for reducing *energy*. There is less challenge in reducing power, since power doesn't incorporate any notion of completing a given task – for example, we could minimize the power used by a processor simply by turning it off or running it very slowly. It is more difficult to reduce the *energy* needed for a computation since that reduction assumes the computation must still be completed (and, typically, completed in a timely fashion).

2. Background and Terminology

In current process technologies active power, or the power used by actively charging and discharging capacitance, dominates the total power consumed by processors. As process feature sizes decrease, static power will become a significant problem, but one that we will not discuss.

Since the active power for a CMOS circuit varies $P \propto CFV^2$, it's clear that the power can be reduced by: reducing the capacitance (C); reducing the processor frequency (F); or, reducing the voltage (V). However, we can typically only adjust capacitance and a combination of frequency and voltage to reduce *energy*. Improving program performance by reducing the number of instructions improves energy use because of a decrease in cycles, and thus optimizing algorithms and compilers are the first step to producing a truly energy efficient system [10]. Other works investigating power (and energy) reduction at the architectural level have concentrated mainly upon cache- or buffer-type structures which can be turned off during periods of inactivity [7, 24, 18]. In certain applications that must compute results at a fixed rate, such as media applications, the total execution time is driven by some exogenous factor, such as the duration of a sound or video image. In these systems, reducing the processor frequency can produce a linear reduction in energy. The power and total energy can be further reduced if some form of *voltage scaling* is available.

Speculation is a fundamental tool in computer architecture. It allows an architectural implementation to achieve higher instruction level parallelism, and thus performance, by predicting the outcome of specific events. Most processors currently implement branch prediction to permit speculative changes in the control-flow; other recent work has focused on predicting data values to reduce data dependencies [19]. When used poorly, speculation is a form of wasted work in the processor and can lead to considerable power or energy overhead We'll examine two forms of *speculation control* that attempt to trade off how aggressively speculation should be applied. The first technique uses *confidence estimation* to implement *Pipeline Gating*. Confidence estimation [14, 11] is a technique for assessing the quality of a particular prediction. Pipeline gating [20] is a form of speculation control that uses confidence estimation to reduce speculation in pipelined processors. Pipeline gating is a form of *activity reduction* – by reducing the number of instructions fetched by the processor, the total power can be reduced.

The second form of speculation control relies on *voltage scaling*. Dynamic voltage scaling reduces the power over time by varying the voltage and frequency of the entire processor; this process typically takes 10's of microseconds and is usually controlled by the operating system or BIOS layer. Another alternative is *clustered voltage scaling* [15, 25], where different parts of the processor run at different voltages. This is usually done during design time, and the results can be dramatic – Igarashi *et al* [13] applied this technique to a media processor and saw a 47% power reduction with 10% increase in area using two voltage levels. In that technique, critical path analysis is performed after the circuit is designed [25]. Circuits that are not on the critical path are said to have "scheduling slack" and are candidates for a lower voltage. The lower voltage drives the logic at a slower rate, but the voltage level is chosen to insure that the results are still computed no later than the circuits on the critical path. We will show how the "scheduling slack" can be applied to *instruction scheduling* in a processor as well as the underlying circuits. To controls speculation by slowing down instructions, executing them no earlier than necessary.

3. Pipeline Gating

We have applied confidence estimation to three areas of computer microarchitecture [20, 17, 21]; in this work, we'll focus on using confidence estimation for *pipeline gating*, which reduces the average power demands of an existing microarchitecture without reducing performance by avoiding unnecessary speculation. Before describing the how confidence estimators can be used to implement speculation control, we'll cover the basic properties of confidence estimators.

3.1 Confidence Estimators

It is more difficult to compare two confidence estimators than two branch predictors in part because confidence estimators can be used for a number of purposes while branch predictors are typically only used to predict the outcome of control-dependent instructions. Most architectures are designed to use speculation and the general assumption is that "you might as well be doing something", and thus each branch is predicted. By comparison, confidence estimators will normally be used for *speculation control*. For example, if a particular branch in a multithreading processor is of low confidence, it may be more cost effective to switch threads than speculatively evaluate the branch. A confidence predictor attempts to corroborate or assess the prediction made by a branch predictor. Each branch is eventually determined to have been predicted correctly or incorrectly. For each prediction, the confidence estimator assigns a "high confidence" or "low confidence" to the prediction. We have found that the following notation simplifies the comparison of different confidence estimators. We can draw a 2×2 matrix listing the frequency for each outcome of a test. When we apply this framework to architectural simulation, each of the quadrants can be directly measured during simulation or analysis. Typically, we normalize the values to insure that the sum equals one. Thus, our quadrant table for confidence estimation is:

		Prediction Outcome	
		C	I
Confidence	HC	C_{HC}	I_{HC}
	LC	C_{LC}	I_{LC}

In this table, "C" and "I" refer to "correct" and "incorrect" predictions, respectively, and "HC" refers to "high confidence" and "LC" to "low confidence". During a simulation, we can measure C_{HC}, I_{HC}, C_{LC} and I_{LC} using a branch predictor for each branch and concurrently estimate the confidence in that branch predictor using a specific confidence estimator. When the branch is actually resolved, we classify the branch as belonging to class C_{HC}, I_{HC}, C_{LC} or I_{LC}.

When a confidence estimator is applied, the architectural feature using that confidence estimation will usually be used for "high confidence" *or* "low confidence" branches, but not both. For example, consider a simultaneous multithreading (SMT) processor that uses a confidence estimator to determine if a predicted branch is likely to be mispredicted. If the branch prediction is of "low confidence", the processor may switch to another available thread rather than fetch additional instructions from the current thread. The performance of such a processor is very sensitive to $P[I|LC] = I_{LC}/C_{LC} + I_{LC}$, the probability that the branch is incorrectly predicted if it was low confidence. A high value for $P[I|LC]$ indicates that the processor can switch contexts only when the following instructions will not commit. A low value of $P[I|LC]$ indicates that the SMT processor may needlessly switch threads, reducing the performance of the primary thread. A low value of the $P[LC|I]$ means that the processor will miss some opportunities to improve aggregate performance by switching threads.

Not all uses of confidence estimators will make the same kind of decisions, but we feel it is most useful to compare confidence estimators using metrics that reflect how the confidence estimators are used. For example, our work in pipeline gating relied on confidence estimators with a large $P[I|LC]$ and a large $P[LC|I]$. We have found in our own discussion that terms such as "accuracy" (*e.g.* $P[I|LC]$) and "coverage" (*e.g.* $P[LC|I]$) tend to cause confusion, because accuracy has an inherit implication about the application of a technique; thus, we have introduced a set of metrics that can be used to compare confidence estimation techniques [11]; we've also shown how confidence estimation can use existing branch predictor state and still yield good performance.

3.2 Using Confidence Estimation for Speculation Control

Much low power research has focused on reducing energy in the memory subsystem [8, 16, 10]. In embedded processors, such as the StrongArm [22], the memory subsystem is thought to be the dominant source of area and power because the rest of the processor has been simplified to reduce power. State-of-the-art microprocessors have a high degree of control complexity and a large amount of area dedicated to structures that are essential for high-performance, speculative, out-of-order execution, such as branch prediction units, branch target buffers, TLBs, instruction decoders, integer and floating point queues, register renaming tables, and load-store queues. For example, $\approx 30\%$ of the core die area on the DECchip 21264 is devoted to cache structures, while the StrongARM processor uses $\approx 60\%$ of the core die area for memory. Figure 4.1 shows a distribution of the power dissipated in a PentiumPro processor [12] during a test designed to consume the most power, which is when the processor is committing each instruction that it fetches. The fetch and decode stages, along with components necessary to perform dynamic scheduling and out-of-order execution, account for a significant portion of the power budget. It it clear that pipeline activity is a dominant portion of the overall power dissipation for many microprocessors.

Performance is the primary goal of state-of-the-art microprocessor design. Architectural improvements for performance have centered on increasing the amount of instruction level parallelism through aggressive speculation and out-of-order execution. Although these advances have increased the number of instructions per cycle (IPC), they have come at the cost of wasted work. Most processors use branch prediction for speculative control flow execution. Branch prediction is used to execute beyond the control boundaries in the code. With high branch prediction accuracy, most issued instructions will actually commit. However, many programs have a high branch misprediction rate, and these programs issue many instructions that never commit. Each of those instructions uses many processor resources. If we can decrease the percentage of uncommitted instructions actually issued, we can decrease the power demands of the processor as a whole [10, 4, 23]. This is the goal of pipeline gating.

We will use the schematic of the processor pipeline shown in Figure 4.2 to describe pipeline gating. Like many high-performance processors, such as the DEC AXP-21164 or Intel PentiumPro, our sample pipeline uses two fetch and decode cycles to allow the clock rate to be increased. We assume the fetch stage has a small instruction buffer to allow instruction fetch to run ahead of

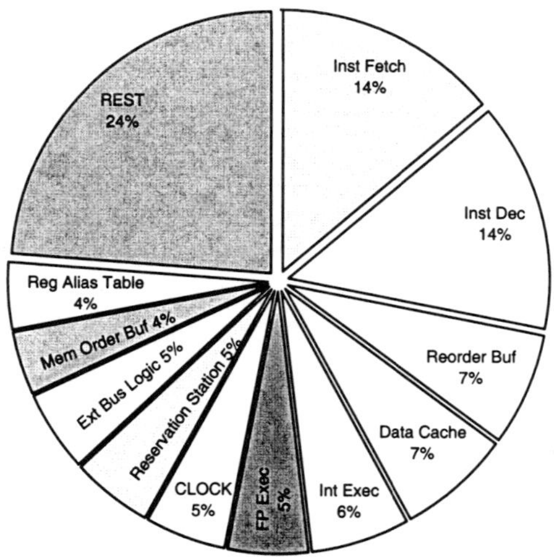

Figure 4.1. Power Consumption for PentiumPro Chip, Broken down by Individual Processor Components

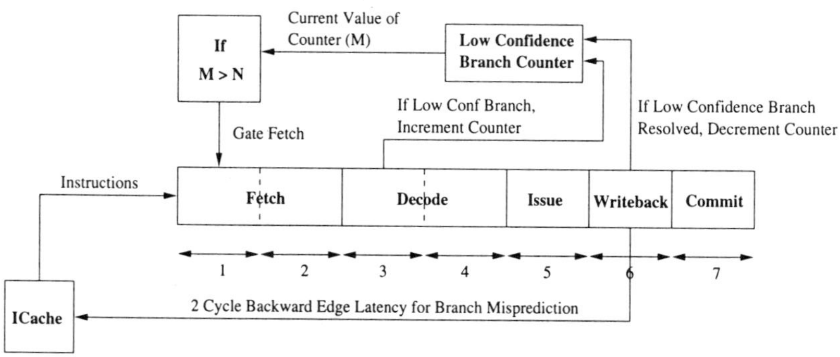

Figure 4.2. Pipeline with a Two Fetch and Decode Cycles Showing Hardware required for pipeline gating. The low-confidence branch counter records the number of unresolved branches that reported as low-confidence. The counter value is compared against a threshold value ("N"). The processor ceases instruction fetch if there are too many unresolved low-confident branches.

decode. Branch prediction occurs when instructions are fetched to reduce the misfetch penalty. The actual instruction type may not be known until the end of decode. Conditional branches are resolved in the execution stage, and branch prediction logic is updated in the commit stage. Since the processor uses out-of-order execution, instructions may sit in the issue window for many cycles, and there may be several unresolved branches in the processor.

We use a *confidence estimator* to assess the quality of each branch prediction. A "high confidence" estimate means we believe the branch predictor is likely to be correct. A "low confidence" estimate means we believe the branch predictor has incorrectly predicted the branch. We use these confidence estimates to decide when the processor is likely to be executing instructions that will not commit; once that decision has been reached, we "gate" the pipeline, stalling specific pipeline stages. In our study [20], we varied a number of system parameters, including the branch predictor, the confidence estimator, the stage at which a gating decision is made, the stage that is actually gated and the number of outstanding low-confident branches needed to engage gating. The decision to gate can occur in the fetch, decode or issue stages. Equally important is the decision about *what* to gate and *how long* to gate. Gating the fetch or decode stages would appear to make the most sense, and we examined both cases. We used the number of unresolved low confidence branches to determine when to gate. For example, if the instruction window includes one low-confident branch, and another low-confident branch exits the fetch (or, alternatively, decode or issue) stage, gating would be engaged until one or the other low-confident branch resolves. Figure 4.2 illustrates this process for a specific configuration. We add a counter that is incremented whenever the decode encounters a low-confident branch and is decremented when a low-confident branch resolves. If the counter exceeds a threshold, the fetch stage is gated. Instructions in the fetch-buffer continue to be decoded and issued, but no new instructions are fetched.

To properly understand the effects of stalling the pipeline, we developed a model of an out-of-order, speculative, wide-issue processor. We have found that gating the processor typically stalls the processor for a very short duration. Figure 4.3 shows the number of times a specific configuration of our pipeline model is stalled while executing different programs. Generally, gating stalls occur for about 2-4 processor cycles. Most processor configurations exhibit a similar distribution, and indicate that our mechanism is exhibiting fine control over the speculative state of the processor.

We explored moving the point of gating to the decode and issue stages rather than simply eliminating fetching. Gating at decode saved less power than gating at the fetch stage. This is reasonable since we are allowing the fetch stage to continue fetching until the fetch buffer is full. Therefore, more instructions will enter the pipeline; this would not happen if gating disabled instruction fetch. We expected to see an improvement in processor performance (CPI) with gating at decode since the recovery penalty for incorrect gating would be less than gating at fetch. It takes only three cycles for an instruction to "catch up" and issue after an incorrect gating event with gating at decode as opposed

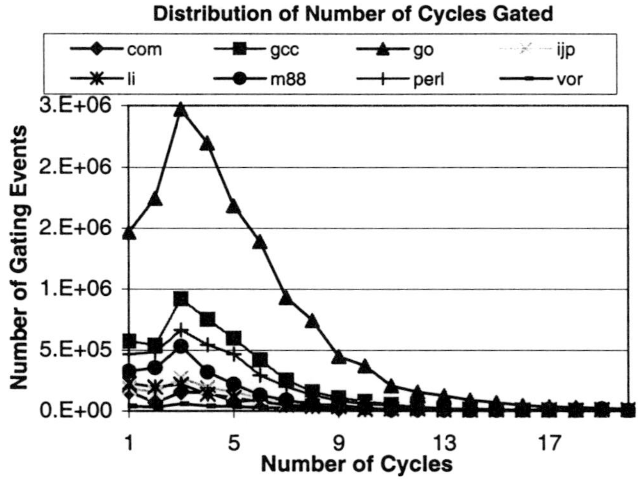

Figure 4.3. Distribution of Gating Events and the Number of Cycles Gated per Event

to five cycles with gating at fetch. However, results showed no real performance benefit from moving the gating point from fetch to decode. As shown in Figure 4.3, the pipeline is generally not gated for more than a few cycles. Our pipeline model used a 64-entry register update unit, and results show that it usually has enough instructions in the issue queue to keep the execution units occupied while the pipeline catches up from gating.

Overall, we found that gating the processor typically stalls the processor for a very short period and that we could vary the design of the confidence estimator to reduce the number of issued instructions by $\approx 20\%$ while reducing performance, measured by instructions per cycle, by typically less than 1%. As with most work on power-efficient microarchitecture, it's difficult to state how much power or energy is saved overall in an actual implementation, since those figures depend on numerous implementation decisions and higher level power analysis tools have poor accuracy [9]. Since the overheads for pipeline gating are so small, it should be effective across a wide range processor designs.

4. Speculation Control by Exploiting Scheduling Slack

We have explored how clustered voltage and frequency scaling techniques can be used to reduce processor energy. We assume the processor has independent clock domains and can, for example, run certain components such as functional units at 600*Mhz* and other functional units at 300*Mhz*. There are a number of complications with such a design, since the different clock domains will require DC-DC power conversion and inter-domain communication may

```
add     r0,   r1,    r2;    (A)
sub     r3,   r4,    r5;    (B)
and     r9,   0x1,   r9;    (C)
ornot   r5,   r9,    r10;   (D)
xor     r2,   r10,   r11;   (E)
```

(a) Code Sequence

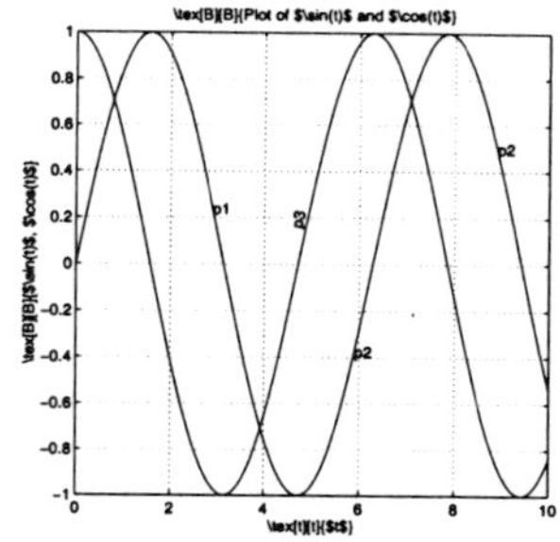

(b) Program Dependence Graph

Figure 4.4. Sample Code Sequence and Program Dependence Graph to Illustrate Scheduling Slack

have some overhead. However, it presents an interesting opportunity for power reduction: *if we are able to dynamically identify instructions that could use the slower components without impacting the overall rate of computation, we may be able to save additional power without impacting performance.* This design technique can be exploited at multiple levels. To date, we have focused on instruction-level scheduling but we are also investigating thread-level scheduling that can be exploited either by multithreaded or chip-multiprocessing designs.

To exploit functional units with different clock speeds, the critical issue is identifying those instructions that can be computed at a slower rate without impacting performance. These instructions must be identified early in the pipeline so that the information can be used at the scheduling/issue stage without impacting performance. Following the terminology of Johnson *et. al* [15], we call this property "scheduling slack". We will formally define scheduling slack shortly, but the simple example shown in Figure 4.4 provides most of the insight.

In Figure 4.4(a), we see a sample dynamic code sequence executing in the microprocessor pipeline with each instruction labeled *A* through *E*. Figure 4.4(b) shows the data dependences for the code sequence. In this example, instructions *A*, *B*, and *C* are in the ready queue of the processor's issue logic, and will issue in the current cycle; instructions *D* and *E* are decoded but are not part of the ready queue since they must wait for instructions *A* through *C* to complete before they can be issued.

This example assumes each instruction has an operational latency of one cycle and that three functional units are available. Figure 4.5(a) shows a possible

1	A	B	C
2			D
3			E

(a) Standard Schedule

1	A_1	B	C
2	A_2		D
3			E

(b) Schedule Exploiting Slack

Figure 4.5. Possible Execution Schedules

execution schedule for this code sequence; instruction A can be executed either at cycle 1 or 2 and still produce a useful result by cycle 3, when the result of A is needed by instruction D.

In this example, instruction A has one cycle of slack; on a conventional architecture, it could be issued one cycle later (*i.e.* either at cycle 1 or 2) and not impact the performance of the processor. If it is possible to design hardware that employs multiple clock domains, this extra degree of slack also implies that instruction A could be run at $1/2$ speed (or, in general, at $1/(slack+1)$ speed) and achieve the same performance. For example, instruction A could take two cycles to actually execute, which we denote in Figure 4.5(b) by the functional unit being filled with A_1 and A_2. Alternatively, the instruction could be scheduled at cycle 2 to reduce the total number of concurrently executing instructions, possibly reducing peak power demands.

4.1 Theoretical Underpinnings of Slack Scheduling

The principles behind "slack scheduling" can be precisely defined using a conventional graphic-theoretic model. However, this module is impractical to implement in hardware. We first describe the formal model and then the realizable implementation.

We use conventional dependence graph notation. To compute scheduling slack for a set of instructions (for either statically or dynamically scheduled processors) each cycle, we construct a dependence graph $G = (N, E)$ of all available instructions. We assume we know the operational latency of each instruction (Lat_j). The "start time" for an instruction (S_j) is the earliest that a given instruction can issue; the "finish time" (F_j) is when that instruction will finish execution. We can recursively define these terms as:

$$\begin{aligned} \mathrm{F}_j &= \mathrm{S}_j + \mathrm{Lat}_j \\ \mathrm{S}_j &= \max_{k \in \mathrm{pred}_j} \mathrm{F}_k \end{aligned}$$

with $\mathrm{S}_j = 0$ when node j has no predecessors ($\mathrm{pred}_j = \emptyset$). We will also need to know the maximum delay for the critical path, $\mathrm{MaxDelay} = \max_{k \in N} \mathrm{F}_k$.

The "slack" (Sl_j) in an instruction is the difference between when that instruction can finish and the earliest time that the results are needed. To compute this, we need to introduce the notion of a "latest time" (L_j) for an instruction - this is the latest time by which an instruction must finish to meet the critical path delay for the instructions. For a single dependence graph, the latest time and finish time are identical; in practice, the dependence relations form a multi-graph and portions of the graph may not be on the critical path and thus may not need to be executed quickly. This leads to the following:

$$\begin{aligned}
\mathrm{L}_j &= \begin{cases} \mathrm{MaxDelay} & \text{if } \mathrm{succ}_j = \emptyset \\ \max_{k \in \mathrm{succ}_j} (\mathrm{L}_k - \mathrm{Lat}_k) & \text{if } \mathrm{succ}_j \neq \emptyset \end{cases} \\
\mathrm{Sl}_j &= \mathrm{L}_j - \min_{k \in \mathrm{succ}_j} \mathrm{S}_k
\end{aligned}$$

It is easiest to explain these equations by extending our earlier example to include multiple dependence graphs, as shown in Figure 4.6; again, we assume all operational latencies are a single cycle. Each node in Figure 4.6 is labeled with the corresponding start, finish and late times as well as the computed slack. The critical path ($B \rightarrow D \rightarrow E$) delay is 3 cycles. Instructions X and Y are not on the critical path. Since no instructions depend on Y, it could execute in either cycle 2 or 3 with no impact on the total execution time, *assuming that preceding initiate at their start time*. Likewise, instruction X can execute at either cycle 1 or 2 without affecting the critical path.

Up to now, we have only addressed the *innate scheduling slack* that depends purely on dependence relations. Actual programs have considerable innate slack available even when considering just the issuable instructions [2]; however, processors can also exploit *runtime slack*. This additional slack arises either due to mandatory delays such as cache misses or due to advisable delays such as those caused by pipeline gating. For example, a load instruction normally has an operational latency of a single cycle, unless the load misses in the cache, at which time the load may have a latency of 10's or 100's of cycles. Likewise, a branch instruction normally has a single cycle of operational latency, but pipeline gating may delay instruction fetch following that branch for speculation control and thus increase the effective operational latency beyond that measured in this paper.

To underscore this point, refer to Figure 4.7, which shows two load instructions (`ld A` and `ld D`). Even if we know the expected load delay of the `ld A` instruction, we can not use it to any advantage. But, if we know that `ld D` is likely to miss, the start time for `Y = X + D` is lengthened, providing more slack for the `X = A + 10` instruction and `ld A`. We would expect this would introduce slack in most programs.

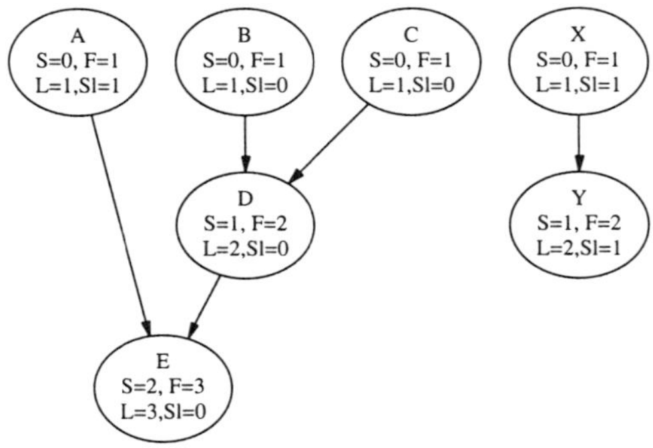

Figure 4.6. More Complex Example Containing a Multi-Graph to Explain Flow Equations

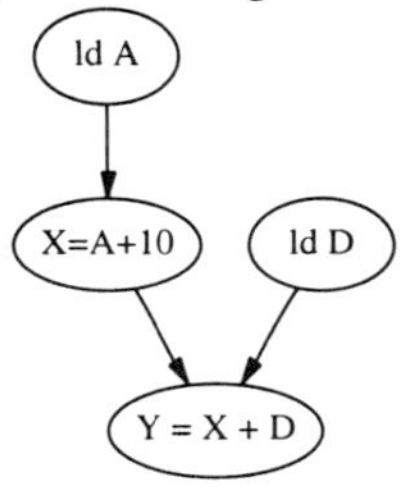

Figure 4.7. Simple Example Showing the Potential Influence of Cache Misses on Schedule Slack

In any given cycle, multiple instructions may have innate slack, but it may be that only a subset of the instructions can exercise that slack. This may affect the efficacy of on-line methods to determine instruction slack.

Scheduling slack can be used for a number of purposes – it may be possible to issue instructions to different clusters based on slack or prioritize dependence resolution based on slack. We have initially used slack to schedule instructions on "low power" functional units. This means we are mainly interested in the scheduling slack available for instructions that are *issuable* (*e.g.* instructions A, B, C and X in Figure 4.6).

4.2 Implementing Slack Detection

We've conducted limit studies to examine the potential scheduling slack in applications [2]; our preliminary findings were encouraging because we found that most execution cycles contain at least one instruction that can be delayed by at least a single cycle. The remaining challenge is to find a practical method to identify slack. The graph-theoretic model we described is impractical, because it examines all instructions every processor cycle; this would lead to excessive overhead.

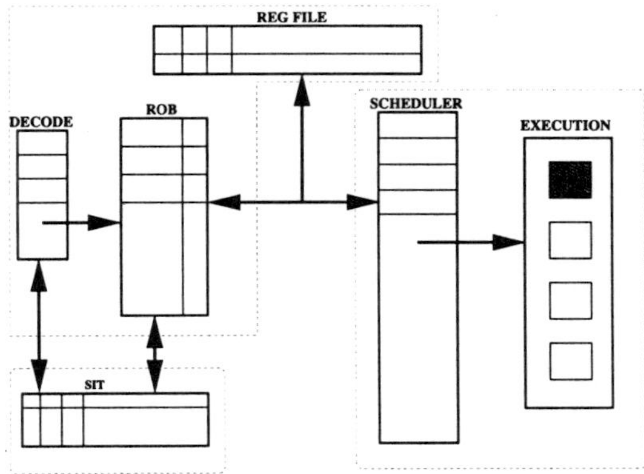

Figure 4.8. Major Architecture Components Involving SIT

We've focused on a technique that "learns" which instructions have scheduling slack using some simple hardware, and have conducted in-depth power and performance analysis for this implementation. The proposed hardware support for the dynamic power reduction system uses three components to detect and exploit instruction dependency slack. The elements perform detection, state tracking, and execution of slackful instructions. First, the instruction must have a minimal amount of *slack*, allowing slow execution without impacting performance. Second, the slack state must be maintained across execution iterations so that the slack may be utilized. Finally, lower power integer execution units exploit the detected slack, saving power at the expense of increased latency. The following components function as a system to accurately detect slackful operations within dynamic instruction sequences and reduce integer execution core power without significantly impacting performance:

Slack Indicator Table (SIT) is a simple structure that is the center of this approach. This tracks instruction slack behavior across execution iterations. In this way, it serves as a history-based prediction mechanism for the scheduling of tolerant operations on slow execution units.

Re-Order Buffer, and Register File extensions are necessary for communicating slack state between both in-flight and completed operations. Further, they identify the *first* consumer of an operations output; consumer ordering is imperative to the computation and back-propagation of slack from a consumer (child) to its producers (parents).

Scheduler Modifications and Low Power Execution Units complete the system with the units which actually save power and the means to utilize them appropriately.

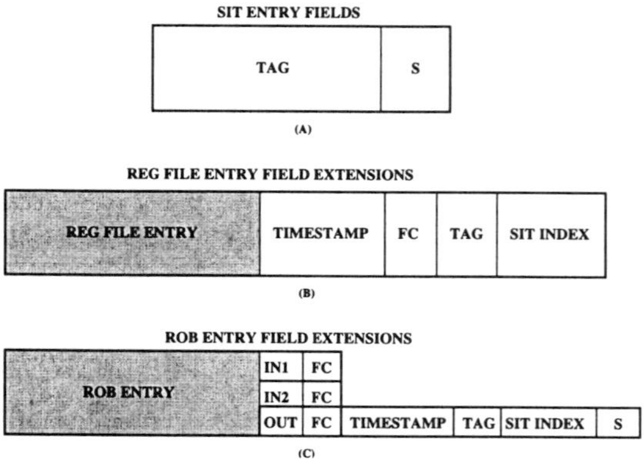

Figure 4.9. Additional Fields Required for SIT

As an instruction is decoded, SIT table is probed to determine if it has had slack in the past. If the operation has had slack, it is preferably dispatched to a slow unit if available; if a fast unit is ready, it may execute there or lose its slack and wait. As instructions complete, they normally announce their completion to all dependent instructions; in our implementation, dependent instructions determine if an input dependence would have scheduling slack. For example, assume an instruction is waiting for two registers, R_s and R_t; if the producer of R_s finishes before the producer for R_t, than the producer of R_s *may* have slack. In fact, the producer of R_s must complete before *any* dependent instruction needs the value of R_s. However, we only need to record when the *first* instruction requests the value of R_s.

4.3 Slack Indicator Table

Figure 4.9(a) is one entry of the Slack Indicator Table. The table is simply an array of such entries, each comprised merely of a *Tag* and *Slack Bit* (S). The tag and index of the SIT are derived and used in the same manner as in a cache. The slack bit indicates if the operation at that index was found to be slackful in its previous execution. The hardware cost of the SIT is proportional the number of entries and the corresponding tag length. The power cost of the SIT can be estimated as an equivalent size cache tag array.

Figure 4.9(b) shows the fields added to the register file entries. The *TIMESTAMP* field is the cycle count value when the result of the operation was written into that register. The timestamp field has to be large enough to capture the largest latency between when a consumer is mapped to a producer and when that producer completes execution (e.g., memory access latency). The *FC* (First Consumer) is a single bit field that is set only by the first operation to read the value after it has been written. Subsequent operations will see that

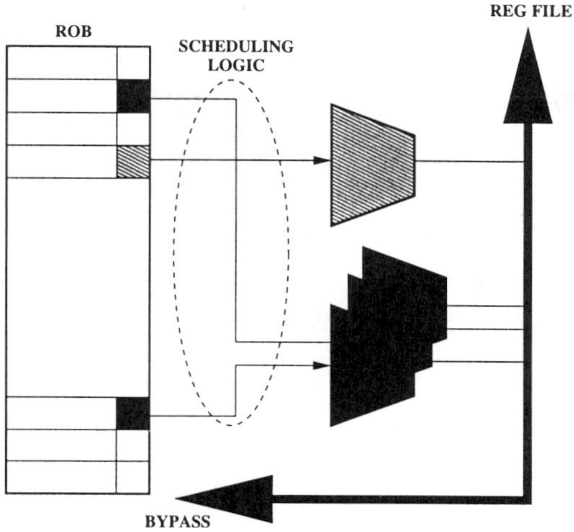

Figure 4.10. Scheduling of Slackful Operations to Slow and Fast Execution Units

this value has been consumed by a previous operation in execution order. The instruction writing the register copies its tag into the *TAG* field. Likewise, the *SIT INDEX* is from the writing instruction's PC, and points to its entry in the SIT.

The ROB entry extensions are shown in Figure 4.9(c). There are two sets of extensions: the inputs set tracks the inputs for which this operation is the first consumer; the output set contains the state for the operation in that ROB entry (and the output it will produce). The FC flags are associated with the input dependence pointers. The *FC*, *TIMESTAMP*, *TAG*, and *SIT INDEX* fields for the ROB output are set and used the same as those in the register file. The output of an operation in the ROB, however, may or may not be written into the register file. Hence, the state must be maintained in that ROB entry until it is either squashed (the state is also squashed) or written into the register file (the state is transferred into the matching fields of the register). The slack bit *S* is read from the SIT and written into the ROB as the instruction is being decoded.

4.3.1 Using the Indicated Slack. The scheduler treats a ROB entry slack bit as a priority bit for selecting an execution resource, as shown in Figure 4.10. If the slack bit is set, the operation will be scheduled to the slow unit, if possible. If the bit is clear, the scheduler attempts to send the operation to a fast unit.

There are several avenues to creating a low power resource. The specific path chosen depends upon the cost-benefit tradeoffs considered. Power can be traded for area or complexity – the circuit design techniques used to achieve a single-cycle functional unit becomes increasingly complex as processors enter multi-Ghz frequency domains. Scaling back the aggressiveness of a functional unit design can significantly reduce the energy needed by that functional unit. Still another option is to design a resource that operates at a lower frequency and/or voltage. The type of tradeoff in part determines the amount of power the device can conserve, as well as the nature and degree of the performance impacts it will incur. The selection of solution is largely implementation specific.

The low power resources used in this work are restricted to integer execution units (IEU). However, the principles for use can be extended to any resource type without loss of generality. The integer units are partitioned into two clusters, one which runs at the standard voltage (V_{dd}) and speed (F), and the other running at one half the standard voltage and one half the standard frequency. This approach leverages the technology of multiple voltage and frequency circuit design. From the Power relation of $P = V_{dd}^2 * F * C$, there is an approximate $3/4$ power savings by halving both the voltage and frequency. However, due to the 2X increase in operation latency (albeit at $1/4$ power) will result in a net energy savings of $1/2$. In this work, the standard IEU latency is a single cycle, with the slow units having a latency of two cycles. Hence, the slack threshold is one cycle.

4.3.2 Costs and Benefits of the SIT Slack Detection.

The power model for this study is derived from actual Pentium III processor power values [1]. Where there is no direct correlation between the hardware which the power is based upon and the architecture modeled by the simulator, the power model was modified to fit. The base power values did not employ any form of clock gating. Inactive and static power values were derived from [5], representing an aggressive clock gating infrastructure which is captured by the TEMPEST power model [6]. The inactive power of each component was taken as 10% of the active power.

Figure 4.11 shows the IPC degradation (higher is worse) for an infinite SIT across the benchmarks. We examined a number of mechanisms to maintain the SIT state, including decay (*Decay*), self-identified slack (*Self-ID*), slack identification stage (*Stage*), and the default (*Default*), which employs none of these state update techniques. Decay allows the in-flight-instruction to clear its slack bit (but leave the SIT state untouched) if it misses a dispatch opportunity. Self-ID marks completing instructions with no dependents as slackful, rather than waiting for a child to propagate slack to it. Stage performs the producer slack determination by operations as they are about to issue, rather than when they are awakened. This variation attempts to capture more of the resource constraints of the machine by making the timestamp comparison closer reflect the actual completion to use time delta. There is no single technique that is

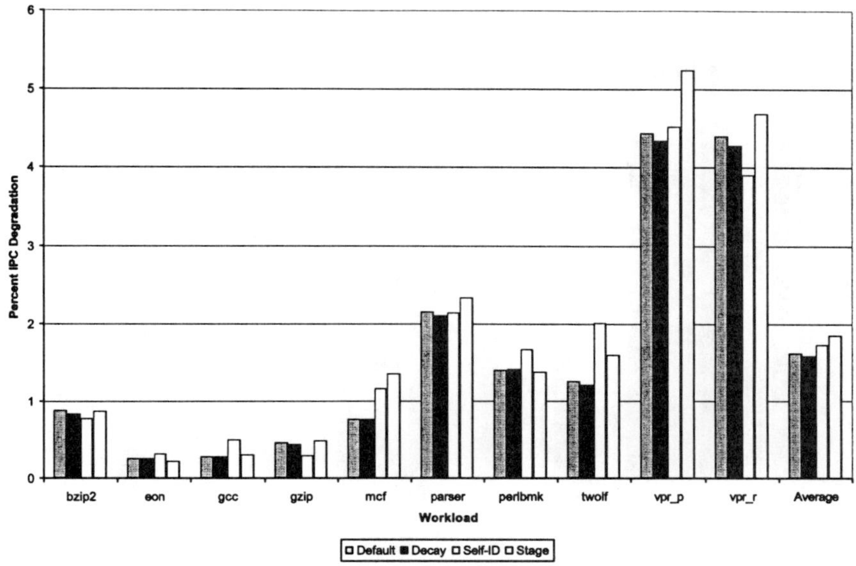

Figure 4.11. IPC Degradation when Using Various State Update Techniques and an Infinite SIT

preferable across all workloads, and the IPC impact varies by workload. It appears that using the default mechanism performs as well as the other techniques, on average. Note that most workloads suffer an IPC degradation of under 2%. The *vpr_p* application suffers the most, using Stage, at 5.3% IPC decrease. There is an inherent lowering of IPC using this scheme, even under ideal conditions, due to the effective integer instruction throughput being reduced to 3.5 per cycle (maximum) from the default of four.

The integer execution unit power reduction is shown in Figure 4.12 (higher is better). The average IEU power is reduced by 11%, with slightly better results from the Self-ID and Stage techniques. These two techniques are more aggressive towards marking instructions as slackful. This accounts for both their resultant decrease in IPC and increase in IEU power savings versus the default mechanism.

Our work on detecting and exploiting slack is only beginning, but it's clear from these results that there are practical mechanisms that can identify scheduling slack and that scheduling slack can be profitably employed. Although our first application of scheduling slack targeted instruction execution, we think it can be applied to processor components that consume more energy, such as the instruction issue and retirement logic. Moreover, we are examining mechanisms other than the SIT tables that may be more accurate in identifying slack.

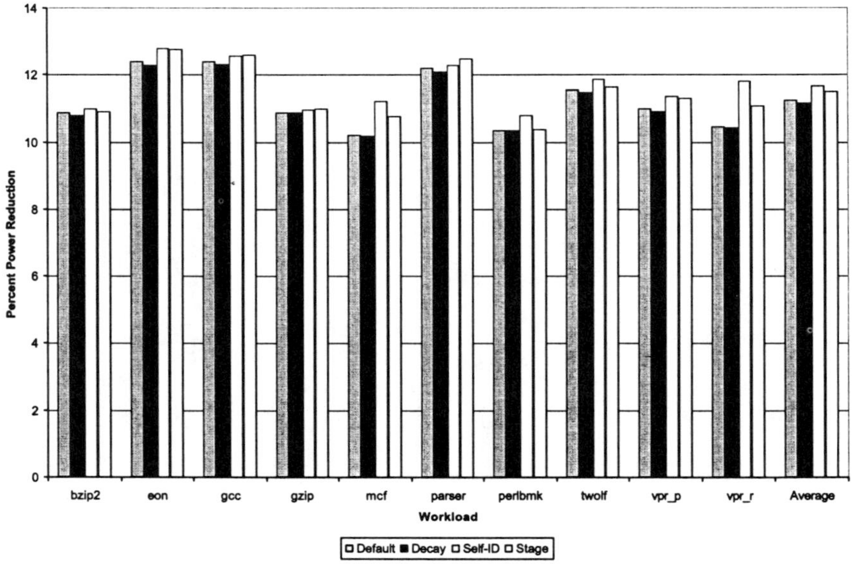

Figure 4.12. IEU Power Reduction when using Various State Update Techniques and an Infinite SIT

5. Summary

The thesis of our research group when we started examining micro-architectural techniques to reduce processor power was that there were mechanisms that could reduce power but that wouldn't normally be considered because they didn't increase performance. Our efforts have focused on speculation control, or attempting to reduce the waste caused by the speculative execution needed to yield high performance processors. Pipeline gating controls speculation by reducing instruction fetch when it is unlikely to be productive. Slack scheduling delays instruction execution to occur no earlier then necessary to save power and energy. We think that the techniques described here and other techniques we've examined show that speculation control can reduce power; in turn, this allows processors to consume less energy and also to run at faster clock speeds.

Acknowledgments

I would like to thank the doctoral students with whom I've had the good fortune to work on issues of power-aware microarchitecture: Jason Casmira Soraya Ghiasi, Artur Klauser and Srilatha ('bobbie') Manne. In particular, the work on confidence estimation was part of the thesis work of Artur Klauser and Bobbie Manne; Bobbie also explored issues in pipeline gating. Jason Casmira has been exploring how "slack scheduling" can be used to reduce processor power. I would also like to thank Steve Gunther of Intel for power data from Pentium processors.

References

[1] D. Carmean. Power, a perspective from the desktop. In *Kool Chips Workshop, invited talk*. ACM, Dec 2000.

[2] J. Casmira and D. Grunwald. Dynamic instruction scheduling slack. In *Kool Chips Workshop*. ACM, Dec 2000.

[3] A. Chandrakasan and R. Brodersen (Eds.). *Low-Power CMOS Design*. IEEE Press, 1998.

[4] A. P. Chandrakasan, S. Sheng, and R. W. Brodersen. Low-power cmos digital design. *IEEE Journal of Solid-State Circuits*, 27(4):473–484, 1992.

[5] Intel Corporation. Pentium 4 Processor Datasheet. http://developer.intel.com/design/Pentium4/datashts/.

[6] A. Dhodapkar, C. Lim, G. Cai, and W. Daasch. Tempest: A thermal enabled multi-model power/performance estimator. In *Workshop on Power Aware Computer Sytstems*. ACM, Nov 2000.

[7] D. Folegnani and A. Gonzalez. Energy-effective issue logic. In *Proc. of 28th. Int. Symposium on Computer Architecture*, June 2001.

[8] Richard Fromm, Stylianos Perissakis, Neal Cardwell, Christoforos Kozyrakis, Bruce McGaughy, and David Patterson. The Energy Efficiency of IRAM Architectures. Technical report, May 1997.

[9] Soraya Ghiasi and Dirk Grunwald. A comparison of two architectural power models. In *Workshop on Power Aware Computer Sytstems*. ACM, Nov 2000.

[10] Ricardo Gonzalez and Mark Horowitz. Energy Dissipation in General Purpose Microprocessors. *IEEE Journal of Solid-State Circuits*, 31(9):1277–1284, September 1996.

[11] Dirk Grunwald, Artur Klauser, Srilatha Manne, and Andrew Pleszkun. Confidence estimation for speculation control. In *Proceedings 25th Annual Annual International Symposium on Computer Architecture*, Barcelona, Spain, June 1998. ACM.

[12] Steve Gunther and Intel Corp. Suresh Rajgopal. Personal communication.

[13] Mutsunori Igarashi, Kimiyoshi Usami, Kazutaka Nogami, Fumihiro Minami, Yukio Kawasaki, Takahiro Aoki, Midori Takano, Chiharu Mizuno adn Takashi Ishikara, Masahiro Kanazawa, Shinji Sonoda, Makoto Ichida, and Naoyuki Hatanaka. Clustered voltage scaling techniques for low-power design. In *IEEE Symposium on Low Power Electronics*, 1997.

[14] Erik Jacobsen, Eric Rotenberg, and J. E. Smith. Assigning confidence to conditional branch predictions. In *Proceedings of the 29th Annual Inter-*

national Symposium on Microarchitecture, pages 142–152, Paris, France, December 2–4, 1996.

[15] M.C. Johnson and K. Roy. Optimal selection of supply voltages and level conversions during data path scheduling under resource constraints. In *Proceedings of the International Conference on Computer Design*, 1996.

[16] J. Kin, M. Gupta, and W. Mangione-Smith. The Filter Cache: An Energy Efficient Memory Structure. *IEEE Micro*, December 1997.

[17] Artur Klauser, Abhijit Paithankar, and Dirk Grunwald. Selective eager execution on the polypath architecture. In *Proceedings 25th Annual Annual International Symposium on Computer Architecture*, Barcelona, Spain, June 1998. ACM.

[18] U. Ko, P. T. Balsara, and A. K. Nanda. Energy optimization of multilevel processor cache architectures. In *Proceedings of the International Symposium on Low Power Design*, 1995.

[19] Mikko Lipasti. *Value Locality and Speculative Execution.* PhD thesis, Carnegie Mellon University, April 1997.

[20] Srilatha Manne, Artur Klauser, and Dirk Grunwald. Pipeline gating: Speculation control for energy reduction. In *Proceedings 25th Annual Annual International Symposium on Computer Architecture*, Barcelona, Spain, June 1998. ACM.

[21] Srilatha Manne, Artur Klauser, and Dirk Grunwald. Branch prediction using selective branch inversion. In *Proceedings of Parallel Architectures and Compilation Techniques PACT'99*, Oct 1999.

[22] J. Montanaro and *et. al.* A 160-MHz, 32-b, 0.5-W CMOS RISC Microprocessor. In *Digital Technical Journal*, volume 9. Digital Equipment Corporation, 1997.

[23] M. Srivastava, A. Chandrakasan, and R. Broderen. Predictive system shutdown and other architectural techniques or energy efficient programmable computation. *IEEE Transactions on Very Large Scale Integration Systems*, 4(1):42–55, March 1996.

[24] C. Su and A. Despain. Cache design trade-offs for power and performance optimization: A case study. In *Proceedings of the International Symposium on Low Power Design*, 1995.

[25] Kimiyoshi Usami and Mark Horowitz. Clustered voltage scaling techniques for low-power design. In *IEEE Symposium on Low Power Electronics*, 1995.

Chapter 5

ENERGY-EXPOSED INSTRUCTION SETS

Krste Asanović, Mark Hampton, Ronny Krashinsky, Emmett Witchel
MIT Laboratory for Computer Science
200 Technology Square
*Cambridge, MA 02139, USA**
{krste, mhampton, ronny, witchel}@lcs.mit.edu

Abstract Modern *performance-oriented* ISAs, such as RISC and VLIW, only expose to software features that impact the critical path through computation. Pipelined microprocessor implementations hide most of the microarchitectural work performed in executing instructions. Therefore, there is no incentive to expose these micro-operations, and their energy consumption is hidden from software.

This work presents *energy-exposed* hardware-software interfaces to give software more fine-grain control over energy-consuming microarchitectural operations. We introduce *software restart markers* to make temporary processor state visible to software without complicating hardware exception management. This technique can enable a wide variety of energy optimizations. We implement *exposed bypass latches* which allow the compiler to eliminate register file traffic by directly targeting the processor bypass latches. Another technique, *tag-unchecked loads and stores*, allows software to access cache data without a hardware tag check when the compiler can guarantee an access will be to the same line as an earlier access.

1. Introduction

Power consumption is emerging as a key factor limiting computational performance in both mobile and tethered systems. Although there has been significant progress in low-power circuit design and low-power CAD and some work in low-power microarchitectures, there has been little work to date at the level of instruction set architecture (ISA) design for low power computing.

*This work was partly funded by DARPA PAC/C award F30602-00-2-0562 and by NSF CAREER award CCR-0093354.

Modern ISAs such as RISC or VLIW are based on extensive research into the effects of instruction set design on performance, and provide a purely performance-oriented hardware-software interface. These instruction sets avoid providing alternate ways to perform the same task unless it will increase performance significantly. Implementations of these ISAs perform many energy-consuming microarchitectural operations during execution of each user level instruction and these dominate total power dissipation. For example, when executing an integer add instruction on a simple RISC processor only around 5% of the total energy consumption is due to the adder circuitry itself. The rest is dissipated by structures such as cache tag and data arrays, TLBs, register files, pipeline registers, and pipeline control logic. Modern machine pipelines have been refined to the point where most of the additional microarchitectural work is performed in a pipelined or parallel manner that does not affect the throughput or user-visible latency of a "simple" add instruction. Because their performance effects can be hidden, there is no incentive to expose these constituent micro-operations in a purely performance-oriented hardware-software interface — their energy consumption is hidden from software.

In this chapter, we present new energy-exposed hardware-software interfaces that give software fine-grain control over energy consumption. The key idea is to reward compile-time analysis with run-time energy savings. Instruction set enhancements enable this goal by providing software with alternative methods of executing an operation; performance is unchanged, but greater compile-time knowledge can be used to deactivate unnecessary portions of the machine microarchitecture. Our primary focus is on integer applications with complex control flow. This type of code will likely become the energy bottleneck in future embedded systems, as more regular computations can be mapped to energy-efficient vector instructions or custom hardware accelerators. We modify a RISC microprocessor architecture to support three energy-exposed techniques and develop compiler algorithms to target the enhanced instruction set.

The first technique is *software restart markers*, which reduce the energy expended in exception state management. Current pipelined machines invest significant energy in preserving precise exception semantics. Instruction results are buffered before being committed in order, requiring register rename logic to find the correct value for new instructions. Even a simple five-stage RISC pipeline has a bypass network that effectively performs these functions. In addition, other information such as PC and faulting memory addresses must be preserved in the pipeline until the exception can be serviced. Software restart markers reduce energy by allowing the compiler to annotate at which points it requires precise exception behavior. Initial results show that the number of precise exception points can be reduced by a factor of three. More importantly, this technique allows additional machine state to be made visible between restart points, enabling the introduction of more energy-exposed features without incurring additional exception management costs.

The second technique is *exposing bypass latches* with a hybrid RISC-accumulator architecture that uses compile-time register lifetime information to reduce the number of register file reads and writes performed at run time. Many register values in a computation are short-lived, being produced by one instruction, consumed by the next instruction, and then never used again. This register lifetime information can be encoded by adding accumulator registers to a general-purpose register (GPR) RISC architecture, which allow software to pass values directly from one instruction to the next without accessing the GPRs. The accumulator registers can be mapped to the bypass latches that are already present in a CPU datapath. To avoid the hardware and energy overhead which would be necessary to preserve this bypass latch state during exceptions, the bypass latch accumulator registers are treated as temporary state which is recreated when a region is re-executed after any trap. This technique can remove a third of all register file writes.

The final technique is *tag-unchecked loads and stores.* Tag accesses consume over half the energy of a data cache access in a low-power microprocessor. In cases where compile-time analysis can guarantee that two accesses will be to the same cache line, tag-unchecked loads and stores allow the hardware to avoid performing a tag-check on the second access. Initial results indicate up to 70% of tag checks in SPECint95 and Mediabench programs can be removed at compile time.

1. Baseline Processor

For mobile and embedded processors, both energy and performance are of interest, and a traditional pipelined datapath is preferable to a more complex superscalar design [8]. An energy-efficient five-stage pipelined MIPS RISC microprocessor based on that presented in [4] was adopted as a baseline with which to evaluate the three energy-exposed instruction set techniques. The basic pipeline structure of this design is shown in Figure 5.1. The design has split 16 KB instruction and data caches that are organized as 64-way set-associative CAM-tag caches with 32-byte lines. This pipeline and cache configuration is designed to be similar to the popular StrongARM-1 [6] low-power microprocessor.

2. Software Restart Regions

Machines that support an operating system with preemptive context switching or demand-paged virtual memory must provide some mechanism to manage exceptions. If precise exceptions are supported in a pipelined machine, hardware must either buffer state updates in some form of future file until all

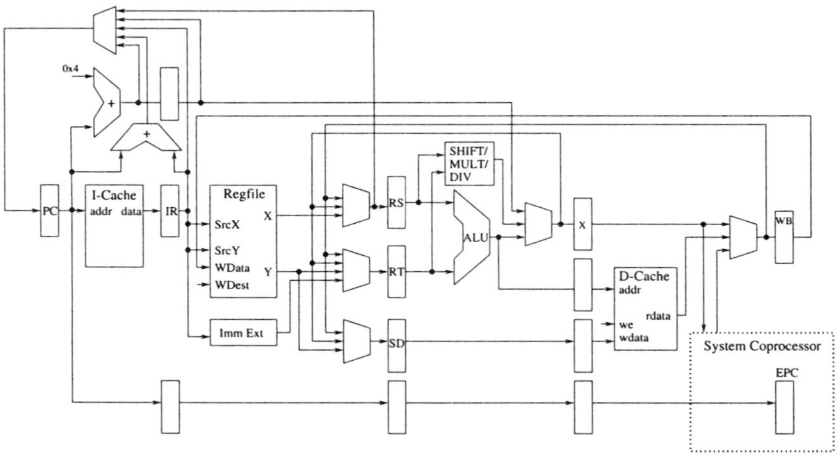

Figure 5.1. Baseline Pipeline Design

possible exceptions have cleared, or alternatively save old machine state in history buffers so that it can be recalled when an exception is detected [9]. Even if precise exceptions are not implemented, hardware must provide access paths to allow software to save and restore all machine pipeline state across exceptions to allow restart [11]. Note that these schemes add additional exception state management energy overhead to the execution of all instructions. An energy-exposed instruction set requires that internal machine state is made visible to software. However, if making this additional machine state visible to software incurs additional exception management overhead, much of the potential energy savings could be lost.

Most current ISAs have sequential instruction semantics and implementations are usually required to provide precise exceptions. That is, exceptions must be taken in program order, and on recognizing an exception the machine must provide the program counter (PC) of the faulting instruction and ensure that all earlier instructions have committed state updates and no later instructions have affected architectural state [9]. This provides a simple model for user code, which is given the illusion of uninterrupted execution; and for system software, which only need save and restore programmer-visible state including the faulting PC to swap contexts. However, supporting these semantics incurs hardware exception management overhead for every instruction executed. In practice, exceptions occur rarely and usually do not require full precision in exception reporting. For example, timer interrupts and page faults only require that the process be restartable.

Software restart markers reduce the energy cost of exception management by requiring software to explicitly divide the instruction stream into restartable regions. After handling a trap, the OS resumes execution at the beginning of the restart region for the associated instruction. A conventional architecture with precise exceptions is equivalent to placing every instruction into its own restart region. A simple local analysis can remove many of these implicit restart points by placing multiple instructions into one region.

2.1 Restart Marker Implementation

Restart points are encoded by marking the last instruction in a restart region. This instruction is called the *barrier* instruction because it acts as a trap barrier that will commit and irrevocably update machine state only if it is guaranteed that it will not raise an exception and that any preceding instruction will not raise an exception. Also, the barrier instruction ensures that if an exception does occur before it commits, the effects of following instructions will not be visible. When the barrier instruction commits it will update a kernel visible register, the *restart program counter*, to point to the next instruction to be executed; this instruction is the beginning of the next restart region. Note that marking every instruction as a barrier instruction is equivalent to conventional precise exception semantics.

The compiler must ensure that the code in a region is such that the operating system kernel can restart the process after an exception by simply jumping to the restart PC. This requires that software ensure the code in each restart region (except for the final barrier instruction) is idempotent, i.e., that it can be re-executed multiple times without changing the result. This restriction still allows for the creation of large restart regions. For example, many of the functions in the standard C library can each be entirely contained within one restart region. To illustrate this point, consider the `sprintf` function, for which the prototype is given below.

```
int sprintf(char *s, const char *format, ...);
```

This function uses the `format` string as an input argument to write to the string pointed to by `s`. As long as the input argument to the function is passed in stack memory and not altered by the routine (meaning that the input arguments and output arguments cannot overlap in memory), the function can be restarted multiple times and still produce the same result. For the `sprintf` function, the `format` argument is declared to be `const`, so it is not modified. Thus, the function satisfies the criteria for idempotency. In general, any function that does not modify its arguments can use an arbitrary amount of local read/write workspace and still be idempotent.

2.2 Categories of Machine State

The use of software restart markers introduces three different types of machine state: checkpointed, stable, and temporary. Checkpointed state is copied into checkpoint registers each time a barrier instruction commits, and is recoverable if an exception occurs. In the simple scheme used in this chapter, the only checkpointed state is the restart PC. Stable state is preserved across an exception by the kernel: conventional registers and memory fall into this category. Finally, temporary state is only valid within a restart region, and is not preserved across an exception. The bypass latches discussed in the next section are an example of temporary state.

The power of software restart markers is that a great deal of internal machine pipeline state can be exposed by mapping it to temporary state without needing hardware support to preserve this state across exceptions. In fact, the temporary state can be ignored by the operating system. As long as the kernel saves the visible stable user state and the restart PC, the process can be restarted after each exception.

It is interesting to note that the MIPS instruction set [3] already incorporates a limited form of restart regions to support delayed branches, i.e., the next PC is user-visible temporary state that is updated by branch instructions but which is never saved and restored by the kernel. Branches are always idempotent and never trap barriers so that any trap on a delay slot instruction restarts at the branch itself to recreate the next PC.

2.3 Example Use

Figure 5.2 presents some example code showing how the restart barriers are used. The code is similar to conventional MIPS code, except that barrier instructions are marked with a `.bar` suffix. The barrier instructions split the code into four regions labeled A, B, C, D. If an exception occurs within a region, the code can be restarted from the beginning of the region. For example, if the store in region A encounters a write fault (perhaps because the page was write protected as part of a copy-on-write protocol), the OS can replay the code starting at the initial load and obtain the same result. The store in region A must be marked as a barrier because its update of memory would otherwise cause the region to be non-idempotent.

Region B contains a store that is not marked as a barrier. If the final load in region B takes a page fault, this store will have likely changed memory irrevocably. Nevertheless, the OS can restart region B from the initial load instruction because the region is idempotent. The final load is marked as the barrier because it overwrites the original value of the pointer `r4` which would be needed to restart the region.

```
A   lw r1, (r2)
    addiu r1, r1, 1
    sw.bar r1, (r2)
B   lw r2, 0(r4)
    lw r3, 4(r4)
    addu r2, r2, r3
    sw r2, 16(r4)
    lw.bar r4, 12(r4)
C   addu.bar r5, r5, 4
D   lw r1, (r4)
    andi r1, r1, 3
    sw r1, (r4)
    lw r3, (r5)
    bnez r3, loop
    addu.bar r5, r5, 4
```

Figure 5.2. Code Example Showing Restart Regions. Instructions with a `.bar` Suffix are the Barrier Instructions at the End of Each Region

Region C contains a single instruction, and shows how a simple compiler can fall back to marking all instructions as barrier instructions to replicate conventional precise exception semantics.

One sufficient, but not necessary, condition for idempotency is that the set of all external sources (registers and memory) read by the region is disjoint from the set of destinations written by the region (note that it is acceptable to overwrite a value produced within the region). This is not a necessary condition as shown by the example in region D. Here, the store to memory changes input source data, but with an idempotent operation (masking out the bottom two bits). The barrier for this region is placed on the delay slot of the branch instruction which will record the branch target in the restart PC when it commits.

2.4 Compiler Analysis

Restart analysis was implemented as a pass within the assembler that performs a purely local optimization at the basic block level after instruction scheduling and register allocation. The analysis begins a new restart region at the start of a basic block, then scans the sequence of instructions, updating the set of external values read and the internal set of values written. When a conflict is detected, the conflicting instruction is marked as a barrier instruction, then the read and write sets are cleared and a new region is started. Barriers are also placed in front of system calls and any other instructions that will likely cause a trap. Although we use a single bit in each set to represent memory, we also incorporate a limited form of memory analysis by separately tracking the base register and offset for each memory instruction. When the first memory instruction in a restart region is encountered, we store the base register, and we create a linked list to hold the offset. For each subsequent memory instruction, if the same base register is used, we can look at the offset to determine whether the access is to a distinct memory location from all previous memory operations. If the base register is modified, or a different base register is used, we revert to treating all of memory as a single location.

One concern with the restart scheme is ensuring forward progress in the face of finite resources. If TLB misses are handled by software exception handlers, then the number of memory operations in a region must be restricted to be less than the number of available TLB entries to ensure that the region can run from start to finish without incurring a TLB fault. Similarly for demand-paging, the number of physical pages must be greater than the number of memory operations allowed in a region. These restrictions can be enforced by the compiler and checked by the operating system, which can abort a process if it fails to progress through a region.

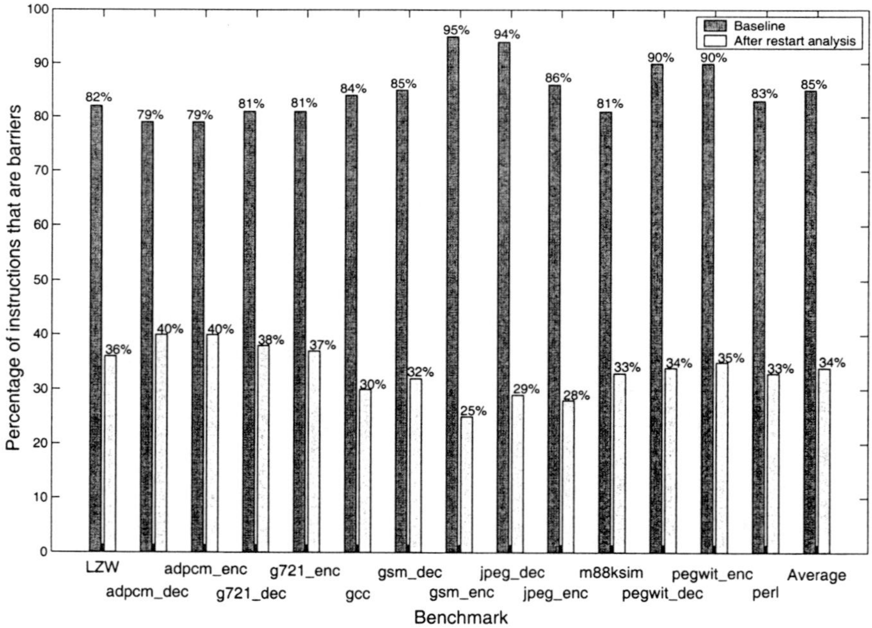

Figure 5.3. Percentage Of Dynamic Barrier Instructions For Baseline MIPS Code And Code After Restart Analysis. For Baseline MIPS, Only Branch And Jump Instructions Exclude Barriers

2.5 Evaluation

The results of this restart analysis are shown in Figure 5.3 for SPECint95 and Mediabench benchmarks. The Figure shows the number of dynamic instructions that are restart points for both baseline MIPS code and for code after the restart analysis. For baseline MIPS code, only branches and jumps do not have barriers and so around 79–95% of all instructions have barriers. After the simple local restart analysis, only 25–40% of instructions are barriers with an average of around 3 instructions in each restart region. More aggressive compiler analysis should generate even larger regions, and allow entire functions to be placed into a single restart region.

For the simple five-stage pipeline, restart analysis by itself only results in a minor energy saving in the exception PC pipeline. The instruction pipeline tags each instruction with its PC as it moves down the pipeline to identify the faulting instruction on an exception. The PC is latched into the EPC register in the system coprocessor if an exception occurs (Figure 5.1). With the restart analysis only the barrier instructions cause an exception PC to shift down the pipeline, allowing the PC pipeline to be gated off in other cases.

The primary advantage of software restart markers is that they make it possible to expose the internal details of a processor to the compiler as temporary state in between restart points. The next section illustrates one use of temporary state to save energy for register file traffic.

3. Exposing Bypass Latches with a Hybrid RISC-Accumulator Architecture

Simulations of the MediaBench and SPECint95 benchmarks reveal that about half of the values written to the register file are used exactly once, usually by the instruction executed immediately after the one producing the value. For example, in the code sequence to increment a memory variable,

```
lw  r1, (r3)     # Load value.
add r1,  r1, 1  # Increment.
sw  r1, (r3)     # Update memory.
```

the result of the load and add are only used once by the subsequent instruction and are normally read from the bypass network rather than the register file. A conventional processor must assume that the register could be read at any arbitrary point in the future or that there could be an exception right after the instruction and hence must always write the value into the register file. The compiler must already have calculated register lifetime information to perform register allocation but has no convenient mechanism to communicate this information to hardware through a standard instruction set.

3.1 ISA Enhancements

By giving software explicit control of the bypass latches, it is possible to reduce the register file traffic considerably. For example, the above code can be rewritten as:

```
lw      RS, (r3)      # Load RS latch.
add     SD,  RS, 1    # Increment and store result into SD latch.
sw.bar SD, (r3)       # Update with barrier.
```

where the RS operand specifies the use of the bypass latch in front of one input to the ALU and the SD operand specifies the use of the bypass latch that holds data being stored to memory (Figure 5.1). Note that this sequence has the same performance as the previous sequence but now two writes and two reads of the register file have been avoided and replaced with accesses to the bypass latches. In effect, when using the bypass latches, software turns off the register fetch and write back stages of the machine pipeline, and thereby removes microarchitectural energy overhead.

The final store is marked as a barrier instruction, because it performs a non-idempotent memory update. If an exception is taken on any instruction in the sequence, the code can be restarted from the load instruction. The bypass latch does not have to be saved and restored by the operating system because the bypass latch state will be recreated when this region is restarted.

This modification creates a hybrid RISC-accumulator architecture, but without the need to preserve accumulator contents around exceptions. This allows the accumulator registers to map directly to the bypass latches that were present in the original pipeline design, with no area, delay, or energy penalty for additional backup state or access paths to support exceptions.

Exposing bypass latches can eliminate register file reads when a temporary value is obtained from a bypass latch instead of the register file. Additionally, register file reads can be eliminated by another method which is referred to as *read caching* [10]. For example, when a procedure is called, it may save registers on the stack before using them, as shown in the following code segment:

```
sw r3, 8(sp)
sw r2, 4(sp)
sw r1, 0(sp)
```

The value in the stack pointer register does not change in the above sequence, yet it is read from the register file and clocked into the RS bypass latch for each instruction. Software can eliminate these extraneous reads as well as the unnecessary clocking of the RS latch by the use of explicit bypass latches, as shown in the following rewritten code segment:

```
sw r3, 8(sp)
sw r2, 4(RS)
sw r1, 0(RS)
```

3.2 Compiler Analysis

Our implementation of the exposed bypass latch code takes advantage of the static liveness information that is already maintained by the compiler. When the compiler determines that a value read by an instruction is being referenced for the last time—i.e. the value will be dead after the instruction executes—it appends a "`.l`" suffix to the assembly opcode with a corresponding operand number to indicate the last use of the value.

The liveness information generated for each instruction is then used by the scheduler that we added to the assembler. The scheduler reorders instructions within a basic block. It performs several passes on the code. First, it attempts to maximize performance by reordering instructions to mask latencies that can cause pipeline stalls–in particular, it tries to fill load-use delay slots with independent instructions. It also attempts to fill the architected branch delay slot. Next, the scheduler uses the lifetime information generated by the compiler to determine if bypass latches can be used in place of general-purpose registers to statically bypass a value. In the subsequent pass, the scheduler creates the restart regions discussed in the previous section. It then looks for read caching opportunities, and finally tries to perform additional static bypassing from the memory stage of the pipeline.

Note that static bypassing from the memory stage raises additional constraints not required for bypassing from one instruction to a subsequent instruction. Consider the following example:

```
add r1, r2, r3
sub r4, r5, r6
and r7, r1, r4
```

In the above code segment, `r1` is read for the last time by the `and` instruction. This would appear to provide an opportunity for static bypassing from the memory stage by having the first `add` instruction target the X latch (Figure 5.1). However, in this scenario, if there is an instruction cache miss for the `and` instruction, the `sub` instruction will overwrite the value in the X latch as it proceeds through the pipeline. To avoid this problem, we must either require strict pipeline sequencing, so that instructions go down the pipeline together with no bubbles between them, or we must not permit an instruction which overwrites the X latch (e.g., the `sub` instruction in the above example) to be the intermediate instruction in a memory stage bypassing sequence. We chose the latter option, as this placed no additional constraints on the hardware implementation. Since they do not write back to the register file, instructions which target the bypass latches are candidates for the intermediate instruction in a memory stage bypassing sequence, for example:

```
add  X, r2, r3
sub RS, r5, r6
```

```
and r7,  X, RS
```

3.3 Evaluation

For our simulations, we modeled the RS, RT, SD, and X bypass latches by reserving four general-purpose registers in the compiler and using their specifiers in the scheduler when modifying an instruction to target a bypass latch. We observed that the loss of these registers in the compiler's register allocator did not have an adverse effect on performance. Ideally, the instruction set encoding would be designed to support bypass latches directly.

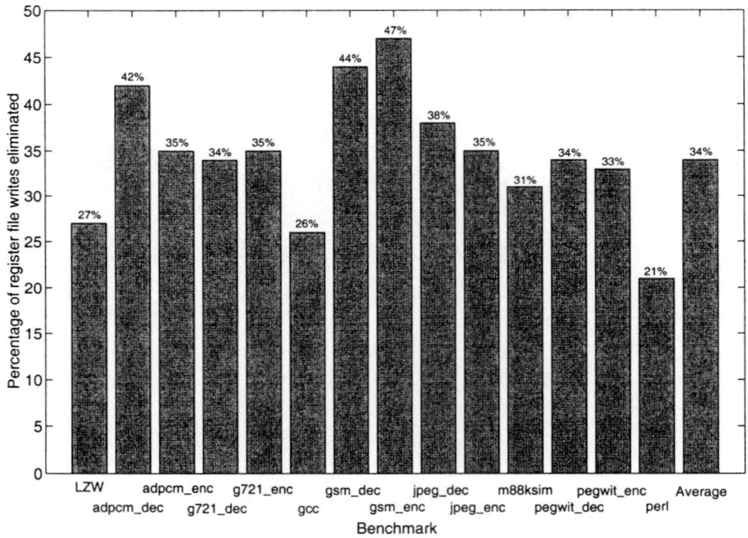

Figure 5.4. Percentage of Register File Writes Eliminated in Energy-Exposed Processor

The reduction in register file writes is shown in Figure 5.4. On average, 34% of all writes are eliminated. The reduction in register file reads is shown in Figure 5.5. On average, 28% of all reads are eliminated.

4. Tag-Unchecked Loads and Stores with Direct Addressing

The memory system, including caches, consumes a significant fraction of total system power. One significant source of energy consumption is the tag check in the primary data cache. *Direct addressing* allows software to access cache data without the hardware performing a cache tag check. These *tag-unchecked loads and stores* save the energy of performing a tag check when the compiler can guarantee an access will be to the same line as an earlier access.

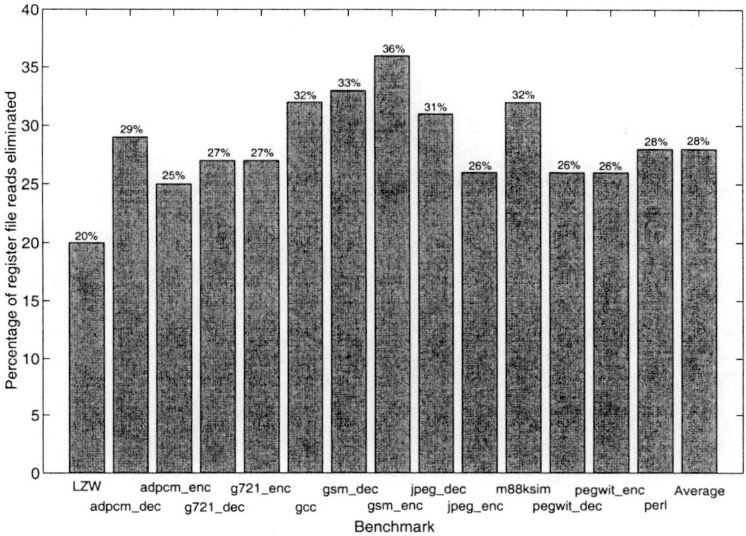

Figure 5.5. Percentage of Register File Reads Eliminated in Energy-Exposed Processor

If the compiler cannot determine this information, or if cache lines are evicted due to interrupts or cache invalidations, direct addressing gracefully degrades back to conventional tag-checked accesses. Direct addressing is only used for data accesses since instruction caches, while they dissipate considerable energy, have very regular access patterns and are only accessed via the program counter. Hence they are amenable to software-invisible micro-architectural techniques for power reduction.

Commercial low-power processors usually employ highly associative CAM-tag caches [1, 7, 6, 2]. Direct-mapped caches, although simpler, are inefficient in terms of energy usage because they experience more misses due to conflicts. Figure 5.6 shows the organization of a CAM-tag cache. Although CAM-tag caches reduce miss rates and hence total access energy, they expend relatively greater energy in tag checks [13]. The tag check for CAM-tag caches is expensive because the tag is broadcast to the CAM in order to find the proper line for the data. If we could shortcut that process—if the software could tell the hardware what line to read, rather than providing a virtual address as a key to the content-addressable memory—then we would save significant amounts of energy. The problem is how to let software directly access cache lines without compromising inter-process protection and while preserving correct operation in the face of cache replacements or other cache coherence actions.

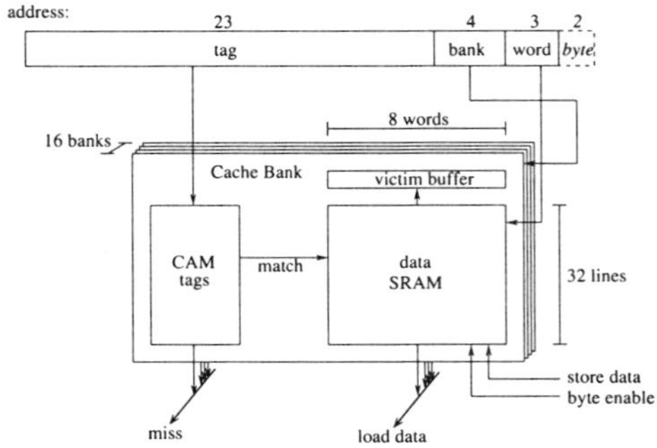

Figure 5.6. The Organization of a Highly-Associative CAM-Tag Cache. The Tag Bits are Broadcast to the CAM, and if There is a Hit, the Word is Read Out

4.1 ISA Enhancements

To eliminate tag checks, the processor-cache interface is enhanced with a direct access mode that tells the hardware exactly where the data is located. The processor state is augmented with some number of *direct address* (DA) registers. These registers contain enough information to specify the exact location of a cache line in the cache data RAM, and also have a valid bit. The exact width and data layout of the DA register is invisible to software to avoid exposing the implementation-dependent structure of the cache. In particular, software is only made aware of the length of a cache line, but not the total cache capacity or associativity.

Table 5.1 shows the instructions needed by the CPU to use the DA registers for direct addressing of the cache (we show only word accesses, but half-word and byte accesses are handled analogously). Software places values in the DA registers as an optional side-effect of performing a regular load or store. A tag-unchecked load or store specifies a full effective virtual address in addition to a DA register number. If the DA register is valid, its contents are used to avoid a tag search; if it is invalid, hardware falls back to a full tag search using the entire virtual address. The valid bit can be checked early in the processor pipeline, so this check does not introduce any additional latency into the cache access path.

4.2 Example Use

As a concrete example, consider the code in Figure 5.7, common at C function entry, and a transformation of that code which uses direct addressing.

The direct addressed operations use `da0` which is set up by the `swlda` instruction. This allows the compiler to use the `swda` instructions to eliminate cache tag checks on up to 7 stores (the remainder of the cache line started by

Instruction	*Explanation*
`(l|s)wlda rt, off(rs), da`	Load or store word, load direct address. These instructions act like regular loads and stores, but they also set the direct address register da with the location of the referenced line. We use MIPS as the basis of our instruction encoding, so the offset for this instruction is 13 bits signed instead of the regular 16-bit offset since there are 3 bits used as a da specifier.
`(l|s)wda rt, off(rs), da`	Load or store word, using direct address. Cache data from the line pointed to by da, using the line offset bits of `rs + off` is transferred to register `rt` (or the contents of `rt` is stored into the line specified by da). If da is invalid, the instruction acts like `(l|s)wlda`, accessing memory and setting the da register.
`jr.dainv rs, mask`	Jump register and invalidate direct address registers. This acts like a jump register instruction, but it also clears the valid bit on the direct address registers specified in the (little endian) bitmask. This instruction is used at the end of function calls and by the operating system when the DA register lifetime has ended.

Table 5.1. A table of instructions for manipulating and using direct address registers

```
Old Code             New Code
sub $sp, 64          sub   $sp,  64
sw   $ra, 60($sp)    swlda $ra, 60($sp), $da0
sw   $fp, 56($sp)    swda  $fp, 56($sp), $da0
sw   $s0, 52($sp)    swda  $s0, 52($sp), $da0
```

Figure 5.7. Code Common at C Function Entry, and its Transformation. The `swlda` instruction writes `$da0` with the cache location of the data at virtual address `$sp + 60`. The swda instructions operate on that cache line without powering up the cache tags.

the store) without adding additional instructions. The compiler keeps the stack 32-byte aligned to support this transformation.

4.3 DA Register Implementation

At minimum, a DA register needs only to record the matching way within the cache set. In this case, the effective address is used to obtain the subbank number, the set index, and the offset within the cache line. In some implementations, however, it will be advantageous to also record subbank and set index information in the DA registers and to physically distribute the DA registers among the cache subbanks. This avoids recalculating and retransmitting these portions of the virtual address for tag-unchecked accesses. Further implementation details are given in [12].

4.3.1 Keeping DA Registers Coherent.

The DA registers must be kept coherent with the state of the cache. If a line pointed to by a DA register is evicted, the DA register contents are no longer valid and cannot be used in a tag-unchecked access; i.e., the inclusion property between the DA registers and the primary cache must be preserved. Lines may be evicted either as a result of cache line replacement or by external invalidate requests to maintain cache coherence with DMA I/O traffic or other processors.

The primary cache can be used as a filter for snooping invalidations to the DA register, and the DA registers are only searched when a cache eviction occurs. Searching the DA tags consumes some additional energy on each evict, but it is only a small addition to the total cost of the cache line replacement which might involve fetching a line from DRAM.

4.4 Compiler Analysis

The general compiler algorithm that is used to eliminate tag checks is straightforward. Find two references, one of which dominates the other (so all paths that cause the subordinate access to be executed cause the dominant reference to be executed first). If we can prove that the two references always point to the same cache line, the second reference can skip the tag check. It is made tag-unchecked by having the dominant reference write a direct address register that the subordinate register reads.

Code between the two references, including assignments, control flow or even function calls, does not affect correctness because hardware will invalidate the DA register if the line gets evicted between the definition and the use of that DA register (as discussed in section 4.3.1). However, the compiler must analyze pointer assignments to prove that two references always point to the same location.

The compiler controls the stack pointer and so can ensure it is always aligned to a cache boundary. This allows easy transformation of function entry/exit code (like in Figure 5.7), spill code, and parameter passing code. Small automatic variables are never allocated across stack cache line boundaries, so references to local variables and spill code profit from use of the DA registers. Heap and static data require special compiler passes to restructure and analyze loops. Further details of compiler techniques are given in [12].

After all potential pairs are identified, each dominant reference that has a successful subordinate reference is a DA register candidate. The next task is

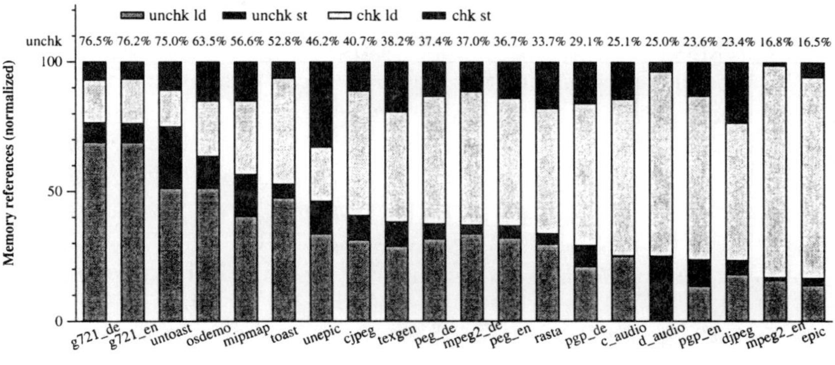

Figure 5.8. Tag Check Elimination for Mediabench Programs Compiled by SUIF Eight direct address registers are used. The lowest part of the bar is tag unchecked loads, then unchecked stores. Over that are tag checked loads and stores. The number on top of each bar (unchk) is the percentage of tag checks eliminated.

to map these candidates onto the limited set of DA registers. This is a standard register allocation problem—DA register candidates that are live at the the same program point interfere and need to be allocated to different registers. The DA register allocation problem is simpler than processor register allocation because a DA register can not be spilled. Instead of spilling, a DA register is simply not allocated to a problematic DA variable.

4.5 Evaluation

The tag-unchecked compiler analysis was implemented for the SUIF compiler [5], which was configured to output instrumented C code. The instrumented code has loops unrolled and is augmented with statistics gathering capability. Figure 5.8 shows how many tag checks were eliminated and whether the elimination was for a load or store. It is important to break these cases out since the tag check is less of the total energy of a store since the value update takes energy.

5. Discussion and Future Work

One concern with exposing detailed internal machine state is that it can seriously constrain the design of future ISA-compatible implementations. In some embedded application areas, it is already accepted practice to change underlying ISA design frequently where there is a significant cost, performance, or power advantage. In other cases, dynamic translation of a virtual machine ISA alleviates the need for native machine compatibility (for example, Transmeta ships two incompatible VLIW ISAs to which x86 code is dynamically translated). Further development should make possible new energy-exposed ISAs

gcc	gsm(encode)	jpeg(decode)	pegwit(decode)
sltiu,beq (9.7)	addiu,subu,sll, addu (30.6)	sll,addu (16.4)	sll,addu (18.8)
lui,lw (9.2)	sll,sra (15.6)	addu,addiu,sra (9.0)	srl,andi (9.0)
addiu,bne (8.2)	addu,sltu,beq (7.2)	addu,addu (8.9)	lui,addu,lw (9.0)
addu,lw (7.3)	mflo,addiu,sll,sra, addu (6.7)	sltu,bne (4.6)	addu,lw (6.8)
sll,addu (4.3)	sll,sra,mult (6.5)	lw,or (4.6)	xor,xor (6.1)
addiu,beq (4.3)	slti,bne (4.2)	lw,addu,addu (4.5)	andi,sltiu,bne (6.0)
slt,bne (3.2)	mflo,addu,sw (3.4)	addu,sra,addu, addu (4.5)	andi,bne (5.5)
lui,addiu (2.9)	addu,addu,addu (2.4)	sll,addu,sll,subu (3.3)	andi,sll (5.4)
slti,bne (2.9)	sll,addu (1.8)	addu,addu,sra,andi, addu (3.2)	andi,addiu (4.9)
slti,beq (2.4)	addu,addu (1.6)	subu,addu,sra,andi, addu (3.2)	andi,sltiu (4.7)

Table 5.2. Most Frequent Instruction Chains for Selected Benchmarks
The numbers in parentheses are the percentages of all chains in the dynamic program execution represented by the corresponding instruction sequences.

that are portable across a range of implementations, and several of these features could be added as extensions to existing ISAs.

5.1 Instruction Chains

Explicitly targeting bypass latches would not work in a processor that reorders instructions, as the latches are typically overwritten every cycle. For example, in the following sequence:

```
add RS, r1, r2
sub r3, RS, r4
slt r5, r6, r7
```

if the `slt` instruction is issued after the `add` instruction but before the `sub` instruction, it will clobber the value in the `RS` bypass latch as it proceeds down the pipeline. One way to make use of exposed temporary state in an out-of-order processor is through the use of *instruction chains*. An instruction chain is a string of instructions where each instruction uses the result of a previous instruction. Data dependencies require that instructions in a chain must be processed in order, but separate chains can be processed out of order. Essentially, the chain is viewed as a single large instruction by the reordering mechanism.

Our simulations show that on average about 41% of all dynamically executed instructions were in chains of at least two instructions; this indicates that the concept of instruction chains warrants further investigation. For example, when designing a new instruction set, chains could be incorporated into the architecture. CISC architectures effectively encode a few types of instruction chains as addressing modes for operands. Table 5.2 lists the frequently occurring chains in programs, and shows the considerable variety in the instruction chains generated for each program. These results indicate that a small number of instruction patterns will not suffice when creating a new instruction set — more generality is needed. Exploiting instruction chains should also allow a more compact instruction set encoding by using implicit chain operands, saving further energy in instruction fetch.

6. Conclusion

Instructions perform many hidden microarchitectural operations as they execute. Compile-time analysis can statically determine that much of this work is unnecessary. By providing an energy-exposed instruction set, this information can be communicated to the hardware to save energy without impacting performance.

One potential difficulty with exposing more machine state to the compiler is that it can potentially increase energy consumption due to exception management overhead. Software restart markers reduce this overhead by enabling the introduction of temporary state that does not have to be saved and restored across exceptions. Exposed bypass latches are an example of allowing software to make use of temporary state to avoid microarchitectural operations at run time; in this case register file reads and writes are statically eliminated. An instruction interface with alternative mechanisms to perform the same task allows a compiler to deactivate unnecessary portions of the machine microarchitecture. Tag-unchecked loads and stores are an example which use compile-time analysis to access the cache with direct address registers instead of costly tag checks.

The three energy-exposed instruction set features presented here will yield greater savings with more sophisticated global compiler analyses, and further energy-exposed instruction set features are under development. These techniques are just a first step towards future energy-efficient instruction set architectures.

References

[1] Furber, S. B. *et al.* ARM3 - 32b RISC processor with 4kbyte on-chip cache. In G. Musgrave and U. Lauther, editors, *Proceedings IFIP TC 10/WG 10.5 International Conference on Very Large Scale Integration (VLSI'89)*, pages 35–44. Elsevier (North Holland), 1989. ISBN 0 444 88344 4.

[2] Intel Corp. *Intel Xscale core developers manual*, order no. 273473-001 edition, December 2000.

[3] G. Kane. *MIPS RISC Architecture (R2000/R3000)*. Prentice Hall, 1989.

[4] R. Krashinsky. Microprocessor energy characterization and optimization through fast, accurate, and flexible simulation. Master's thesis, Massachusetts Institute of Technology, May 2001.

[5] Lam, M. S. *et al.* The SUIF compiler system, 1992–2001. `http://www-suif.stanford.edu`

[6] Montanaro, J. *et al.* A 160-MHz, 32-b, 0.5-W CMOS RISC microprocessor. *IEEE Journal Solid-State Circuits*, 31(11):1703–1714, November 1996.

[7] M. Muller. Power efficiency & low cost: The ARM6 family. In *Hot Chips IV*, August 1992.

[8] V. Oklobdzija. Architectural tradeoffs for low power. In *Power Driven Microarchitecture Workshop at ISCA98*, Barcelona, Spain, June 1998.

[9] J. E. Smith and A. R. Pleszkun. Implementation of precise interrupts in pipelined processors. In *Proc. 12th ISCA*, 1985.

[10] J. Tseng and K. Asanović. Energy-efficient register access. In *Proceedings of the 13th Symposium on Integrated Circuits and System Design*, pages 377–382, Manaus, Amazonas, Brazil, September 2000.

[11] W. Walker and H. G. Cragon. Interrupt processing in concurrent processors. *IEEE Computer*, 28(6):36–46, June 1995.

[12] Witchel, E. *et al.* Direct addressed caches for reduced power consumption. In *MICRO 34*, Dec 2001.

[13] M. Zhang and K. Asanović. Highly-associative caches for low-power processors. In *Kool Chips Workshop, MICRO 33*, December 2000.

III

OPERATING SYSTEM LEVEL POWER MANAGEMENT

Chapter 6

DYNAMIC MANAGEMENT OF POWER CONSUMPTION

Tajana Simunic
HP Labs

Abstract Power consumption of electronic devices has become a serious concern in the recent years. Power efficiency is necessary to lengthen the battery lifetime in the portable systems, as well as to reduce the operational costs and the environmental impact of stationary systems. Two new approaches that enable systems to save power by adapting to changes in environment are proposed: *dynamic power management* and *dynamic voltage scaling*. Dynamic power management (DPM) algorithms aim to reduce the power consumption at the system level by selectively placing components into low-power states. A new event-driven power management algorithm that guarantees globally optimal decisions is presented that is based on Time-Indexed Semi-Markov Decision Process model (TISMDP). TISMDP power management policies have been implemented on four devices: two different hard disks, a laptop WLAN card and a SmartBadge portable system [1]. The measurement results show power savings ranging from a factor of 1.7 up to 5.0 with performance basically unaffected. Dynamic voltage scaling (DVS) algorithms reduce energy consumption by changing processor speed and voltage at run-time depending on the needs of the applications running. This work extends the TISMDP power management model with a DVS algorithm, thus enabling even larger power savings. The measurements of MPEG video and MP3 audio algorithms running on the SmartBadge portable device show savings of a factor of three in energy consumption for combined DVS and DPM approaches.

1. Introduction

Power consumption has become one of the primary concerns in electronic design due to the recent popularity of portable devices and the environmental concerns related to desktops and servers. The battery capacity has improved very slowly (a factor of 2 to 4 over the last 30 years), while the computational demands have drastically increased over the same time frame. Better low-power circuit design techniques have helped to increase battery lifetime

Power Aware Computing
Edited by Graybill and Melhem, Kluwer Academic/Plenum Publishers, 2002

[2,3,4]. On the other hand, managing power dissipation at higher levels can considerably decrease energy requirements and thus increase battery lifetime [5].

System-level *dynamic power management* [6] decreases the energy consumption by selectively placing idle components into lower power states. System resources can be modelled using state-based abstraction where each state trades off performance for power [7]. The transitions between states are controlled by commands issued by a *power manager* (PM) that observes the workload of the system and decides when and how to force power state transitions. The power manager makes state transition decisions according to the *power management policy*. The choice of the policy that minimizes power under performance constraints (or maximizes performance under power constraint) is a constrained optimisation problem. The most common power management policy at the system level is a timeout policy implemented in most operating systems. The drawback of this policy is that it wastes power while waiting for the timeout to expire [8,9]. Predictive policies developed for interactive terminals [10,11] force the transition to a low power state as soon as a component becomes idle if the predictor estimates that the idle period will last long enough. An incorrect estimate can cause both performance and energy penalties. Both timeout and predictive policies are heuristic in nature, and thus do not guarantee optimal results.

In contrast, approaches based on stochastic models can guarantee optimal results. Stochastic models use distributions to describe the times between arrivals of user requests (*interarrival times*), the length of time it takes for a device to service a user request, and the time it takes for the device to transition between its power states. The system model for stochastic optimisation can be described either with memoryless distributions (exponential or geometric) [12,13,14] or with general distributions [15]. Power management policies can be classified into two categories by the manner in which decisions are made: *discrete time* (or clock based) [12,13] and *event driven* [14,15]. In addition, policies can be *stationary* (the same policy applies at any point in time) or *non-stationary* (the policy changes over time). All stochastic approaches except for the discrete adaptive approach presented in [13] are stationary. The optimality of stochastic approaches depends on the accuracy of the system model and the algorithm used to compute the solution. In both discrete and event-driven approaches optimality of the algorithm can be guaranteed since the underlying theoretical model is based on Markov chains. Approaches based on the discrete time setting require policy evaluation even when in low-power state [12,13], thus wasting energy. On the other hand, event-driven models based on only exponential distributions show little or no power savings when implemented in real systems since the exponential model does not describe well the request interarrival times of users [15].

Dynamic voltage scaling (DVS) algorithms reduce energy consumption by changing processor speed and voltage at run-time depending on the needs of the applications running. Early DVS algorithms set processor speed based on the processor utilization of fixed intervals and did not consider the individual requirements of the tasks running. There has been a number of voltage scaling techniques proposed for real-time systems. The approaches presented in [16,17,18,19] assume that all tasks run at their worst case execution time (WCET). The workload variation slack times are exploited on task-by-task basis in [20], and are fully utilized in [21]. Work presented in [22] introduces a voltage scheduler that determines the operating voltage by analysing application requirements. The scheduling is done at task level, by setting processor frequency to the minimum value needed to complete all tasks. For applications with high frame-to-frame variance, such as MPEG video, schedule smoothing is done by scheduling tasks to complete twice the amount of work in twice the allocated time. In all DVS approaches presented in the past, scheduling was done at the task level, assuming multiple threads. The prediction of task execution times was done either using worst case execution times, or heuristics. Such approaches neglect that DVS can be done within a task or for single-application devices. For, instance, in MPEG decoding, the variance in execution time on frame basis can be very large: a factor of three in the number of cycles [23], or range between 1 and 2000 IDCTs per frame [24] for MPEG video.

This chapter introduces a new model that combines dynamic power management and dynamic voltage scaling. The new approach is based on Time-Indexed Semi-Markov Decision Process model (TISMDP). The power management policy optimisation problem is solved *exactly and in polynomial time with guaranteed optimal results*. Large savings are measured with TISMDP model as it handles general user request interarrival distributions and make decisions in the event-driven manner. The dynamic voltage scaling enables even larger savings than are possible with only DPM. DVS represents the active state as a series of states characterized by varying degrees of performance and energy consumption. The system's processor runs at the minimum frequency and voltage required to sustain the performance level required by an application and thus saves power when the system is active, in addition to the savings obtained by DPM during idle periods. Combined savings of DPM and DVS on the SmartBadge are as high as a factor of 3.

The remainder of the chapter is organized as follows. Section 1 describes the stochastic models of the system components based on the experimental data collected. Time-Indexed Semi-Markov Decision Process model for the dynamic power management policy optimisation problem is discussed in Section 2, followed by the expansion of TISMDP model to support DVS in Section 3. Measured results for power managing two hard disks and a WLAN card are discussed in Section 4, in addition to the DPM and DVS simulation results for the SmartBadge. Finally, this chapter is summarized in Section 5.

1. System Model

The system model consists of three components: the user, the device and the queue as shown in Figure 6.1. Each system component can be described probabilistically. The user, or the application that accesses each device by sending requests via operating system, is modelled by a request interarrival time distribution. When one or more requests arrive, the user is said to be in the active state, otherwise it is in the idle state. Figure 6.1 shows three different power states for the device: active, idle and sleep. Service time distribution describes the behaviour of the device in the active state. When the device is in either the idle or the sleep state, it does not service any requests. Typically, the transition to the active state is shorter from the idle state, but the sleep state has a lower power consumption. The transition distribution models the time taken by the device to transition between its power states. The queue models a buffer associated with each device. The combination of interarrival time distribution (incoming jobs to the queue) and service time distribution (jobs leaving the queue) appropriately characterizes the behaviour of the queue. Each system component models are described next.

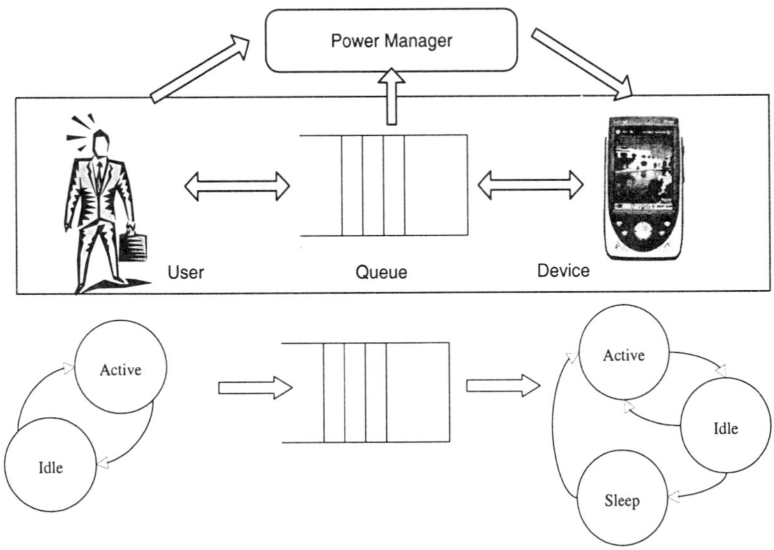

Figure 6.1. System Model

1.1 User Model

The request interarrival times in the active state (the state where at least one user's request is queued up) for all devices are exponentially distributed. Figure 6.2 shows the exponential cumulative distribution fitted to the measurements of 11hr user trace accessing the hard disk of the PC running Windows OS with standard software (e.g Excel, Word, MS VC++). Similar results have been observed for the other devices in the active state [15]. Thus, the user in active state can be modelled with rate λ_U and the mean request interarrival time, $1/\lambda_U$, where the probability of a device receiving a user request within time interval t follows the cumulative probability distribution shown below.

$$F_U(t) = 1 - e^{-\lambda_U t} \tag{6.1}$$

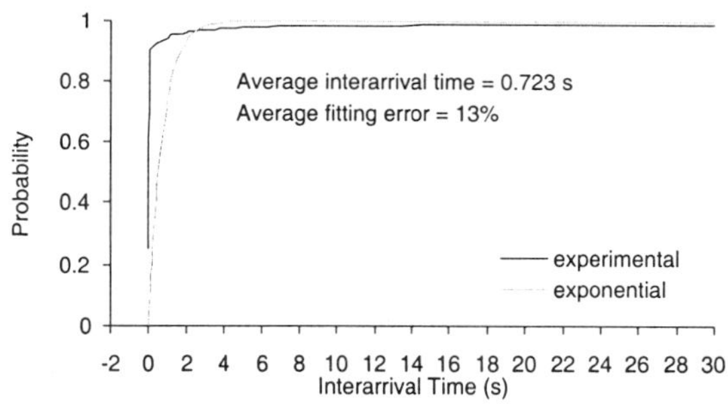

Figure 6.2. User Request Arrivals in Active State for Hard Disk

The exponential distribution does not model well the first request arrival in the idle state. The tail distribution highlights the probability of longer idle times that are of interest for power management. Figure 6.3 shows the measurements of idle period tail distributions fitted with the Pareto and the exponential distributions for the hard disk and Figure 6.4 shows the same measurements for the WLAN card. The Pareto distribution shows a much better fit for the long idle times as compared to the exponential distribution. The Pareto cumulative distribution is defined in Equation 6.2. The Pareto parameters are a=0.9 and b=0.65 for the hard disk, a=0.7 and b=0.02 for WLAN web requests and a=0.7 and b=0.06 for WLAN telnet requests. SmartBadge arrivals behave the same way as the WLAN arrivals.

$$F_U(t) = 1 - at^{-b} \tag{6.2}$$

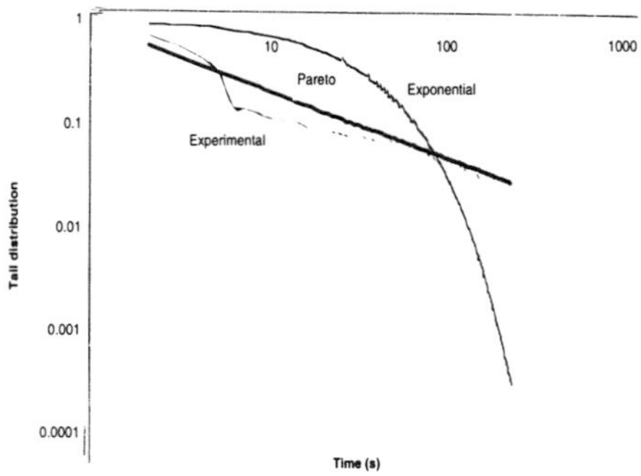

Figure 6.3. Hard Disk Idle State Arrival Tail Distribution

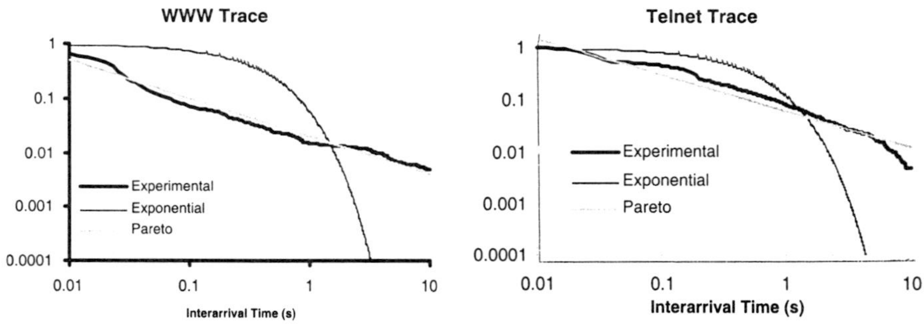

Figure 6.4. WLAN Idle State Arrival Tail Distribution

1.2 Portable Devices

A power managed device often has one or more active states in which it services user requests, and one or more low-power states. The power manager can trade off power for performance by placing the device into low-power states or by adjusting the level of service in the active states. Each low power state can be characterized by the power consumption and the performance penalty incurred during the transition to or from that state. Usually higher performance penalty corresponds to lower power states. While the methods presented in this work are general, the optimization of energy consumption under performance constraints (or vice versa) is applied to and measured on the following devices: WLAN card [25] on the laptop, the SmartBadge [1] and laptop and desktop hard disks. More details about the device models are presented below.

1.2.1 SmartBadge.

The SmartBadge is used as a *personal digital assistants* (PDAs) that operates as a part of a client-server system. The SmartBadge's components, the power states and the transition times of each component from standby (t_{sby}) and off (t_{off}) state into active state, and the transition times between standby and off states (t_{so}) are shown in Table 6.1.

Table 6.1. SmartBadge Components

Component	Active Pwr (mW)	Idle Pwr (mW)	Standby Pwr (mW)	t_{sby} (ms)	t_{off} (ms)	t_{so} (ms)
Display	1000	1000	100	100	240	110
RF Link	1500	1000	100	40	80	20
SA-1100	400	170	0.1	10	35	10
FLASH	75	5	0.023	0.6	160	150
SRAM	115	17	0.13	5.0	100	90
DRAM	400	10	0.4	4.0	90	75
Total	3.5 W	2.2 W	200 mW	110 ms	250 ms	160 ms

The StrongARM processor on the SmartBadge can be configured at run-time by a simple write to a hardware register to execute at one of 11 different frequencies. The number of frequencies is predefined by the design of the StrongARM processor. The transition between two different frequency settings takes 150 microseconds. For each frequency, there is a minimum voltage the SA-1100 needs in order to run correctly, but with lower energy consumption. Figure 6.5 shows the frequency-voltage tradeoff.

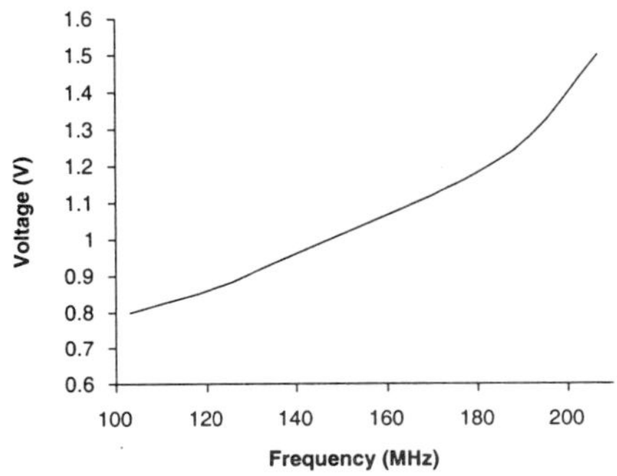

Figure 6.5. Frequency vs. Voltage for SA-1100

1.2.2 Hard Disks. Two different hard disks whose characteristics are shown in Table 6.2 have been used in experiments: the Fujitsu MHF 2043AT hard disk in a laptop and the IBM hard disk in a desktop. Although the power savings in the sleep state are large (P_{active} versus P_{sleep}), the disks should be placed into the sleep state only when the idle period between successive accesses is long enough to justify the performance (T_{active} and T_{sleep}) and power overhead incurred during the transition.

Table 6.2. Disk Parameters

Model	P_{sleep} (W)	P_{active} (W)	T_{sleep} (sec)	T_{active} (sec)
IBM	0.75	3.48	0.51	6.97
Fujitsu	0.13	0.95	0.67	1.61

Hard disk service times in the active state follow an exponential distribution as shown in Figure 6.6. The SmartBadge and the WLAN card measurements have similar patterns. Equation 6.3 defines the cumulative probability of the device servicing a user request within time interval t with the average service rate λ_D.

$$F_D(t) = 1 - e^{-\lambda_D t} \tag{6.3}$$

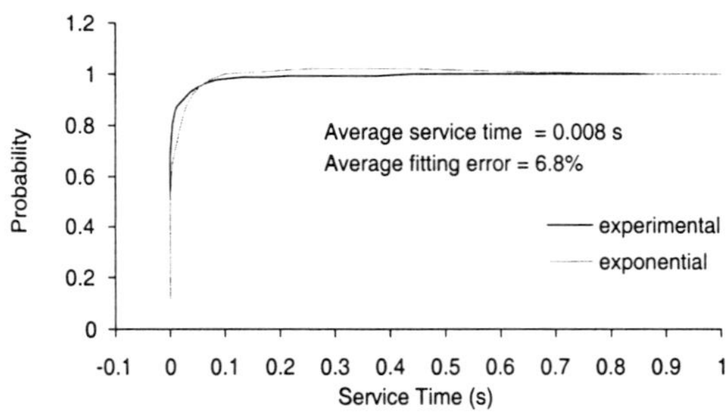

Figure 6.6. Hard Disk Service Time Distribution

The transition from sleep to active state requires a spin-up of the hard disk, which is very power intensive. This transition, in addition to the transition from idle to the sleep state, is best described using the uniform distribution shown in Equation below, where t_0 and t_1 can be defined as $t_{ave} - \Delta t$ and $t_{ave} + \Delta t$

respectively. Figure 6.7 shows the measurement results for the transition of Fujitsu hard disk from sleep to active state. Similar results can be obtained for the other devices.

$$F_D(t) = \begin{cases} 0 & t \le t_0 \\ \frac{t-t_0}{t_1-t_0} & t_0 < t \le t_1 \\ 1 & t > t_1 \end{cases} \tag{6.4}$$

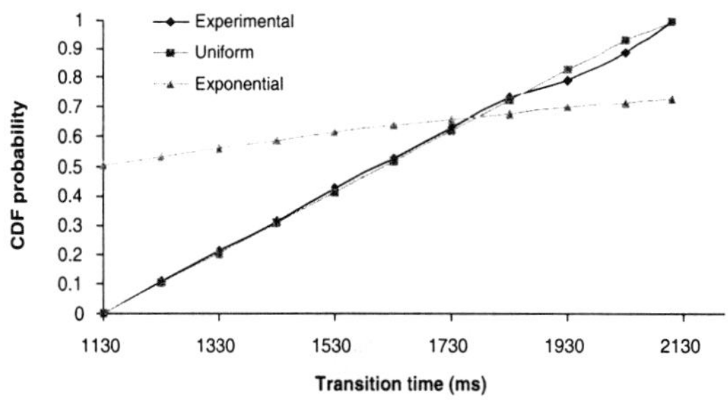

Figure 6.7. Hard Disk Transition from Sleep to Active State

1.2.3 WLAN card. The wireless local area network (WLAN) card has multiple power states: two active states, *transmitting, receiving*, and two inactive states, *doze* and *off*. Transmission power is 1.65W, receiving 1.4W, the power consumption in the doze state is 0.045W [25] and in the off state it is 0W. Once both receiving and transmission are done, the card automatically enters the doze state. Unfortunately, savings of only 5-10% in power have been measured with this approach, due to the overhead of having to be awake every 100ms to find out if any communication needs to take place. In client-server systems, such as the laptop used in this work, it is clear when communication is finished on the client side. Thus, the power manager can turn the card off once the communication is finished, and turn in back on when the client wishes to resume communication. Once in the off state, the card waits for the first user request arrival before returning back to the doze state. The transitions between the doze and the off states have been measured and are best described using the uniform distribution. The transition from the doze state into the off state takes on average $t_{ave} = 62ms$ with variance of $t_{var} = 31ms$. The transition back takes $t_{ave} = 34ms$ with $t_{var} = 21ms$ variance.

1.3 Queue

Portable devices have a buffer for storing requests that have not been serviced yet. This buffer is modelled as a queue. An experiment on a hard disk using a typical user trace measured a maximum queue size of 10 jobs. Because the service rate in the SmartBadge and WLAN card is higher, and the request arrival rate is comparable, the same maximum queue size can be used. As there is no priority associated with requests coming into the queue, active and low-power states are differentiated only by the number of jobs pending for service.

1.4 Model Overview

Table 6.3 shows the probability distributions used to describe each system component derived from the experimental results. User request interarrival times with at least one job in the queue are best modelled with the exponential distribution. Pareto distribution best models the arrival of the first user's request when the queue is empty. The service times in the active state follow the exponential distribution. The transitions to and from the low power states are uniformly distributed. The combination of these distributions is used to derive the state of the queue. Although the experimental section of this Chapter utilizes the fact that the non-exponential user and device distributions can be described with well-known functions (Pareto or uniform), the models presented are general in nature and thus can give optimal results with both experimental distributions obtained at run time.

Table 6.3. System Model Overview

System Component	Component State	Distribution
User	Queue not empty	Exponential
	Queue empty	Pareto
Device	Active	Exponential
	Transition	Uniform

2. Dynamic Power Management

Dynamic power management (DPM) techniques selectively place system components into low-power states when they are idle. A power managed system can be modelled as a *power state machine*, where each state is characterized by the power consumption and the performance. In addition, state transitions have power and delay cost. Usually, lower power consumption also implies lower performance and longer transition delay. When a component is placed into a low-power state, such as a sleep state, it is unavailable for the time period spent there, in addition to the transition time between the states. The break-even time, T_{be}, is the minimum time a device should spend in the low-power state to compensate for the transition cost. The break-even time can be calculated directly from the power state machine of the device as shown in Equation 6.5, where P_{as}, T_{as} and P_{sa}, T_{sa} are the power consumption and the time to transition between the active and the sleep states, and P_a, P_s are the power consumptions in the active and the sleep states. With devices where the transition cost into inactive state is minimal, the power management policy is trivial (once in the idle state, shut off). In all other situations it is critical to determine the most appropriate policy that the PM should implement.

$$T_{be} = \frac{P_{as}T_{as} + P_{sa}T_{sa} - P_a(T_{as} + T_{sa})}{P_a - P_s} \tag{6.5}$$

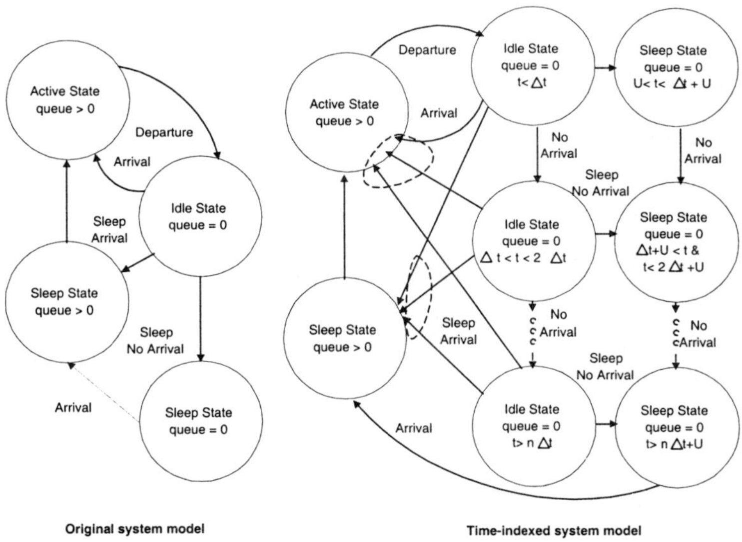

Figure 6.8. Time-Indexed SMDP States

This section presents the power management optimisation problem formulation based on *Time-Indexed Semi-Markov Decision Processes* (TISMDP). TISMDP model is a generalization of *Semi-Markov decision processes* (SMDP) model. SMDP allows for at most one non-exponentially distributed transition at a time and enables the power manager to respond on event occurrences instead of being tied to a system clock as in traditional *Discrete-Time Markov Decision Processes*. *Continuous-time Markov decision processes* (CTMDP) [14] can be viewed as a special case of SMDP model in which all transitions are exponentially distributed. TISMDP model introduced in this work, in contrast to SMDP and CTMDP, uses general distributions to describe system transitions, is still event driven and guarantees optimal results. On transitions where none of the processes are exponential (e.g. transition from idle to sleep state, where user's requests are governed with Pareto distribution and the transition time follows the uniform distribution), time-indexed Markov chain formulation is used to keep the history information [27].

Time-indexing is done by dividing the time line into a set of intervals of equal length Δt. The original state space is expanded by replacing one idle and one queue empty low-power state in SMDP model with a series of time-indexed idle and low-power empty states in TISMDP model as shown in Figure 6.8. The expansion of idle and low-power states into time-indexed states is done only to aid in deriving in the optimal policy. Once the policy is obtained, the actual implementation is completely event-driven in contrast to the policies based on discrete-time Markov decision processes.

The *interevent time set* is defined as $\mathcal{T} = \{t_i, \{s.t.\} i = 0, 1, 2, \ldots, i_{max}\}$ where each t_i is the time between the two successive event arrivals and i_{max} is the index of the maximum time horizon. We denote by $s_i \in \mathcal{S}_i$ the system state at decision epoch i. At each event occurrence, the power manager issues a *command* or an *action* that decides the next state to which the system should transition. An action that is issued at decision epoch i is denoted by $a_i \in \mathcal{A}$. In general, commands given are functions of the state history and the policy. Commands can be either deterministic or randomised. In the former case, a decision implies issuing a command, in the later case it gives the probability of issuing a command. The actions taken by the PM form a sequence which completely describes the optimal power management policy.

The goal of TISMDP optimisation is to minimize the performance penalty under an energy consumption constraint (or vice versa). The linear program minimizing energy consumption under performance constraint is shown in Equation 6.6. Additional constraints can be added easily. Because the problem can be formulated as a linear program, globally optimal solution can be obtained that is *stationary* (the functional dependency of PM actions on the states does not change with time) and *randomised* under the presence of constraints. The problem can be solved in polynomial time as a function of number of states and actions.

$$\textbf{LPD:} \min \sum_{s \in S} \sum_{a \in A} cost_{energy}(s,a) f(s,a) \tag{6.6}$$
$$\text{s.t.} \sum_{a \in A} f(s,a) - \sum_{s' \in S} \sum_{a \in A} m(s'|s,a) f(s',a) = 0$$
$$\sum_{s \in S} \sum_{a \in A} y(s,a) f(s,a) = 1$$
$$\sum_{s \in S} \sum_{a \in A} cost_{perf}(s,a) f(s,a) < Constraint$$

The linear program minimizes the cost in energy under a set of constraints. The first constraint is really a set of constraints for each system state. This constraint formulation is called a *balance equation*, because it specifies that the number of ways a system can transition into any given state has to equal the number of ways it transitions out of that state. The second constraint defines the expected time spent in each state when action a is given. The last constraint is the performance constraint. The next set of equations defines all variables in the optimisation problem shown above for both the states that do not need time indexing and for those that need it.

The average cost incurred between two successive decision epochs (events) for states that are not time indexed is defined in Equation 6.7 as a sum of the lump sum cost $k(s_i,a_i)$ incurred when action a_i is chosen in state s_i, in addition to the cost in state s_{i+1} incured at rate $c(s_{i+1},s_i,a_i)$ after choosing action a_i in state s_i. We define $\mathcal{S}_{i+1}$ as the set of all possible states that may follow s_i. Equation 6.8 defines the same cost for the time-indexed states. $\mathcal{S}_{i+1}$ as the set of all possible states that may follow s_i. When action a_i is chosen in system state s_i, the probability that the next event will occur by time t_i is defined by the cumulative probability distribution $F(t_i|s_i,a_i)$. The probability that the system transitions to state s_{i+1} at or before the next decision epoch t_i is defined by $p(s_{i+1}|t_i,s_i,a_i)$.

$$cost(s_i,a_i) = k(s_i,a_i) + \tag{6.7}$$
$$\int_0^\infty [E(du|s_i,a_i) \sum_{s_{i+1}} \int_0^u c(s_{i+1},s_i,a_i) p(s_{i+1}|t_i,s_i,a_i)]dt$$

$$cost(s_i,a_i) = k(s_i,a_i) + \sum_{s_{i+1} \in \mathcal{S}_{i+1}} c(s_{i+1},s_i,a_i) y(s_i,a_i) \tag{6.8}$$

The probability of arriving to state s_{i+1} given that the action a_i was taken in state s_i is defined by $m(s_{i+1}|s_i,a_i)$ for states that do not need time indexing as shown in Equation 6.9.

$$m(s_{i+1}|s_i,a_i) = \int_0^{\infty} p(s_{i+1}|t_i,s_i,a_i)F(dt|s_i,a_i) \tag{6.9}$$

For time indexed state, the probability of transition to the next idle state is defined to be $m(s_{i+1}|s_i,a_i) = 1 - p(s_{i+1}|t_i,s_i,a_i)$ and of transition back into the active state is $m(s_{i+1}|s_i,a_i) = p(s_{i+1}|t_i,s_i,a_i)$. The general cumulative distribution of event occurrences is given by $F(t_i)$, while the probability of getting an arrival is defined using the time indices for the system state where $t_i \leq t \leq t_i + \Delta t$:

$$p(s_{i+1}|t_i,s_i,a_i) = \frac{F(t_i+\Delta t) - F(t_i)}{1-F(t_i)} \tag{6.10}$$

Expected time spent in each state for the states that are not time indexed is given in Equation 6.11, while the expected time for the time indexed states is given in Equation 6.12:

$$y(s_i,a_i) = \int_0^{\infty} t \sum_{s_i \in S} p(s_{i+1}|t_i,s_i,a_i)F(dt|s_i,a_i) \tag{6.11}$$

$$y(s_i,a_i) = \int_{t_i}^{t_i+\Delta t} \frac{(1-F(t))dt}{1-F(t_i)} \tag{6.12}$$

The $A \cdot S$ unknowns in the LPD, $f(s,a)$, called *state-action frequencies*, are the expected number of times that the system is in state s and command a is issued. The exact and the optimal solution to the TISMDP policy optimization problem belongs to the set of *Markovian randomized stationary policies* [26] that can be compactly represented by associating a value $x(s,a) \leq 1$ with each state and action pair in the TISMDP, as defined in Equation 6.13.

$$x(s_i,a_i) = \frac{f(s_i,a_i)}{\sum_{a_i \in A} f(s_i,a_i)} \tag{6.13}$$

2.1 Policy Implementation

The optimal policy obtained by solving the linear program given in the previous Section can be presented as a table of cumulative probabilities $P(s,a)$ calculated based on the probability distribution described with $x(s,a)$. Once the system enters a decision state (e.g. idle state), a pseudo-random number

RND is generated. The device stays in the decision state until either the transition to the low-power state as given by RND and the policy, or until a request arrival forces the transition into the active state. The time interval for which the policy gives the cumulative probability $P(s,a)$ of going to the low-power state greater than RND is the time when the device will start that transition. Thus the policy works like a randomised timeout. Once the device is in the low-power state, it stays there until the first request arrives, at which point it transitions back into the active state.

Table 6.4. Sample Policy

Idle Time (ms)	Idle to Standby Probability	Idle to Off Probability	Standby to Off Probability
0	0	0	0
50	0.4	0.2	0
100	0.1	0.9	0.8

Example

The SmartBadge has two decision states: idle and standby. From the idle state, it is possible to transition to the standby or to the off state. From standby, only a transition to the off state is possible. The optimal policy (sample is shown in Table 6.4) gives a table of probabilities determining when the transition between the idle, the standby and the off states should occur. A sample policy may specify that if the system has been idle for 50ms, the transition to the standby state occurs with probability of 0.4, the transition to the off state with probability of 0.2 and otherwise the device stays idle. If the SmartBadge was placed into standby state at time 50ms, then the probability to transition into the off state at 100ms would be 0.8, and otherwise the device stays in the standby state. When a user request arrives, the SmartBadge transitions back into the active state.

3. Dynamic Voltage Scaling

Dynamic Voltage Scaling (DVS) algorithms adjust the device speed and voltage according to the workload at run-time. Since most systems do not need peak performance at all times, decreasing the device speed and voltage during less busy periods increases energy efficiency. Implementing a DVS algorithm for a processor requires both hardware and software support that is not commonly available yet, even though there have been a few examples of DVS implementation such as in [28].

TISMDP policy presented in Section 2 only decides when to transition the device into one of the low-power states. The addition of DVS algorithm enables power manager to also make decisions on the CPU frequency and voltage setting while in the active state. Thus, instead of having only one active state, now there is a set of active states, each characterized by different performance (CPU frequency) and power consumption (CPU voltage) as shown in Figure 6.9. The transformation from one active into multiple active states is completely compatible with the rest of the original model, as both TISMDP and DVS use exponential distributions to describe the behaviour in the active states.

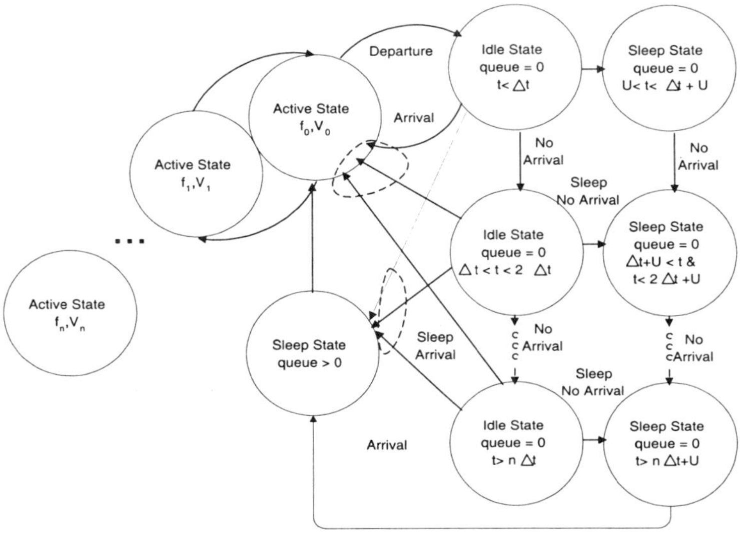

Figure 6.9. Expansion of the Active State

The newly extended power manager works as follows. At run-time, it observes user request arrivals and service completion times (e.g. frame arrivals and decoding times), the number of jobs in the queue (e.g. number of frames in the buffer) and the time elapsed since last entry into idle state. When in the active state, the power manager checks if the rate of incoming or decoding frames has changed, and then adjusts the CPU frequency and voltage accordingly. Once the decoding is completed, the system enters idle state. At this point the power manager observes the time spent in the idle state, and depending on the TISMDP policy, it decides when to transition into one of the sleep states. When an additional processing request arrives from the user, the power manager transitions the system back into the active state and starts the processing requests. The DVS algorithm consists of two main portions: detection of the change in request arrival or servicing rate, and the policy that adjusts the CPU frequency and voltage. The detection is done using maximum likelihood ratio that guarantees optimal results. Policy is implemented based on M/M/1 queue results to ensure constant average delay. Detecting the change in rate is a critical part of optimally matching CPU frequency and voltage to the requirements of the user. For example, the rate of MP3 audio frames coming via RF link can change drastically due to the changes in the environment. The servicing rate can change due to variance in computation needed between MPEG frames [23,24], or just by changing the MP3 audio source. The request (frame) interarrival times and servicing (decoding) times follow the exponential distribution as discussed in Section 1. The two distributions are characterized by the user arrival rate, λ_U, and the device servicing rate, λ_D.

The change point detection is performed using maximum likelihood ratio, P_{max}, as shown in Equation 6.14. Maximum likelihood ratio computes the ratio between the probability that a change in rate did occur (numerator in Equation 6.14) and the probability that rate did not change (denominator). The probability that the rate changed is computed by fitting the exponential distribution with an old rate, λ_o, to the first $k-1$ interarrival or servicing times (t_i), and another exponential distribution with a new rate, λ_n, to the rest of the points observed in window of size w (which contains the last w interarrival times of user requests). The probability that the rate did not change is computed by fitting the interarrival or decoding times with the exponential distribution characterized by the current (or old) rate, λ_o.

$$P_{max} = \frac{\prod_{i=1}^{k-1} \lambda_o e^{-\lambda_o t_i} \prod_{i=k}^{w} \lambda_n e^{-\lambda_n t_i}}{\prod_{i=1}^{w} \lambda_o e^{-\lambda_o t_i}} \tag{6.14}$$

An efficient way to compute the maximum likelihood ratio, P_{max}, is to calculate the natural log of P_{max} as shown below:

$$ln(P_{max}) = (w-k+1) ln \frac{\lambda_n}{\lambda_o} - (\lambda_n - \lambda_o) \sum_{i=k}^{w} t_i \tag{6.15}$$

The advantage of using $ln(P_{max})$ is that only the sum of interarrival (or decoding) times needs to be updated upon every arrival (or service completion). A set of possible rates, Λ, where $\lambda_o, \lambda_n \in \Lambda$, is predefined, as well as the size of the window w. Variable k is used to locate the point in time when the rate has changed.

The change point detection algorithm consists of two major tasks: off-line characterization and on-line threshold detection. Off-line characterization is done using stochastic simulation of a set of possible rates to obtain the value of $ln(P_{max})$ that is sufficient to detect the change in rate. The results are accumulated in a histogram, and then the value of maximum likelihood ratio that gives very high probability that the rate has changed is chosen for every pair of rates under consideration. In this work 99.5% likelihood is selected.

On-line detection collects the interarrival time sums at run time and calculates the maximum likelihood ratio. If the maximum likelihood ratio computed is greater than the one obtained from the histogram, then there is 99.5% likelihood that the rate change occurred, and thus the CPU frequency and the voltage are adjusted. We found that a window of $w = 100$ is large enough. Larger windows cause longer execution times, while much shorter windows do not contain statistically large enough sample and thus give unstable results. The change point can be checked every $k = 10$ points. Larger values of k interval mean that the changed rate is detected later, while with very small values the detection is quicker, but also causes extra computation.

The adjustment of frequency and voltage is done using M/M/1 queue model [27,29]. The goal is to keep the average processing queue delay constant:

$$Delay = \frac{\lambda_D}{\lambda_U(\lambda_U - \lambda_D)} \tag{6.16}$$

When either interarrival rate, λ_U, or the servicing rate, λ_D, change, the delay is evaluated and the new frequency and voltage are selected that will keep the delay constant. For example, if the arrival rate for MP3 audio frames changes, the equation shown above is used to calculate the required decoding rate in order to keep the frame delay (and thus performance) constant. If a different frame decoding rate is detected while processor is set to the same frequency, then piece-wise linear approximation based on the application frequency-performance tradeoff curve is used to obtain the new processor frequency setting. In either case, when CPU frequency is set to a new value, the CPU voltage is always adjusted according to Figure 6.5.

4. Results

4.1 Dynamic Power Management

Policy optimisation is performed with a linear program solver [30] in just under 1 minute on a 300MHz Pentium processor. Large savings are measured on three different devices: laptop and desktop hard disks and the WLAN card. All policies implemented are compared with two bounds: *always-on* and *oracle* policies. Always-on policy leaves a device in the idle state, and thus does not save any power. Oracle policy gives the lowest possible power consumption, as it transitions the device into sleep state with the perfect knowledge of the future. It is computed off-line using a previously collected trace. Obviously, the oracle policy is an abstraction that cannot be used in run-time DPM.

Hard disk measurements have the power manager as a part of a filter driver template attached to the vendor-specific device driver [31]. Application programs such as word processors or spreadsheets send requests to the OS. When any event occurs that concerns the hard disk, power manager is notified. When the PM issues a command, the filter driver creates a power transition call and sends it to the device which implements the power transition using Advanced Configuration and Power Interface standard [7]. The change in power state is also detected with the digital multimeter that measures current consumption of the hard disk.

Table 6.5. Hard Disk Measurement Comparison

Algorithm	Laptop Pwr (W)	N_{sd}	N_{wd}	$T_{ss}(s)$	Desktop Pwr (W)	N_{sd}	N_{wd}	$T_{ss}(s)$
Oracle	0.33	250	0	118	1.64	164	0	166
TISMDP	0.40	326	76	81	1.92	156	25	147
Adaptive	0.43	191	28	127	1.97	168	26	134
Karlin's	0.44	323	64	79	1.94	160	15	142
30s timeout	0.51	147	18	142	2.05	147	18	142
DTMDP	0.62	173	54	102	2.60	105	39	130
120s timeout	0.67	55	3	238	2.52	55	3	238
Always on	0.95	0	0	0	3.48	0	0	0

Comparison of power and performance penalty for all policies measured on the laptop and the desktop is shown in Table 6.5. Performance of the policies is compared using three different measures. N_{sd} is defined as the number of times the policy issued sleep command. N_{wd} gives the number of times the sleep command was issued and the hard disk was asleep for shorter than the time needed to recover the cost of spinning down and spinning up the disk. Clearly, it is important to minimize N_{wd} while maximizing N_{sd}. The average length of time spent in the sleep state (T_{ss}) should be as large as possible while still keeping the power consumption down. From our experience with the user interaction with the hard disk, our algorithm performs well, thus giving us low-power consumption with still good performance. In comparison, Karlin's policy consumes 10% more power and has worse performance. Karlin's algorithm [8] guarantees to yield a policy that consumes at worst twice the minimum amount of power consumed by the policy computed with perfect knowledge of the user behaviour. In addition, our policy consumes 1.7 times less power than the default Windows timeout policy of 120s and 1.4 times less power than the 30s timeout policy on the laptop. TISMDP policy performs better than the adaptive model [13], and significantly better than the policy based on discrete-time Markov decision processes (DTMDP). The event-driven nature of TISMDP algorithm, as compared to algorithms based on discrete time intervals, saves considerable amount of power while in sleep state as it does not require policy evaluation until an event occurs. Similar results can be observed for the desktops.

In addition to the hard disks, the measurements have been performed also on Lucent's WLAN 2Mb/s card [25] connected to a Linux laptop. When comparing different policies, a LAN-attached host reads the 2.5 hr WWW and 2hr telnet traces collected by tcpdump [32] utility and delays or drops packets accordingly. Three different versions of TISMDP algorithm (labelled TISMDP a,b,c) with different power and performance penalty are implemented for each application. Since web and telnet arrivals behave differently (see Figure 6.4),

Table 6.6. WLAN Measurement Comparison

	WWW				Telnet			
Algorithm	N_{sd}	N_{wd}	$T_p(s)$	$P_{ave}(W)$	N_{sd}	N_{wd}	$T_p(s)$	$P_{ave}(W)$
Oracle	395	0	0	0.467	766	0	0	0.220
TISMDP(a)	363	96	6.90	0.474	798	21	2.75	0.269
TISMDP(b)	267	14	1.43	0.477	782	33	2.91	0.296
Karlin's	623	296	23.8	0.479	780	40	3.81	0.302
TISMDP(c)	219	9	0.80	0.485	778	38	3.80	0.310
CTMDP	3424	2866	253.7	0.539	943	233	20.53	0.361
Default	0	0	0	1.410	0	0	0	1.410

the OS observes what application is currently actively communicating and informs the power manager. Performance penalty is determined with three different measures. Delay penalty, T_p, is the time the system had to wait to service a request since the card was in the sleep state when it should not have been. The total number of shutdowns, N_{sd} and the number of shutdowns where sleep time is too short to make up for the total cost of transition, N_{wd} are measured as well.

The power management policy results presented in Table 6.6, TISMDP a,b,c, show, on average, a factor of three in power savings with a low performance penalty for the WWW application. Karlin's algorithm [8] has low power consumption, but its performance penalty is an order of magnitude larger than for TISMDP policy. A policy that models all stochastic processes with the exponential arrivals only, CTMDP, has a larger performance penalty because its decision is based only on the current system state and not on the previous history. TISMDP policy applied to Telnet trace shows a factor of five in power savings. Telnet allows larger power savings because on average it transmits and receives much less data then the WWW browser, thus giving more chances to shut down the card.

4.2 Dynamic Voltage Scaling

DVS algorithm is implemented as a part of a power manager on the SmartBadge for two different applications: MPEG video decoder and MP3 audio decoder. During the times that the system is idle, the TISMDP power management policy described in this Chapter decides when to transition the SmartBadge into a sleep state. When it is in the active state (the state where audio and video decoding occur), the power manager (PM) observes changes in the frame arrival and decoding rates using change point detection algorithm described previously. Once a change is detected, the PM evaluates the required value of the processor frequency that keeps the frame delay constant. The CPU voltage is set using results shown in Figure 6.5.

Rate change detection algorithm is compared to the ideal detection and to the exponential moving average algorithm. Ideal detection assumes knowledge of the future; thus the system detects the change in rate exactly when the change occurs. The exponential moving average can be defined as follows:

$$\lambda_n = (1-g)\lambda_o + g\lambda_{cur} \tag{6.17}$$

where λ_n is the new average rate, λ_o is the old average, λ_{cur} is the current measured rate and g is the gain. Figure 6.10 shows the comparison results for detecting a change from 10 fr/sec to 60 fr/sec. Change point detection algorithm presented in this Chapter detects the correct rate within 10 frames and is more stable than either of the two the exponential moving average algorithms (with different gain values).

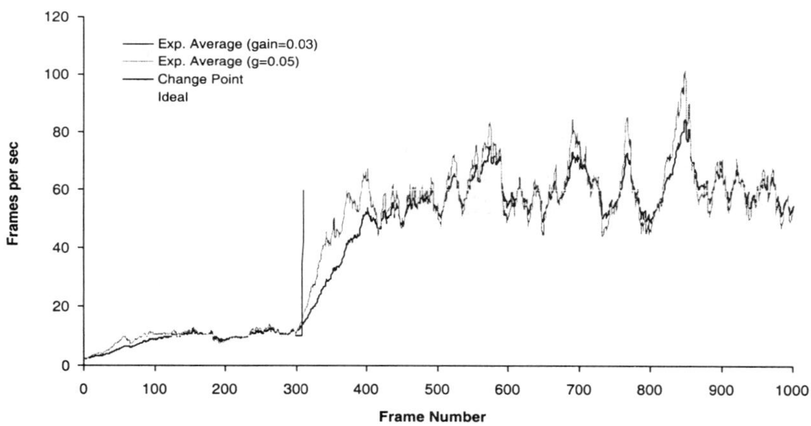

Figure 6.10. Rate Change Detection Algorithms

Three different audio clips totaling 653 seconds of audio, each running at a different set of bit (16,32,64 Kb/s) and sample rates (16 or 32 KHz) have been used to test the DVS algorithm. The ideal detection algorithm, the exponential average approximation used in previous work and the maximum processor performance are compared to the change point algorithm presented in this Chapter. During decoding, the DVS algorithm detects changes in both arrival and decoding rates for the MP3 audio sequences. The resulting energy (kJ) and average total frame delay (s) are displayed in Table 6.7. The change point algorithm presented in this work performs well, its results are very close to the ideal, with no performance loss as compared to the ideal detection algorithm that allows an average 0.1 s total frame delay (corresponding to 6 extra frames of audio in the buffer).

Table 6.7. MP3 Audio DVS

MP3 Audio Sequence	Result	Ideal	Change Point	Exp. Ave.	Max
Clip 1	Energy	196	217	225	316
	Fr.Delay	0.1	0.09	0.1	0
Clip 2	Energy	189	199	231	316
	Fr.Delay	0.1	0.09	0.1	0
Clip 3	Energy	190	214	232	316
	Fr.Delay	0.1	0.04	0.1	0

The next set of results are for decoding two different video clips. Results are reported for the ideal detection, the exponential average, the maximum processor performance and change point algorithm presented in this work. The ideal detection algorithm allows for 0.1s average total frame delay equivalent to 2 extra frames of video in the buffer. The arrival rate varies between 9 and 32 frames/second. Energy (kJ) and average total frame delay (s) are shown in Table 6.8. The results are similar to MP3 audio. The exponential average shows poor performance and higher energy consumption due to its instability (see Figure 6.10). The change point algorithm performs well, with significant savings in energy and a very small performance penalty (0.11s frame delay instead of allowed 0.1s).

Table 6.8. MPEG Video DVS

MPEG Video Clip	Result	Ideal	Change Point	Exp. Ave.	Max
Football	Energy	214	218	300	426
(875s)	Fr.Delay	0.1	0.11	0.16	0
Terminator2	Energy	280	294	385	570
(1200s)	Fr.Delay	0.1	0.11	0.16	0

Finally, the dynamic voltage scaling is combined with power management algorithm. This experiment uses a sequence of audio and video clips, separated by some idle time. During longer idle times, the power manager has the opportunity to place the SmartBadge in the sleep state. Table 6.9 shows the energy savings with only dynamic voltage scaling, only power management and finally also for the combination of the two approaches. Combined savings are as high as a factor of three.

Table 6.9. DPM and DVS

Algorithm	Energy (kJ)	Factor
None	4260	1.0
DVS	3142	1.4
DPM	2460	1.7
Both	1342	3.1

5. Summary

As most systems do not need peak performance at all times, it is possible to transition system components into low-power states when they are idle (dynamic power management) and to adjust frequency and voltage of operation to the workload (dynamic voltage scaling). This chapter presents an event-driven power management algorithm based on Time-Indexed Semi-Markov Decision Process (SMDP) model that guarantees globally optimal results for systems modeled with general distributions. Large power savings, ranging from a factor of 1.7 to a factor of 5 have been observed when using TISMDP policy on four different portable devices: the laptop and the desktop hard disks, the WLAN card and the SmartBadge portable device.

TISMDP model has been extended to enable dynamic voltage scaling. The dynamic voltage scaling algorithm consists of two tasks: (i)change point detection to recognize the change in arrival or decoding rates, and (ii) the frequency setting policy that sets the processor frequency and voltage based on the current arrival and decoding rates in order to keep constant performance. The new DVS algorithm gives a factor of 1.4 to 2.2 savings in energy at a small performance penalty for MP3 audio and MPEG video applications running on the SmartBadge. DPM and DVS algorithms implemented together on the SmartBadge have a factor of 3 energy savings.

Acknowledgments

The author wishes to thank Dr. Giovanni De Micheli and Dr. Luca Benini for their input into this work. In addition, this work would not have been possible without the help and support of colleagues at HP Labs and Stanford University.

References

[1] G. Q. Maguire, M. Smith and H. W. Peter Beadle "SmartBadges: a wearable computer and communication system", *6th International Workshop on Hardware/Software Codesign*, 1998.

[2] A. Chandrakasan, R. Brodersen, *Low power digital CMOS design*, Kluwer, 1995.

[3] J. Rabaey, M. Pedram (Editors), *Low power design methodologies*, Kluwer, 1996.

[4] W. Nabel, J. Mermet (Editors), *Lower power design in deep submicron electronics*, Kluwer, 1997.

[5] C. Ellis, "The case for higher-level power management", *7th IEEE Workshop on Hot Topics in Operating Systems*, pp.162–167, 1999.

[6] L. Benini and G. De Micheli, *Dynamic Power Management: design techniques and CAD tools*, Kluwer, 1997.

[7] Intel, Microsoft and Toshiba, "Advanced Configuration and Power Interface specification", 1996.

[8] A. Karlin, M. Manesse, L. McGeoch and S. Owicki, "Competitive Randomized Algorithms for Nonuniform Problems", *Algorithmica*, pp. 542–571, 1994.

[9] D. Ramanathan, R. Gupta, "System Level Online Power Management Algorithms", *Design, Automation and Test in Europe*, pp. 606–611, 2000.

[10] M. Srivastava, A. Chandrakasan. R. Brodersen, "Predictive system shutdown and other architectural techniques for energy efficient programmable computation," *IEEE Transactions on VLSI Systems*, vol. 4, no. 1, pp. 42–55, March 1996.

[11] C.-H. Hwang and A. Wu, "A Predictive System Shutdown Method for Energy Saving of Event-Driven Computation", in *International Conference on Computer Aided Design*, pp. 28–32, 1997.

[12] L. Benini, G. Paleologo, A. Bogliolo and G. De Micheli, "Policy Optimization for Dynamic Power Management", in *IEEE Transactions on Computer-Aided Design*, vol. 18, no. 6, pp. 813–833, June 1999.

[13] E. Chung, L. Benini and G. De Micheli, "Dynamic Power Management for non-stationary service requests", *Design, Automation and Test in Europe*, pp. 77–81, 1999.

[14] Q. Qiu and M. Pedram, "Dynamic power management based on continuous-time Markov decision processes", *Design Automation Conference*, pp. 555–561, 1999.

[15] T. Simunic, L. Benini, P. Glynn, G. De Micheli, Event-driven Power Management, *IEEE Transactions on CAD*, pp.840–857, July 2001.

[16] I. Hong, D. Kirovski, G. Qu, M. Potkonjak, M. Srivastava, "Power optimization of variable voltage-core based systems," *Proceedings of Design Automation Conference*, pp.176–181, 1998.

[17] I. Hong, M. Potkonjak, M. Srivastava, "On-line Scheduling of Hard Real-time Tasks on Variable Voltage Processor," *Proceedings of International Conference on Computer-Aided Design*, Nov. 1998.

[18] T. Ishihara, H. Yasuura, "Voltage Scheduling Problem for dynamically variable voltage processors," *Proceedings of IEEE International Symposium on Low Power Electronics and Design*, pp.197–202, 1998.

[19] F. Yao, A. Demers, S. Shenker, "A scheduling model for reduced CPU energy," *IEEE Annual foundations of computer sciend*, pp.374–382, 1995.

[20] Y. Shin, K. Choi, "Power conscious fixed priority scheduling for hard real-time systems," *Proceedings of Design Automation Conference*, pp.134–139, 1999.

[21] S. Lee, T. Sakurai, "Run-time voltage hopping for low-power real-time systems," *Proceedings of IEEE International Symposium on Low Power Electronics and Design*, pp.806–809, 2000.

[22] T. Pering, T. Burd, R. Brodersen, "The simulation and evaluation of Dynamic Voltage Scaling Algorithms" *Proceedings of IEEE International Symposium on Low Power Electronics and Design*, 1998.

[23] A. Bavier, A. Montz, L. Peterson, "Predicting MPEG Execution Times," *Proceedings of SIGMETRICS*, pp.131–140, 1998.

[24] A. Chandrakasan, V. Gutnik, T. Xanthopoulos, "Data Driven Signal Processing: An Approach for Energy Efficient Computing," *Proceedings of IEEE International Symposium on Low Power Electronics and Design*, pp.347–352, 1996.

[25] Lucent, *IEEE 802.11 WaveLAN PC Card - User's Guide*, p.A-1.

[26] M. Puterman, *Finite Markov Decision Processes*, John Wiley and Sons, 1994.

[27] H. Taylor and S. Karlin, *An Introduction to Stochastic Modeling*, Academic Press, 1998.

[28] L. Geppert, T. Perry, "Transmeta's magic show," *IEEE Spectrum*, vol. 37, pp.26–33, May 2000.

[29] S. Ross, *Stochastic Processes*, Wiley Press, 1996.

[30] S. Skiena, *The Algorithm Design Manual* , 1997.

[31] Y. Lu, T. Šimunić and G. De Micheli, "Software Controlled Power Management", *7th International Workshop on Hardware/Software Codesign*, pp. 157–161, 1999.

[32] V. Jacobson, C. Leres, S. McCanne, *The "tcpdump" Mannual Page*, Lawrence Berkeley Laboratory.

Chapter 7

POWER MANAGEMENT POINTS IN POWER-AWARE REAL-TIME SYSTEMS

Rami Melhem, Nevine AbouGhazaleh, Hakan Aydin, Daniel Mossé
Department of Computer Science
University of Pittsburgh
*Pittsburgh, PA 15260**
{melhem, mosse}@cs.pitt.edu

Abstract Managing power consumption while simultaneously delivering acceptable levels of performance is becoming a critical issue in several application domains such as wireless computing. We integrate compiler-assisted techniques with power-aware operating system services and present scheduling techniques to reduce energy consumption of applications that have deadlines. We show by simulation that our dynamic power management schemes dramatically decrease energy consumption.

Keywords: Processor Power, Dynamic Speed Setting, Speculation, Voltage Scaling

1. Introduction

As the remarkable advances in VLSI and communication technologies have culminated in a proliferation of mobile, embedded and wireless computers in the last decade, system designers are faced with a relatively new and difficult resource management problem. Most of these devices usually have to rely on *battery* power, which is usually rather scarce. Moreover, many popular and emerging applications designed for these platforms, such as wireless communication, and image/audio/video processing, tend to consume considerably higher power than more traditional tasks. Some recent studies even advocate the replacement of CPU-centric operating system design view by the energy-centric view [37]. In short, the *power bottleneck* has to be addressed in an

*This work has been partially supported by the Defense Advanced Research Projects Agency under contract F33615-00-C-1736

efficient way to guarantee the functionality in the upcoming *pervasive computing* era.

The Engineering and Computer Science communities at large confronted the low power system design problems with a multi-dimensional effort [12, 28]. Hardware and software manufacturers have agreed to introduce standards such as the ACPI (Advanced Configuration and Power Interface) [16] for power management of laptop computers that allows several modes of operation, such as predictive system shutdown [34]. Such on-going low-power research has important implications for real-time systems design, simply because most of the applications running on power-limited systems inherently impose temporal constraints on the response time (such as real-time communication in satellites).

An increasingly popular technique for saving power, *Dynamic Voltage Scaling* (DVS) [38], is based on exploiting the convex (usually quadratic) relation between the supply voltage and the CPU power consumption. In addition, it has been long recognized that the CPU clock frequency (hence, the speed) should be reduced in parallel with the supply voltage [8]. In this case, it is possible to obtain striking (quadratic) energy savings at the expense of roughly linearly increased response time. One aspect that needs to be carefully taken into consideration is the energy and delay overhead associated with speed/voltage changes. Some studies are optimistic about the overhead imposed by DVS schemes [30]; we examine this issue in more detail in Section 4.

The DVS framework aims at stretching out task executions through speed and voltage reduction. For real-time systems, the proposed DVS schemes focus on minimizing energy consumption in the system, while still meeting the deadlines. The extensive literature on traditional real-time scheduling theory [22, 7] deals with settings where the CPU speed is constant, and hence can not be directly applied.

The principle of slowing down the processor can and should be applied in multiple dimensions. The natural starting point is the *static* DVS dimension, where the aim is to compute the optimal speed assignments for a given real-time task set and a (worst-case) workload.

In one of the earliest studies in this line, Yao et. al [40] provided an optimal static off-line scheduling algorithm to minimize the total energy consumption while meeting all the deadlines, assuming independent aperiodic tasks with release times and timing constraints. The algorithm has $O(n \log^2 n)$ time complexity for a system of n tasks. Heuristics for on-line scheduling of aperiodic tasks while not hurting the feasibility of off-line periodic requests are proposed in [14], which also suggested assigning a uniform speed value to all periodic tasks using the *total task utilization* as a basis. Non-preemptive power aware scheduling is investigated in [13]. Concentrating on a periodic task set with identical periods, the effects of having an upper bound on the voltage change rate are examined in [15]. The authors show that the problem is intractable even with a linear change rate and propose a heuristic to tackle the problem. The static solution for the general periodic model where tasks have potentially different power consumption characteristics is provided in [4]. Aydin et. al recently showed [5] the equivalence of the *static* dynamic voltage scaling problem to the reward-based scheduling problem [3].

Designing a real-time system with worst-case workload assumption in mind is common and often necessary. However, the *actual* workload may be much lower than the worst-case assumption in many real-time applications. Most of the scheduling schemes presented in the above studies, while using exclusively worst-case execution time (WCET) to guarantee the timeliness of the system, lack the ability to dynamically take advantage of unused computation time. In fact, applications usually exhibit a large variation in *actual* execution times; for example, [9] reports that the ratio of the worst-case execution time to the best-case execution time can be as high as 10 in typical applications. Consequently, *dynamically monitoring and reclaiming the 'unused' computation* can be (and, as we show below, is in fact) a powerful approach to obtain considerable power savings and to minimize the effects of conservative predictions of the *actual* execution time by the WCET information. In this line of research, the aim is to dynamically reduce the CPU speed of running task(s) by taking into account the early completions in the history of task executions. The main problem, naturally, is to determine the speed reduction amount that would not compromise any timing constraints, in addition to choosing the tasks that will be executed at the low speed.

One technique for dynamic reclaiming relies on slowing down the processor whenever there is only a single task eligible for execution [33], where a set of periodic tasks scheduled by Rate Monotonic Scheduling [20] is considered. A more general dynamic reclaiming approach, based on comparing the worst-case execution schedule to the actual schedule is proposed in [17]. In that study, a detailed analysis is provided for frame-based tasks; the extension to general Earliest Deadline First (EDF) scheduling of periodic tasks is sketched. One assumption of that work is that there are (only) two discrete speed levels. However, systems which are able to operate on a (more or less) continuous voltage spectrum are rapidly becoming a reality thanks to advances in power-supply electronics and CPU design [10, 27]. For example, the Crusoe processor is able to dynamically adjust clock frequency in 33 MHz steps [36].

Despite the gains due to *static* and *dynamic reclaiming* schemes, there is still room for additional savings provided that we have access to the *statistical* workload information; in this chapter, we present also aggressive schemes where we *anticipate* the early completions of *future* executions and *speculatively* reduce the CPU speed. This approach immediately raises two intertwined questions, namely, (a) the *level* of aggressiveness that justifies speculative speed reductions under a given probability distribution of actual workload; and (b) the issue of guaranteeing the timing constraints.

Note that both dynamic reclaiming and aggressive scheduling techniques can be adopted at the *task level* **and** at the *system level*. In the former, user- or compiler-inserted Power Management Points (PMPs) allow intra-task monitoring of the execution and controlling speed of a given task to improve the energy savings [1, 32]. In the latter, the operating system invokes the PMPs at context switch times, taking advantage of the global knowledge (i.e., system-wide workload). Finally, it is worth mentioning that a whole new line of research that tolerates (and tries to minimize) deadline misses for the sake of energy savings has recently emerged [23, 18, 19, 31, 11]. These *soft* real-time scheduling techniques also make use of the *statistical* workload information while determining the CPU speed assignments.

In this chapter, we summarize the results of our multi-layered research effort in power-aware scheduling for real-time systems. Our solution is based on a three-dimensional approach that can be applied at the task level or at the system level, while taking into account energy and time overheads. Hence, we present:

1. A *static* (off-line) solution to compute the optimal speed, assuming *worst-case workload*,
2. A *dynamic* (on-line) speed adjustment mechanism based on the *actual workload*, used to reclaim unused time/energy, when executions fall short of their worst-case workload, and
3. An adaptive and *speculative* speed adjustment mechanism based on *statistical information about the workload*, used to anticipate and compensate probable early completions of future executions.

We emphasize once again that, in the context of real-time systems, all these components should be designed not to cause any deadlines to be missed even under the worst-case scenario: the aim is to **meet the timing constraints while simultaneously and dynamically reducing power consumption as much as possible**.

2. Real-Time Task and System Models

Typical real-time research assumes that a task, τ_i, has a deadline, D_i, which is derived from the system's real-time requirements. If we assume that a task is ready at time 0, then D_i can be seen as the length of the time interval within which τ_i is allowed to execute. Given that variable voltage CPUs are available, the time to execute a program, P_i, depends on the processor speed. We will characterize a task (we will use task τ_i and program P_i to denote the same entity) by the worst case *number of CPU cycles*, C_i, needed to execute the program.

In order to simplify the analysis and to allow for the derivation of analytical formulas, we would like to assume that C_i is independent of the CPU speed for a given processor architecture. This assumption, however, does not hold if the speed of the memory system is independent of the speed of the CPU, since memory references will consume larger number of cycles when the processor speed is high, thus increasing the total number of cycles needed to execute the program. For this reason, we assume that C_i is the worst case number of CPU cycles needed to execute a program at the maximum processor speed.

We have conducted a number of simulation experiments using SimpleScalar 3.0 [6] (a micro architectural simulator) to determine the degree of pessimism in the definition of C_i. These experiments show that, with on-chip caches and low cache miss rates, C_i does not change substantially with the processor speed. For the Li, Perl, Go and Compress programs from the SPEC benchmarks [35], changing the processor speed from 700 MHz to 300 MHz changed the number of CPU cycles needed to execute the benchmarks by 0.01%, 1.2%, 1.9% and 0.6%, respectively. In all the experiments, the default SimpleScalar configurations for the L1 and L2 caches are used and no disk I/O is performed (typical assumption for real-time systems).

In this chapter, we normalize the units of C_i such that the maximum processor speed is 1. That is, if the maximum processor speed is s CPU cycles per second, then we define a *hypercycle* to consist of s CPU cycles and express C_i in terms of the number of hypercycles. The maximum processor speed is thus normalized to $S_{max} = 1$ hypercycles per second. We will simply refer to hypercycles by "cycles", and thus, at S_{max}, the time for executing C_i cycles is C_i seconds.

2.1 Modeling control flow

We consider a general form of program execution in which a program P_i is divided into n_i segments, $\tau_{i,(j)}, 1 \leq j \leq n_i$, where a segment is a loop, a procedure call or, in general, any subgraph of the control flow graph of P_i (see Figure 7.1). We assume that each segment $\tau_{i,(j)}$ executes a maximum of $C_{i,(j)}$ cycles. Each segment is represented by a circle. A segment that is composed of a loop is represented by a square and a number representing the maximum loop index. Note that a "segment flow graph" is a compact version of the control flow graph of the program, in which subgraphs are replaced by single nodes. Each execution of the program will follow a specific path from the start node to an end node.

For any given node, j, in the segment flow graph, let $\Pi\_wc_{i,(j)}$ denote the maximum number of cycles to complete the execution of the program, starting at the beginning of segment j. Clearly, $\Pi\_wc_{i,(j)}$, $1 \leq j \leq n_i$ can be computed recursively from

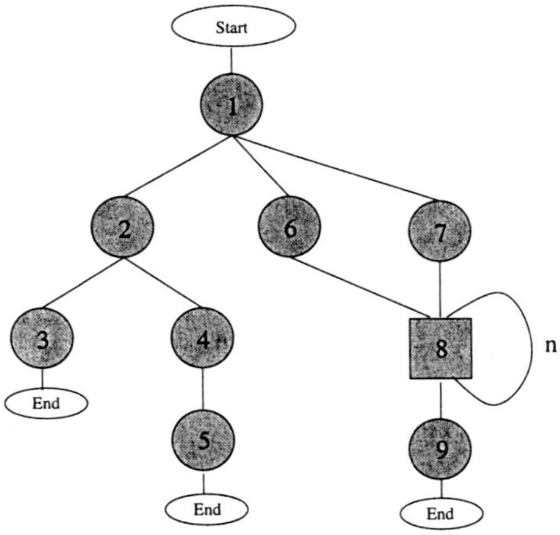

Figure 7.1. Schematic of a Segment Flow Graph

$$\Pi\_wc_{i,(j)} = C_{i,(j)} + \max_{k \in B(j)} \{\Pi\_wc_{i,(k)}\} \tag{7.1}$$

where $B(j)$ is the set of children of node j in the segment flow graph. If node 1 represents the first segment in the program, then $C_i = C_{i,(1)}$ represents the worst case execution of the entire program P_i.

Although knowing the worst case execution scenario of a task is essential to guarantee that a task meets its deadline, usually a program, P_i, executes for much less than its worst case estimate, C_i [9]. This is because the input data and system architecture (e.g., the amount of cache in the system) determine not only the actual number of cycles the segment executes (typically less than C_i), but also determines the execution path of P_i. If we denote by $C\_avg_{i,(j)}$ the average number of cycles consumed in the execution of segment j of P_i, then the average number of cycles to complete the execution of P_i starting at the beginning of segment j, denoted by $\Pi\_avg_{i,(j)}$, can be computed recursively from

$$\Pi\_avg_{i,(j)} = C\_avg_{i,(j)} + \sum_{k \in B(j)} Prob_k \cdot \Pi\_avg_{i,(k)} \tag{7.2}$$

where $Prob_k$ is the probability that execution proceeds from segment j to segment k. Clearly, execution should proceed from segment j to one of the segments in $B(j)$, and thus $\sum_{k \in B(j)} Prob_k = 1$.

2.2 Periodic task model

Typically, real-time systems execute periodic tasks, where each task τ_i has associated with it a period, T_i, which represents the interarrival time of consecutive instances of the task. We will assume that a task τ_i is ready for execution at the beginning of its period and should complete execution by the end of its period. *Frame-based systems* are special periodic real-time systems in which a sequence of frames is repeatedly executed, and all tasks in a frame have the same period, T, and the same initial phasing. In practice, many real-time systems are frame-based, since designing and verifying the correctness for such systems is much simpler than for more general real-time systems.

Given a set of tasks, $\{\tau_1, \ldots, \tau_N\}$, let $U = \sum_{i=1}^{N} \frac{C_i}{T_i}$ be the total utilization of the task set under the maximum processor speed (recall that we normalized $S_{max} = 1$). The utilization U can also be seen as the load imposed on the system by the task set under consideration. It is well known that if $U \leq 1$, $T_i = D_i$, and EDF scheduling is used, then each instance of every task will meet its deadline [20]. In this chapter, we will assume that there are no precedence constraints among the N periodic tasks and that EDF is used to schedule these tasks.

2.3 Power consumption model

Variable-voltage CPUs can reduce power consumption *quadratically* or *cubically* at the expense of *linearly* increased delay (reduced speed) [15]. Thus, any effective DVS scheme should be able to vary the voltage fed to the system component and the frequency of the system clock. The power consumption of the processor under the speed S is given by $g(S)$, which is assumed to be a strictly increasing convex function, represented by a polynomial of at least the second degree [15]. If task τ_i occupies the processor during the time interval $[t_1, t_2]$, then the *energy* consumed during this interval is $\int_{t_1}^{t_2} g(S(t))dt$, which is equal to $g(S)(t_2 - t_1)$ if S is constant during the period $[t_1, t_2]$. Unless stated otherwise, we assume that the CPU speed can be changed continuously between a minimum speed, S_{min} (minimum supply voltage necessary to keep the system functional), and the maximum speed S_{max}. The idle processor power consumption, g_{idle}, is usually less than $g(S_{min})$ but larger than $g(0)$.

The importance of the speed management is derived from the convexity of the power function $g(S)$. Specifically, if τ_i is allotted Y_i time units to execute, then $S_i = \frac{C_i}{Y_i}$ is the speed that will execute C_i in exactly Y_i time units. Because the function g is convex, we have

$$g(S_i)Y_i \leq g(S')x + g(S'')(Y_i - x)$$

for any $S' \neq S''$ and x such that $S'x + S''(Y_i - x) = C_i$. This means that the total energy consumption is reduced when the processor speed is uniform during the Y_i time units. Note that in case of a single task, $Y_i = D$, but when there are several tasks, their allocations typically fall short of their deadline.

3. PMPs: Power Management Points

In the context of power management through CPU speed adjustment, a *power management point*, PMP, is an abstraction that lets us reason about and influence power management at specific points in time. Typically, a PMP will have associated with it a piece of code that manages information about the execution of the tasks in the system and decides about changing the CPU speed. A PMP code can be part of the user's program (executes in user space) or can be part of the operating system (executes in kernel space). In either case, after making a decision to change the CPU speed, a PMP typically makes the appropriate system calls to change both the CPU clock frequency and CPU voltage. We distinguish between two types of PMPs.

- A *task-level PMP* is invoked during the execution of a given task, τ_i, and uses information only about τ_i to make decisions about τ_i's execution speed. Task-level PMPs may be inserted by the user or by the compiler in the program. For instance, the user or the compiler may insert a PMP at the beginning of each segment of the program.
- A *system-level PMP* takes a more global view of the system and uses information about all the tasks in the system to make speed adjustment decisions. For instance, after determining the next task to be dispatched, the scheduler in an operating system may execute a PMP to determine the execution speed of the next task.

To make a decision, a PMP uses *task profile information*, such as worst case executions, average case executions and timing or performance constraints, and *execution progress information* such as CPU time consumption and early termination of tasks. The task profile information can be precomputed and stored, while the execution progress information must be collected at run time with help from the hardware and/or the operating system. Different speed adjustment schemes can be designed based on how a PMP computes the slack existing in the system and on how to use this slack to carry out power (via CPU speed) management.

3.1 Static (off-line) power management

In static power management, a PMP computes the processor speed based on the assumption that each task, τ_i, will execute the maximum number of cycles, C_i. If executing C_i cycles at speed S_{max} does not consume the entire time allocated to τ_i, then it is possible to reduce the speed of executing τ_i thus reducing the consumed energy, while still meeting the timing constraints.

This power management scheme is called *static* because only worst-case task profile information and no execution progress information is used in speed adjustments.

For task-level power management, if D_i is the time allocated to execute τ_i, then the speed during the execution of τ_i can be safely set to $S_i = \frac{C_i}{D_i}$. Clearly, if $S_i > S_{max}$, then it is not possible to guarantee completion of τ_i's execution within the given time constraints. Also, if $S_i < S_{min}$, we should set $S_i = S_{min}$. A single PMP at the beginning of task's execution is needed to calculate and set the processor speed to S_i.

For system-level power management of N periodic tasks scheduled using EDF, the convexity of the power function, $g(S)$, implies that all deadlines can be met and that the total energy is minimized when the speed of the processor is the same for all the tasks. Hence, when executing a periodic task set with utilization $U \leq 1$, the energy consumption is minimized if the CPU speed is set uniformly to $max\{S_{min}, US_{max}\}$.

In order to appreciate the energy savings resulting from static power management, assume conservatively that the idle power, g_{idle}, is equal to zero and assume that $g(S) = \alpha S^3$, for some constant α. If T_{lcm} is the least common multiple of the periods $T_1, \ldots, T_N$, then, executing at $S_{max} = 1$ during T_{lcm} will result in an energy consumption equal to $\sum_{i=1}^{N} g_i(S_{max}) \frac{C_i}{S_{max}} \frac{T_{lcm}}{T_i} = \alpha U T_{lcm}$. If the speed is set to US_{max}, then the energy consumption during T_{lcm} reduces to $\sum_{i=1}^{N} g_i(U) \frac{C_i}{US_{max}} \frac{T_{lcm}}{T_i} = \alpha U^3 T_{lcm}$, which is a factor of U^2 lower than $\alpha U T_{lcm}$. For example, if $U = 0.5$, then static power management consumes only 25% of the energy that is consumed without power management.

The optimality of the uniform speed based on the load assumes that the power functions, $g_i(S)$, are the same for all tasks. However, due to the fact that different tasks may use different hardware units and have different patterns of memory and cache usage, we may have a different power consumption function, $g_i()$, for each task τ_i. (The different power functions could refer to different tasks or different segments of the same task.) In this case, the energy consumption is minimized when each task τ_i executes at a speed S_i derived from solving the following optimization problem:

$$minimize\{\sum_{i=1}^{N} g_i(S_i) \frac{C_i}{T_i S_i}\}$$

such that

$$\sum_{i=1}^{N} \frac{C_i}{T_i S_i} \leq 1$$

$$S_{min} \leq S_i \leq S_{max}$$

The above formulation is obtained by noting that when the speed of executing τ_i is set to S_i, then each instance of τ_i executes for a time $\frac{C_i}{S_i}$, and thus the fraction of time allocated to τ_i (the time utilization) is increased from $\frac{C_i}{T_i}$ to $\frac{C_i}{T_i S_i}$. EDF can always meet the deadlines if the sum of utilizations of the tasks in the system is less than one.

After solving the above minimization problem (see [4] for solution techniques), the speed S_i of each task, τ_i, can be stored in its process control block. Before the scheduler dispatches τ_i, it calls a PMP to set the processor speed to S_i. That is, the processor speed change becomes part of the context switch operation.

In summary, static power management for both task-level and system-level aim at the same goal, namely fully utilizing the CPU in the system, assuming worst case execution scenarios. When a single task is involved, the task-level management is sufficient to determine the uniform speed of that task, but when several tasks are involved, the operating system must intervene because it is the only entity that knows about all tasks and their characteristics.

3.2 Dynamic (on-line) power management

Dynamic power management is based on the observation that tasks usually do not execute their worst case scenarios, and thus by using execution progress information, the processor speed can be adjusted during the execution to increase the energy saving beyond that achieved by static power management. To simplify the discussion, we assume that the power consumption function is the same for all the tasks, and that S_s is the optimal speed obtained by static power management.

When tasks do not execute their worst case scenario, slack (unused computation time) is generated dynamically in the system; the first task of a PMP is to compute this slack. One way of estimating this slack is to determine whether a task or task segment is running earlier than predicted by the worst-case scenario. For instance, if the worst case scenario indicates that a PMP should be reached at time t_{wc} and this PMP is actually reached at time t_{ac} ($t_{ac} \leq t_{wc}$), then the difference of $slack_{early} = t_{wc} - t_{ac}$ can be considered as an *earliness slack* which can be used to slow down the processor.

An alternate way of looking at the *earliness slack* is to consider the work (in terms of number of CPU cycles) that remains to be executed until the next deadline, D. Let Π_{wc} and Π_{av} be the worst case estimate and the average case estimate of that work, respectively. Given that static power management fully utilizes the processor assuming worst case execution scenarios, the speed S_s guarantees that the period from t_{wc} to the deadline D is exactly equal to $\frac{\Pi_{wc}}{S_s}$, which is the time needed to execute Π_{wc} at speed S_s. In other words, the slack

$D - t_{ac} - \frac{\Pi_{wc}}{S_s} = D - t_{ac} - (D - t_{wc}) = t_{wc} - t_{ac}$ is the slack time between the current time and the deadline, after accounting for the execution of Π_{wc} at speed S_s.

Given that the worst case execution scenario occurs very rarely, it is reasonable to assume that a more useful estimate of the slack is obtained by assuming the average case scenario for the remaining work, rather than the worst case scenario. For example, one may speculate that the usable slack is equal to $slack_{speculate} = slack_{early} + \frac{\Pi_{wc} - \Pi_{av}}{S_s}$. This speculation is based on the assumption that only the average case scenario for the remaining work will be executed, and that this work will execute at speed S_s. The rationale behind using $slack_{speculate}$ is to consider the average case behavior, which is more common than the worst case behavior.

After a PMP computes *slack*, it uses this slack to slowdown the execution of the next task or task segment. Assuming that the worst case estimate of the number of cycles in this next task or task segment is C, and that S_s is the execution speed calculated from static power management, then the time allocated to the execution of the C cycles is $\frac{C}{S_s}$. Adding the slack to this time, the new execution speed is computed as:

$$S_{next} = \frac{C}{\frac{C}{S_s} + slack} \tag{7.3}$$

However, it should be noted that there is a lower bound on the processor speed in order to guarantee that, in the worst case, the remaining work will be completed by the deadline D. Specifically, if the speed for executing the next task or task segment is S_{next}, then this execution will consume at most a time equal to $\frac{C}{S_{next}}$. Hence, the remaining time until the deadline, $D - t_{ac} - \frac{C}{S_{next}}$ should be at least large enough to guarantee that the worst case scenario of the remaining work, $\Pi_{wc} - C$, can finish by the deadline. Given that the maximum speed is S_{max}, this remaining work will require at least $\frac{\Pi_{wc} - C}{S_{max}}$ to execute. Hence,

$$D - t_{ac} - \frac{C}{S_{next}} \geq \frac{\Pi_{wc} - C}{S_{max}}$$

should always hold, which means that

$$S_{next} \geq \frac{C}{D - t_{ac} - \frac{\Pi_{wc} - C}{S_{max}}} = S_{feasible} \tag{7.4}$$

From Equations (7.3) and (7.4), we can put an upper bound on the amount of slack that can be used to compute the next speed. Namely,

$$slack \leq slack_{max} = (D - t_{ac} - \frac{\Pi_{wc} - C)}{S_{max}}) - \frac{C}{S_s} \tag{7.5}$$

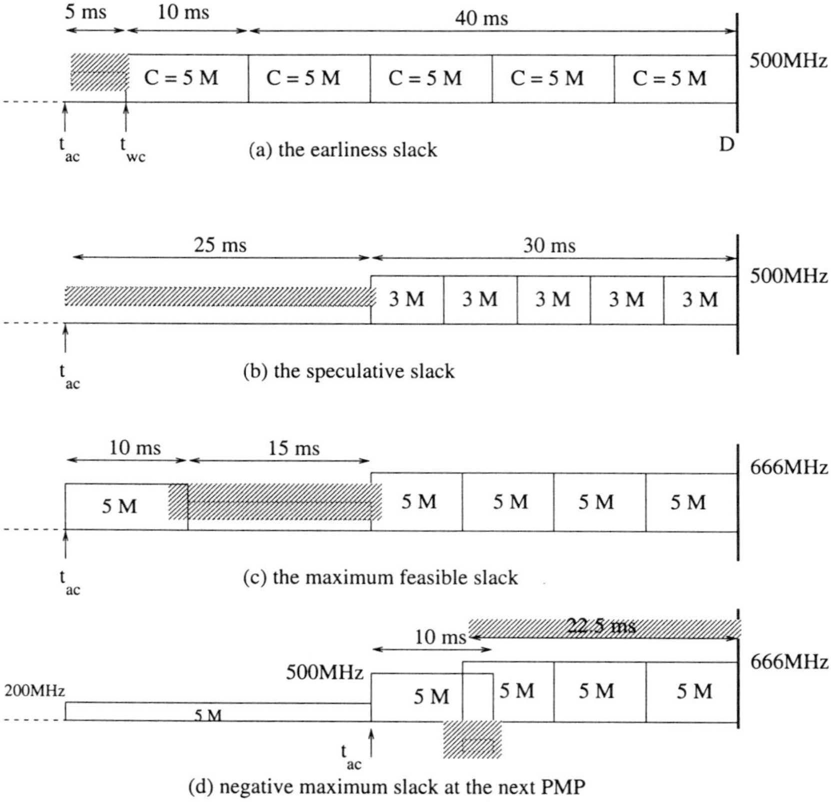

Figure 7.2. An Example of Slack Computation; Height of Box Is CPU Speed for Task

In Figure 7.2 we show an example to clarify the above slacks. The deadline D is $50ms$ later than t_{wc} and the remaining Π_{wc} = 25 million cycles should execute at a speed $S_s = 500$ MHz to meet the deadline (in the figure, M denotes a million cycles). It is assumed that a PMP is placed every 5 million cycles of the remaining work. In the figure, the execution of C at speed S cycles is represented by a rectangle whose area is C. The height of the rectangle is S and its width, $\frac{C}{S}$, is the execution time. Assume that a PMP is reached at time $t_{ac} = t_{wc} - 5ms$ (recall that t_{wc} is reached in the worst case scenario). Figure 7.2(a) shows the earliness slack, while Figure 7.2(b) shows the speculative slack assuming that Π_{av} = 15 million cycles. Figure 7.2(c) shows the maximum slack assuming that the next task consumes 5 million cycles and that $S_{max} = 666MHz$.

In general, the PMP can select any value for the $slack$ between zero and $slack_{max}$. In fact, a zero slack means that the PMP does not perform any dynamic speed adjustment and selects the speed computed by static power management. A slack equal to $slack_{max}$ is a very aggressive scheme which gives all the available slack to the next task or task segment. This scheme may slow down the next task or task segment too much without taking into consideration the remaining computations in the system, perhaps causing the CPU speed to be raised to S_{max}.

If a PMP is very aggressive in speculatively reducing the processor speed, the next PMP may be reached at a time, t'_{ac}, which is later than the time, t'_{wc}, predicted by the worst case scenario, and the maximum slack, $slack_{max}$, may be negative. The consequence of negative slack (too much aggressiveness) is the setting of speed above S_s in order to meet the deadlines. As shown in Figure 7.2(d), if $slack_{max} = 15ms$ were used, $S_{next} = 200$ MHz; if the next task executes its worst case scenario, then at the next PMP, $slack_{max}$ is equal to $-2.5ms$, which means that the speed has to be raised to 666 MHz in order to guarantee that the deadline will not be missed.

4. The Overhead of Speed Management

Changing the speed of a CPU takes time and energy, an overhead that was ignored in the above estimation of the usable slack. In this section we study how long and how much energy is spent in each PMP, and later compare this overhead with the actual gains from DVS.

4.1 Time Overhead

At every PMP, a time overhead is incurred for *computing* the new speed, S_{next}, and for *changing* the speed from the current speed, $S_{current}$, to S_{next} through a voltage transition in the processor's DC-DC switching regulator (resulting in a processor frequency change) and the clock generator. We denote changing both voltage and frequency by the term *speed change*. In order to guarantee that the deadlines are met, the above time overheads must to be considered.

When a PMP is added at the beginning of a segment $\tau_{i,(j)}$, the estimates $C_{i,(j)}$ and $C\_avg_{i,(j)}$ should be modified to include the number of cycles, F_c, needed to run the PMP code. From experiments with SimpleScalar 3.0, where we implemented speed setting and inserted PMPs in different applications, we observed that the overhead, F_c, of computing the new speed varied between 280 and 320 CPU cycles. Therefore, we consider F_c to be constant.

To change voltage, a DC-DC switching regulator is employed. This regulator cannot make an instantaneous transition from one voltage to another [15]. When setting a new speed, both the CPU clock and the voltage fed to the CPU need to be changed, incurring a wide range of delays. For example, the *Strong Arm SA-1100* is capable of on-the-fly clock frequency changes in the range of

59MHz to 206MHz where each speed and voltage change incurs a latency of up to 150 μsec [25], while the $lpARM$ processor [29] (a low-power implementation of the $ARM8$ architecture) takes 15 μs for a full swing from 10 MHz to 100 MHz. Another example is the Transmeta TM5400, which is specifically designed for DVS [36]. Some systems can continue operation while changing speed and voltage [29, 15], but the frequency continues to vary during the transition period. A conservative approach, which we adopt, is to assume that the processor cannot execute application code during this period.

Hence, we assume that a fixed time, F_t, is needed for each speed step transition. That is, the time overhead for speed changes is $F_t \cdot d(S_{current}, S_{next})$, where $d(S_i, S_j)$ is a function that returns the number of speed steps needed to make a transition between S_i and S_j. In the Transmeta model, this function returns how many multiples of 33MHz is the difference between S_i and S_j.

The overhead of changing the speed should be accounted for in order to guarantee that deadlines are met. Specifically, the maximum feasible slack, $slack_{max}$, should be adjusted by subtracting this overhead before computing the new speed. Moreover, because $slack_{max}$ is computed assuming that the processor runs at S_{max}, we should allow the time for a future PMP to switch the processor speed to S_{max} to meet the deadline. This will require $F_t d(S_{next}, S_{max})$ to switch the speed to S_{max}. In other words, $slack_{max}$ computed by Equation (7.5) should be reduced by $F_t d(S_{current}, S_{next}) + F_t d(S_{next}, S_{max})$ to guarantee that deadlines are met.

4.2 Energy Overhead

Given that a PMP executes code which is not part of the application code, the energy consumed during the execution of a PMP is an overhead that exists only because of the power management. In addition to the energy consumed in executing a PMP, there is an energy overhead associated with the change in voltage and frequency to change the CPU speed from $S_{current}$ to S_{next}. This overhead is proportional to the number of speed steps, $d(S_{current}, S_{next})$. In the simulations presented below, we will assume that the energy overhead for changing the speed is equal to $F_t \cdot d(S_{current}, S_{next}) \cdot g(S_{current})$. This means that during the speed change, the CPU consumes power at a rate equal to its consumption, while executing at speed $S_{current}$.

In the next two sections, we compare different power management schemes that use different methods for estimating the slack and different schemes for using the slack. These methods can take into consideration the energy and time overheads, as discussed above.

5. Task-Level Dynamic Power Management

Consider a task, τ_i, which is allotted a time D_i to execute, and assume that a PMP is executed at the beginning of segment j of this task. Moreover, assume that the PMP is actually invoked at time t_{ac}, even though this PMP is supposed to be called at time t_{wc}, according to the worst case scenario executing at speed S_s. In order to compute the speed, $S_{i,(j)}$, for executing segment $\tau_{i,(j)}$, the PMP needs to estimate the slack that it will use for speed adjustment. This slack can vary between zero and $slack_{max}$, which, due to the time and energy overheads should be modified from Equation (7.5) to

$$slack_{max} = \frac{C}{S_{feasible}} - \frac{C}{S_s} - F_t \cdot (d(S_{current}, S_{i,(j)}) + d(S_{i,(j)}, S_{max}))$$

where $S_{current}$ is the currently executing speed. Given that $S_{i,(j)}$ is not known, it can be safely and conservatively approximated by S_{min} to obtain

$$slack_{max} = \frac{C}{S_{feasible}} - \frac{C}{S_s} - F_t(d(S_{current}, S_{min}) + d(S_{min}, S_{max}))$$

As described in Section 3, a PMP can use all or part of $slack_{max}$ to compute the speed of the segment. Then, for $0 \leq slack \leq slack_{max}$,

$$S_{i,(j)} = \frac{C_{i,(j)}}{\frac{C_{i,(j)}}{S_s} + slack} \tag{7.6}$$

In [26], three specific schemes were described for the computation of the slack. The first scheme, called **Greedy**, uses $slack = slack_{early}$. That is, it uses the *earliness slack* to adjust the speed of the next task segment. The second scheme, called **Proportional**, uses $slack = slack_{early}\frac{C_{i,(j)}}{\Pi\_wc_{i,(j)}}$. That is, it distributes the earliness slack to all the future segments and gives to the next segment only a proportional amount of that slack. The third scheme, called **Statistical**, uses $slack = min\{slack_{max}, slack_{speculate}\frac{C_{i,(j)}}{\Pi\_avg_{i,(j)}}\}$. That is, it assumes an average case scenario for the computation of the slack and distributes this slack to all the remaining segments in a proportional fashion.

We implemented a simulator to experiment with the different power management schemes. Inputs to the simulator are a segment flow graph, the ratio of the worst case execution time to the best case execution time (β), and the system utilization or load (U). In the graphs below, the energy consumption values are normalized to the energy consumed by the Static scheme. We ran experiments using a synthetic program that has a segment flow graph similar to one shown in Figure 7.1 where C_i values, the loop indexes, and the actual execution times for each segment are drawn from a normal distribution.

From our experiments with SimpleScalar, we extracted $F_c = 300$ cycles and $F_t = 320$ cycles for a single step of 33 MHz. These values are used in the calculation of the total overhead associated with each PMP, and consequently augmented the C_i of each task i.

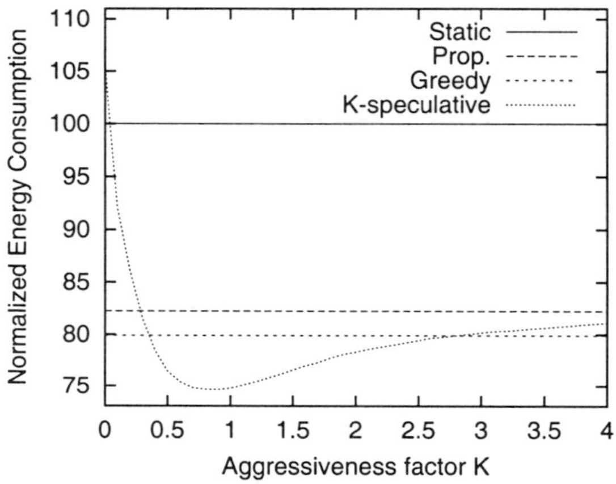

Figure 7.3. Simulation Results for $\beta = 3$ and Load = 0.5.

In Figure 7.3, we show the results of simulating task-level dynamic power management for the Greedy and the Proportional schemes mentioned above, as well as for a scheme which uses $slack = min\{slack_{max}, Kslack_{speculate}\frac{C_{i,(j)}}{\Pi\_avg_{i,(j)}}\}$, where K represents the aggressiveness in using the slack for slowing down the next segment. We call this scheme, which is more flexible, **K-speculative**. Note that the Statistical scheme is the K-speculative scheme with $K = 1$. As shown in the figure, Proportional and Greedy consume less energy than Static, because of the dynamic slack reclaiming. In addition, varying the aggressiveness factor K dramatically affects the K-speculative scheme: for $K = 0$, Static outperforms K-speculative since the latter does not take any advantage of the reclaimed slack, *and* it has to pay the overhead cost at each PMP, while Static has overhead only once, at the beginning of execution. At the other end of the spectrum, it is clear that K-speculative tends to Greedy when K increases. K-speculative reaches its minimum consumption around $K = 1$, that is, approximately the average behavior of the system.

To study the effect of the variability of the workload, we experimented with the same program at different values of $\beta = \frac{C_i}{C_i - 2(C_i - C\_avg_i)}$ (see Figure 7.4). We found that the larger variability, the more dynamic slack to reclaim, and thus the smaller energy consumption. However, the pattern is very similar for all values of β.

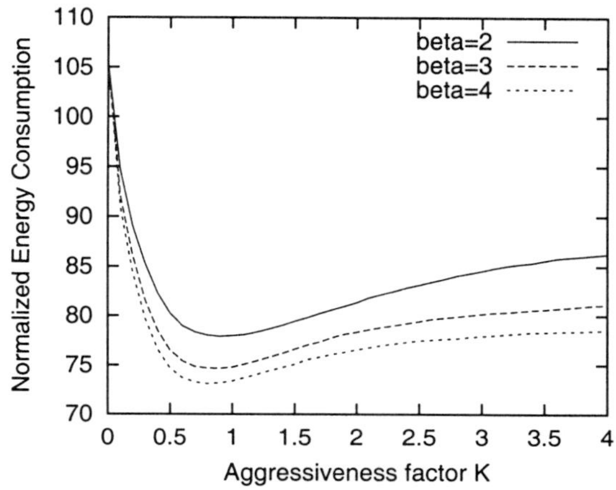

Figure 7.4. Simulation Results for Different Values of β with Load = 0.5.

6. System-Level Dynamic Power Management

The first observation that we make in this section is that the speed adjustment schemes used in system-level dynamic power management for frame-based periodic tasks are similar to the schemes described in the previous section for task-level power management. The difference is that task-level management adjusts the speed for n segments of the same task with a common deadline, D, while system-level management adjusts the speed for N frame-based tasks with a common deadline, T.

The second observation is that it is a good idea to combine task-level and system-level dynamic management, since the former carries out adjustments within the task's allotted time, while the latter attempts to adjust speeds between tasks (when context switches occur). This way, system-level complements the task-level dynamic management, allowing for benefits stemming from both compilers and operating systems.

Thirdly, we note that the case of general periodic tasks is more complicated than frame-based systems because each task has a different deadline. For simplicity of presentation, we do not take the PMP overhead (of computing or changing speeds) into account in this section. The principles are the same as outlined in Section 4. To detect earliness slack, we perform comparisons between the actual execution history and the canonical schedule $\mathcal{S}^{static}$, which is the static optimal schedule on which every instance presents its worst-case workload to the processor and instances run at the constant speed S_s. The CPU speed is adjusted (i.e., a PMP is executed) only at task dispatch times.

Due to the periodic nature of the tasks we consider, it is impractical to produce and keep the entire static optimal schedule $\mathcal{S}^{static}$ (length of $\mathcal{S}^{static}$ could be T_{lcm}) during the execution. To address feasibility and efficiency while tasks execute, complete, and re-arrive dynamically, we choose to construct and update a data structure (called **α-queue**) that helps to compute the earliness slack at dispatch time. At any time t during actual execution, the α-queue contains information about the (up to) N tasks that would be active (i.e., running or ready) at time t in the worst-case static optimal schedule $\mathcal{S}^{static}$. That is, α-queue is the ready queue of $\mathcal{S}^{static}$ at time t and it is constructed and maintained so that the remaining execution time, $rem_i(t)$, of τ_i at time t in $\mathcal{S}^{static}$, under the static optimal speed S_s, is available (note that $rem_i(t) > 0$).

In this chapter, we assume that tasks are scheduled and dispatched according to EDF* policy, which is the same as EDF [20], but if deadlines are the same, the task with the earliest arrival time and then lowest task id has the highest priority. This EDF* *priority ordering* is essential in our approach because it provides a total order on the priorities. (Any scheduling policy that provides total order on priorities will work as well.) We denote the EDF* priority-level of the task i by d_i^*; low values denote high priorities.

To relate the α-queue with the computation of earliness slack, let $w_i^S(t)$ denote the remaining worst-case execution time of task τ_i under the speed S at time t. Note that when task τ_x is being dispatched, tasks with higher priority that are still in the α-queue must be already finished in the actual schedule (since τ_x currently has the highest EDF* priority), but they would have not yet finished in $\mathcal{S}^{static}$.

Therefore, for any task τ_x which is about to execute, any unused computation time (slack) of any task in the α-queue having strictly higher priority than τ_x will contribute to the earliness of τ_x along with the already-finished portion of τ_x in the actual schedule. That is, *total earliness* of τ_x is no less than $\varepsilon_x(t) = \sum_{i|d_i^* < d_x^*} rem_i(t) + rem_x(t) - w_x^{S_s}(t) = \sum_{i|d_i^* \leq d_x^*} rem_i(t) - w_x^{S_s}(t)$.

The α-queue can be easily implemented using the following rules:

R1. Initially the α-queue is empty.

R2. Upon arrival, each task τ_i "pushes" its worst-case execution time under speed S_s to the α-queue in the correct EDF* priority position (this happens only once for each arrival, no re-push at 'return from preemptions').

R3. As time elapses, the elements in the α-queue are updated (consumed) accordingly: the rem_i field at the head of α-queue is decreased with a rate equal to that of the passage of time. Whenever the rem_i field of the head reaches zero, that element is removed from α-queue and the update continues with the next element. No update is done when the α-queue is empty.

Note that, *at time t, the α-queue, updated according to the rules R1, R2 and R3, contains only the tasks that would be ready at time t in the static optimal schedule* $\mathcal{S}^{static}$. This observation stems from the following: (a) α-queue is ordered according to EDF* order, (b) every arriving task pushes its remaining worst-case execution time according to S_s into the α-queue only once, (c) the queue is updated only at the head, reflecting the fact that only the task with the highest EDF* priority would be running in $\mathcal{S}^{static}$, and (d) a task that would have finished in $\mathcal{S}^{static}$ (i.e., $rem = 0$) is removed from the α-queue. This effectively yields *a dynamic image* of the ready queue in $\mathcal{S}^{static}$ at time t.

Note that the dynamic reduction of rem_i in R3 above does not need to be performed at every clock cycle; instead, for efficiency, we perform the reduction whenever a task is preempted or completes, by taking into account the time elapsed since the last update. The above approach relies on two facts. First, the system-level speed adjustment decision will be taken only at arrival/preemption and completion times, and it is necessary to have an accurate α-queue only at these points (if speeds are to be changed at other points like the task-level PMPs, the update of rem_i must reflect that). Second, between these points, each task is effectively executed non-preemptively.

As mentioned in Section 3 any specific algorithm should specify the *exact* amount of earliness slack. One natural choice is to use $\varepsilon_x(t)$, that is, to reduce the speed so as to profit from the full earliness. We call this simply *Dynamic Reclaiming Algorithm* (DRA).

6.1 Speculative Speed Reduction

Another dimension of the reclaiming process is based on the fact that we may speculate on how early tasks will finish. This speculative move is similar to the one for frame-based systems, but the periodic model has different deadlines and periods per task. With tasks arriving and departing dynamically, we need to make sure that no deadlines are violated while speculatively reducing the speed of the current task.

To simplify this speculative slack management, we restrict the speculative power management to occur only when we can limit their effects upto the next event (NE) that corresponds to the arrival/deadline of any task. When we can ensure that a ready task τ_x will not finish beyond NE, we can guarantee that it will complete before its deadline (because, by definition, it is not later than NE). Even then, we may need to increase the speed of other tasks $\tau_{x+1}, \ldots, \tau_r$ that also complete before NE to guarantee timeliness, since these tasks may be delayed if a worst-case scenario occurs for τ_x. Clearly, all these speed adjustments should adhere to S_{min} and S_{max}.

In addition to tasks that finish before NE, even the highest priority ready task that completes after NE, τ_{r+1}, may provide a portion of its time allocation under certain conditions. For simplicity of presentation, we do not describe this more complex scheme (for details, see [5]).

As in frame-based systems, after computing $slack_{speculate}$ and $slack_{early}$, we can assign all or some of this slack to the next task to be executed. This is controlled by an aggressiveness parameter that controls how much of the slack will be used by the next task and how much will be "saved" for future tasks. The computation of these slacks is very similar to the one in Section 5, where the deadline D is replaced by the next event NE. In fact, if $t_{wc} - t_{ac} > 0$, we will be able to slow down task τ_x. Further, if there are other ready tasks that will complete before NE, we need change the speed settings for these tasks.

6.2 Evaluation of the Dynamic Schemes

In order to experimentally evaluate the performance of DRA, we implemented a periodic scheduling simulator for EDF* policy. We implemented the following schemes: (a) **Static** uses constant speed S_s, and switches to power-down mode (i.e., $S = S_{min}$) whenever there is no ready task; (b) **DRA**, which reflects only using $slack_{early}$, and (c) **SPECULATE**, which uses $slack_{speculate}$. All graphs shown here are normalized to the Static scheme, and use the best available aggressiveness parameter (which, similar to frame-based systems, is around 1).

In our experiments, we investigated the average performance of the schemes over a large spectrum of worst-case utilization U (or load) and variability in actual workload (β). The periods of the tasks were chosen randomly in the interval [1000, 32000], S_{min} is set to 0.1, and S_s is set to U. The results shown here focus on the average energy consumption of task sets containing 30 tasks each and random values from a normal probability distribution function; results with different number of tasks and uniform distribution are rather similar [2]. The mean and the standard deviation for any task, τ_i, are set to C_avg_i and $\frac{C_i - C\_avg_i}{3}$ respectively, for a given β, as suggested in [33]. These choices ensure that, on the average, 99.7% of the execution times fall in the interval $[C_i - 2(C_i - C\_avg_i), C_i]$. For each task set, we measured the energy consumption using a cubic power/speed function [15].

6.2.1 Effect of Utilization.

We observed that the energy consumption has very little variation when the utilization of the task set (i.e., U) is changed. This is because the use of optimal speed S_s results in having very similar *effective* utilization, for any value of U. In other words, when the utilization decreases, the speed decreases making the CPU fully utilized.

6.2.2 Effect of β.

The simulation results confirmed our prediction that the energy consumption would be highly dependent on the variability of actual workload. The average energy consumption of the task sets, as a function of β, with $U = 0.6$, is shown in Figure 7.5.

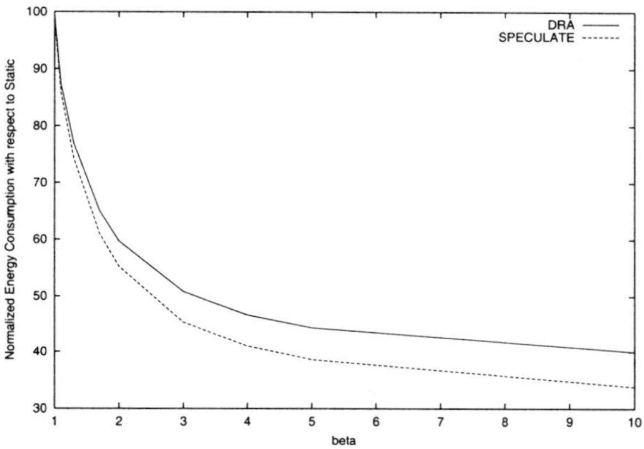

Figure 7.5. Effect of Variability in Actual Workload (Load = 60%)

- When $\beta = 1$, there is no CPU time to reclaim dynamically, and thus the energy consumption is the same for all three techniques, as expected. However, once the actual workload starts decreasing (that is, increasing β), DRA and SPECULATE are able to reclaim unused computation time and are able to save additional energy.
- Once we increase the β beyond 4, power savings of DRA and SPECULATE continue to increase, but the improvement is not as impressive as the case where that ratio is ≤ 4. This is because the expected workload converges rapidly to 50% of the worst-case workload with increasing β (remember that the mean of our probability distribution is C_avg_i.)

7. Maximizing Reward while Satisfying Time and Energy Constraints

So far, we have assumed that the main goal of power management is to minimize the energy consumption while meeting timing constraints. In this section, we will consider a different model in which the goal of the power management is to maximize the system *value* while meeting both timing and energy constraints. In this *reward based* model, each task τ_i has a certain "value", v_i. Given timing and energy constraints, the goal is to select the tasks to execute and the speed at which these tasks should execute such that the total value of the system is maximized. In other words, the goal is to find a subset M of $\{1, \ldots, N\}$, such that when the tasks τ_i, $i \in M$, execute at speed S_i, the timing and energy constraints are satisfied, and the sum of the values of the tasks in M is maximized. To formalize the problem assume that T_{lcm} is the least common multiple of the N periods, $T_1, \ldots, T_N$, and that E_{lcm} is the energy that is available for consumption during one T_{lcm}. Then, the problem is to find the subset of tasks (the set M) and the execution speed S_i for each task τ_i in this subset to

$$maximize \sum_{i \in M} v_i \tag{7.7}$$

subject to

$$\sum_{i \in M} \frac{C_i}{S_i T_i} \leq 1 \tag{7.8}$$

$$\sum_{i \in M} g_i(S_i) \frac{C_i}{S_i T_i} T_{lcm} \leq E_{lcm} \tag{7.9}$$

$$S_{min} \leq S_i \leq S_{max} \tag{7.10}$$

Inequality (7.8) guarantees temporal feasibility (i.e., all deadlines will be met) if EDF is used, inequality (7.9) guarantees that the energy budget will not be exceeded, and inequality (7.10) guarantees that the selected speeds are within the allowable speed bounds. The knapsack problem, which is shown to be NP-hard [24], is a special case of the above problem in which only the first inequality applies with $S_i = S_{max}$. Heuristic search algorithms [24] can be adapted to solve the problem with energy constraints.

A different model is the imprecise computation model [21] in which the value of a task, τ_i, depends on the number of cycles, X_i, that the task actually executes. This dependence is usually expressed in the form of a non-decreasing continuous value function $V_i(X_i)$, in which more reward is given when the task executes more cycles.

Formulated in terms of reward functions, the problem is now to find the allotment X_i, $i = 1, \ldots, N$, and the execution speeds S_i such that to

$$maximize \sum_{i=1}^{N} V_i(X_i) \tag{7.11}$$

subject to

$$\sum_{i=1}^{N} \frac{X_i}{S_i T_i} \leq 1 \tag{7.12}$$

$$\sum_{i=1}^{N} g_i(S_i) \frac{X_i}{S_i T_i} T_{lcm} \leq E_{lcm} \tag{7.13}$$

$$X_i \leq C_i \tag{7.14}$$

$$S_{min} \leq S_i \leq S_{max} \tag{7.15}$$

This is still an open problem and one of our next research targets.

8. Concluding Remarks

In this chapter, we have introduced and described the concept of power management points (PMPs) in real-time systems. The main purpose of a PMP is to manage energy consumption by adjusting the processor speed based on task profiling information (static power management) and/or execution progress information (dynamic power management). Both static and dynamic power management can be done at the task-level, at the system-level, or a combination thereof. Although the static scheme is based on the worst-case workload offered to the system, the dynamic schemes is based on the fact that computations often finish before their predicted worst-case workload. Therefore, the dynamic schemes can take advantage of reclaiming unused computation (based on past actual workload) and speculatively predicting future early completions (based on statistical information of future workload).

The basic concept of reclaiming unused computation time to manage processor speed was described for single processor systems. The extension of these principles to multiprocessor systems depends on the mechanism used for mapping tasks to processors. If such a mapping is done statically, then the extension is rather simple, while if the mapping is done dynamically, then the extension should consider the interaction between the scheduling and the mapping mechanisms. For more details about power management in multiprocessors, see [39, 41].

The overhead of power management is a very important factor that determines the effectiveness of using PMPs. Given that a PMP has to consume some energy to save energy, the issue of the optimal number of PMPs and the placement of these PMPs becomes a crucial factor for the overall energy efficiency of the system. In [1], the tradeoff between the cost and the benefits of the PMP is discussed in a simple computational model. The assessment of this tradeoff in more realistic environments is still open for research.

A basic assumption made in this chapter is that the speed of the processor can be changed continuously within a given range. Current variable speed processors, however, allow the processor speed to be set only to a set of discrete values. For instance, the Transmeta processor allows the speed to be set to increments of 33 MHz within the range from 200 to 700 MHz. The power management technique that we described for the continuous speed range can be safely applied to the discrete speed processors if the speed is set to the closest available speed that is faster than the calculated one. Such a speed setting may not be optimum but guarantees that the deadlines are met.

References

[1] N. AbouGhazaleh, D. Mosse, B. Childers and R. Melhem. Toward The Placement of Power Management Points in Real Time Applications. *Work-*

shop on Compilers and Operating Systems for Low Power (COLP'01), September 2001.

[2] H. Aydin. *Enhancing Performance and Fault Tolerance in Reward-Based Scheduling*. Ph.D. Dissertation, University of Pittsburgh, August 2001.

[3] H. Aydin, R. Melhem, D. Mossé and P.M. Alvarez. Optimal Reward-Based Scheduling for Periodic Real-Time Tasks. *IEEE Transactions on Computers* 50(2): 111-130, February 2001.

[4] H. Aydin, R. Melhem, D. Mossé and P.M. Alvarez. Determining Optimal Processor Speeds for Periodic Real-Time Tasks with Different Power Characteristics. In *Proceedings of the 13th EuroMicro Conference on Real-Time Systems (ECRTS'01)*, Delft, Netherlands, June 2001.

[5] H. Aydin, R. Melhem, D. Mossé and P.M. Alvarez. Dynamic and Aggressive Scheduling Techniques for Power-Aware Real-Time Systems. In *Proceedings of* 22nd *Real-Time Systems Symposium*, December 2001.

[6] D. Burger, and T. Austin. The SimpleScalar Tool Set, Ver 2.0 . *University of Winsconsin-Madison CS Technical report no. 1342*, 1997.

[7] G. Buttazzo. *Hard Real-Time Computing Systems.* Kluwer Academic Publishers, 1997.

[8] E. Chan, K. Govil, and H. Wasserman. Comparing Algorithms for Dynamic Speed-setting of a Low-Power CPU. In *Proceedings of the First ACM International Conference on Mobile Computing and Networking (MOBICOM 95)*, November 1995.

[9] R. Ernst and W. Ye. Embedded Program Timing Analysis based on Path Clustering and Architecture Classification. In *Computer-Aided Design (ICCAD)'97.* pp. 598-604.

[10] V. Gutnik and A. Chandrakasan. An Efficient Controller for Variable Supply Voltage Low Power Processing. *Symposium on VLSI Circuits*, pp.158-159, 1996.

[11] F. Gruian. Hard Real-Time Scheduling for Low-Energy Using Stochastic Data and DVS processors. *International Symposium on Low Power Electronics and Design*, 2001.

[12] P. J. M. Havinga and G. J. M. Smith. Design Techniques for Low-power Systems. *Journal of Systems Architecture*. Vol. 46:1, 2000

[13] I. Hong, D. Kirovski, G. Qu, M. Potkonjak and M. Srivastava. Power optimization of variable voltage core-based systems. In *Proceedings of the 35th Design Automation Conference, DAC'98*

[14] I. Hong, M. Potkonjak and M. B. Srivastava. On-line Scheduling of Hard Real-Time Tasks on Variable Voltage Processor. In *Computer-Aided Design (ICCAD)'98.* pp. 653-656.

[15] I. Hong, G. Qu, M. Potkonjak and M. Srivastava. Synthesis Techniques for Low-Power Hard Real-Time Systems on Variable Voltage Processors.

In *Proceedings of 19th IEEE Real-Time Systems Symposium (RTSS'98)*, Madrid, December 1998.

[16] Intel, Microsoft, and Toshiba. Advanced Configuration and Power Management Interface (ACPI) Specification, 1999. www.intel.com/ial/WfM/design/pmdt/acpidesc.htm.

[17] C. M. Krishna and Y. H. Lee. Voltage Clock Scaling Adaptive Scheduling Techniques for Low Power in Hard Real-Time Systems. In *Proceedings of the 6th IEEE Real-Time Technology and Applications Symposium (RTAS'00)*, Washington D.C., May 2000.

[18] P. Kumar and M. Srivastava. Predictive Strategies for Low-Power RTOS Scheduling. In *Proceedings of International Conference on Computer Design*, 2000.

[19] P. Kumar and M. Srivastava. Power-aware Multimedia Systems Using Run-Time Prediction In *Proceedings of International Conference on VLSI Design*, 2001.

[20] C.L. Liu and J.W.Layland. Scheduling Algorithms for Multiprogramming in Hard Real-time Environment. *J. of ACM* 20(1): pp.46-61, 1973.

[21] J. W.-S. Liu, K.-J. Lin, W.-K. Shih, A. C.-S. Yu, C. Chung, J. Yao and W. Zhao. Algorithms for scheduling imprecise computations. *IEEE Computer,* 24(5): 58-68, May 1991.

[22] J. W. S. Liu. *Real-Time Systems*. Prebtice Hall, 2000.

[23] J. R. Lorch and A. J. Smith. Improving Dynamic Voltage Scaling Algorithms with PACE. In *Proceedings of the ACM SIGMETRICS 2001 Conference*, Cambridge, MA, June 2001.

[24] S. Martello and P. Toth. "Knapsack Problems", *Wiley, InterScience Series in Discrete Mathematics and Optimization*, 1990.

[25] R. Min, T. Furrer, and A. Chandrakasan. Dynamic Voltage Scaling Techniques for Distributed Microsensor Networ ks. *IEEE VLSI Workshop*, 2000.

[26] D. Mossé, H. Aydın, B. Childers and R. Melhem. Compiler-Assisted Dynamic Power-Aware Scheduling for Real-Time Applications. *Workshop on Compilers and Operating Systems for Low-Power (COLP'00)*, Philadelphia, PA, October 2000.

[27] W. Namgoang, M. Yu and T. Meg. A High Efficiency Variable-Voltage CMOS Dynamic DC-DC Switching regulator. *IEEE International Solid-State Circuits Conference*, pp.380-391

[28] M. Pedram. Power Minimization in IC Design: Principles and Applications. *ACM Transactions on Design Automation of Electronics Systems.* 1:1 - pp. 3-56, January 1996.

[29] T. Pering, T. Burd, and R. Brodersen. Voltage Scheduling in the lpARM Microprocessor System. *International Symposium on Low Power Electronics and Design 2000*, pp.96-101, 2000.

[30] J. Pouwelse, K. Langendoen and H. Sips. Dynamic Voltage Scaling on a Low-Power Microprocessor. 7th *International Conference on Mobile Computing and Networking (MOBICOM)*, Rome, Italy, July 2001.

[31] V. Raghunathan, P. Spanos and M. Srivastava. Adaptive Power-Fidelity in Energy-Aware Wireless Embedded Systems. In *Proceedings of* 22nd *Real-Time Systems Symposium (RTSS'2001)*, December 2001.

[32] D. Shin, J. Kim and S. Lee. Intra-task voltage scheduling for low-energy hard real-time applications. *IEEE Design and Test of Computers*, 18:(2), March-April 2001

[33] Y. Shin and K. Choi. Power Conscious Fixed Priority Scheduling for Hard Real-Time Systems. In *Proceedings of the 36th Design Automation Conference, DAC'99*, pp. 134-139.

[34] M. Srivastava, A. P. Chandrakasan and R. W. Brodersen. Predictive System Shutdown and other Architectural Techniques for Energy Efficient Programmable Computation. *IEEE Trans. on VLSI Systems*, 4(1): 42-55, 1996.

[35] http://www.specbench.org

[36] http://www.transmeta.com

[37] A. Vahdat, A. R. Lebeck and C. S. Ellis. Every Joule is Precious: A Case for Revisiting Operating System Design for Energy Efficiency. In *the 9th ACM SIGOPS European Workshop*, September 2000.

[38] M. Weiser, B. Welch, A. Demers, and S. Shenker. Scheduling for Reduced CPU Energy. In *Proceedings of USENIX Symposium on Operating Systems Design and Implementation*, 1994.

[39] P. Yang, C. Wong, P. Marchal, F. Catthoor, D. Desmet, D. Verkest and R. Lauwereins. Energy-Aware Runtime Scheduling for Embedded-Multiprocessor SOCs. *IEEE Design & Test of Computers*, Sep/Oct 2001

[40] F. Yao, A. Demers and S. Shenker. A Scheduling Model for Reduced CPU Energy. *IEEE Annual Foundations of Computer Science*, pp. 374 - 382, 1995.

[41] D. Zhu, R. Melhem, and B. Childers. "Scheduling with Dynamic Voltage/Speed Adjustment Using Slack Reclamation in Multi-Processor Real-Time Systems", RTSS'01 (Real-Time Systems Symposium), London, England, Dec 2001

Chapter 8

A POWER-AWARE API FOR EMBEDDED AND PORTABLE SYSTEMS

Cristiano Pereira*, Rajesh Gupta*, Paleologos Spanos*, Mani Srivastava*

★ *Center for Embedded Computer Systems*

University of California, Irvine

{cpereira,rgupta}@ics.uci.edu

∗ *Networked and Embedded Systems Lab*

University of California, Los Angeles

{paleolog,mbs}@ee.ucla.edu

Abstract Dynamic Power Management (DPM) is an important component of the overall power minimization in embedded computer systems. DPM strategies must often make a tradeoff between real-time performance, computation accuracy and energy needs. However, to make such tradeoffs, we need an efficient way to exchange energy/power and performance related information among the hardware, operating system and the application. In this article, we describe the design and implementation of a power-aware application programming interface (API) that enables the application programmer and OS to implement energy-efficient real-time operating system (RTOS) services and applications. The API provides the capability to specify functionality, precision, and timing contracts between the RTOS and underlying hardware and between the application and the RTOS. Both of these specifications are independent of hardware and RTOS platforms being used. The current implementation provides for services to enable processor voltage and frequency scaling, constraints on frequency-voltage transitions, polling of battery status (if the system is battery based), routines to control energy-speed and energy-accuracy knobs available on IO devices such as network interfaces, display screen, etc. The API is implemented using an open source and object oriented eCos RTOS. The API implementation has been shown to work using the *Compaq iPAQ* (SA 1110) and an Intel XScale based hardware platform.

Keywords: Embedded Systems, Energy-aware operating systems, Power-aware API, Energy-aware applications.

Power Aware Computing
Edited by Graybill and Melhem, Kluwer Academic/Plenum Publishers, 2002

1. Introduction

Dynamic power management (DPM) is an important component of overall power minimization and management strategies in embedded computer systems [3]. Several techniques are being explored from hardware level voltage and frequency scaling [4, 17], to operating system services that use different power modes of resources in an attempt to reduce overall system energy consumption and to modify power profile to match available battery power [1]. While many strategies are specific to the underlying hardware (e.g., voltage-frequency scaling) and software (e.g., task scheduling), the application requirements for functionality and performance delivery under energy constraints can often be specified independent of the platform being used. For instance, the application's requirements may specify a tradeoff between real-time performance, computation accuracy and energy needs. Further, the runtime system often provides a good place to make such tradeoffs. However, to make such tradeoffs, we need an efficient way to exchange energy/power and performance related information among the hardware, operating system and the application. To address this need, we describe here the definition and implementation of a power-aware applications programming interface (API). The API provides means by which an application and the operating system can optimize the power consumption while ensuring satisfaction of application-specific performance requirements. The API is designed to be as independent as possible of the underlying hardware and the specific choice of the operating system. System level power management touches upon almost all aspects of the system architecture from processor, memory, IO to communication subsystems. In this article, we focus on the API aspects related to the processor power management, and specifically inter-task frequency scaling.

Power management has been an important issue for a range of devices. Some of the early works include BIOS-based Advanced Power Management (APM) [11], which provides for CPU and device power management. APM code is executed when CPU is idle or device activity timeout occurs. The principal limitation of APM is that the OS has no knowledge of the APM actions. Its follow on, Advanced Configuration and Power Interface or ACPI [12] allows OS-directed power management by defining hardware registers and BIOS interfaces (tables, control methods). While the actual power management policies are implemented and controlled within the operating system, the user interface allows limited input to the policy use. ACPI is primarily directed at the computer hardware system builders to ensure availability and use of OS-directed power management, giving no support to modify OS services. ACPI also defines standards for configuration and thermal management, which are specific to Intel-based personal computers. ACPI also does not provide any support for real time constraints specification and as a last disadvantage, ACPI software layers are complex, big and not flexible with respect to configurability, characteristics that make it ill suited to be built-in in embedded and portable systems. We propose an API that enables specification of application-specific performance requirements including real-time constraints and the per application tradeoffs against available power versus functionality needs. This API should also enable power-efficient implementation of OS services instead of only making the low-power hardware knobs available to the OS.

By enabling the implementation of different algorithms using a common interface and facilitating their automatic generation in a systematic and orthogonal way, the API provides a framework for comparison of DPM algorithms and application benchmarks.

1. API Requirements

For effective system-level power management, it is important that an application is able to monitor and control both electrical knobs (such as voltage and scheduling), as well as exercise control on the task scheduling and shutdown states of different parts of the system.

Several algorithms in the literature addressing different techniques to both exercise the task scheduling and shutdown states of different parts of the system have been presented in the last few years [25, 22, 23, 18, 19, 13, 20, 14, 26, 6, 2, 24]. Among the algorithms to exercise the task scheduling, [20] presented a framework using a predictive algorithm based on static and dynamic slow down factors. The static slow down factor is derived offline and is based on the schedulability test of a given task set and hence guarantees that the task set is still schedulable, although power is saved. The actual slow down factor depends upon the specific scheduling technique used. The dynamic slow down factor, on the other hand, is derived during run-time and it depends on the time prediction for each task being executed in the system. Consequently, if a task is predicted to spend less time than its worst case execution time in the next execution, the frequency and voltage can be slowed down using the dynamic slow down factor together with the static slow down factor.

To enable the implementation of a DPM technique, the API should provide means to create and instantiate a task, taking into consideration the task real time parameters such as deadline, worst case execution time, period and frequency level. It should also provide code to calculate voltage and frequency scale down factors of task frequencies statically and at run-time, admit the entrance of new tasks in the system and implement the prediction scheduling technique. Further, the API must also support killing a task (The terms task and thread are used interchangeably in the remaining of this text) when its deadline is not met (based on the assumption that the work that would be executed by such task is no longer useful). In addition to OS independence, it is also important to make the API scheduling technique independent as well, so that it can be used with different energy efficient strategies. Therefore, the chief requirements for a power-aware API (w.r.t. task scheduling OS service) are:

1 Application should be able to tell information such as period, worst case execution time, deadline and whether the task is a soft or hard to the OS.

2 Application should be able to create a new instance of a task.

3 Application should be able to tell the OS what the expected remaining run-time for the current instance of the task is. This is application specific and adds more accuracy to the prediction once specific knowledge about the current setting of the task and the workload can be used.

4 OS should be able to tell the application what the OS predicts as the remaining execution time of the current task. A simple example is to predict the execution time by using previous execution times of each task. However, if some input is gotten from requirement 3, this data should be used instead.

5 If the task was hard, the OS must kill the task instance that missed its deadline.

In the following section, we shall show how these requirements have been achieved in our implementation.

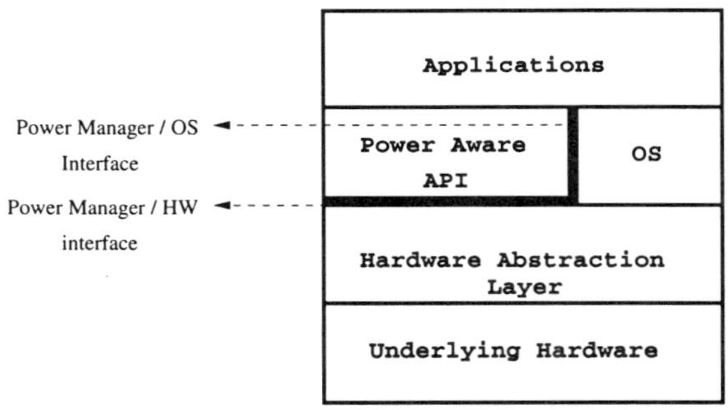

Figure 8.1. Representation of the Power-Aware API in the Context of the Application, OS and Hardware

2. The API

The power-aware API is written with a view towards facilitating a continuous dialogue between the OS and the application. The API enables the interaction between OS and application (in addition to providing an interface to the low-power hardware capabilities) so that the application can become power-aware. Similarly to the work described by [16], although in the power awareness context instead of mobility adaptation, a collaborative partnership between operating system and applications exists. The adaptation of the operating system services is also facilitated by the interface OS/Hardware so that those services can be modified to also become power-aware.

The API described in this document has been implemented within a RedHat object oriented and open source real time operating system named *eCos* [21]. *eCos* is well known by its high level of configurability and by its adequacy to be used in highly embedded systems. Similar to most of the commercial available embedded operating systems, *eCos* consists of a Hardware Abstraction Layer (HAL), which isolates hardware dependent details from the OS services, and the OS services themselves. The OS services are task scheduling, memory management, IO and file system, etc. The end user writes its own application, which may use one or more OS system services, and compiles the OS source code along with the application source code resulting in the embedded application as a final product.

Figure 8.1 shows the API and its relationship to the application, OS and the Hardware Abstraction Layer (HAL). The API must enable the operating system developer as well as the application developer to transform the OS and the application in power aware embedded software, respectively. The operating system developer modifies operating system services using the suitable hardware/software knobs and the application developer uses the modified OS services along with the application specific information to make power efficient decisions. We provide an overview of the application and the HAL interfaces of the API in the following sections.

2.1 Interface between the Power Manager and Hardware

Power management at the processor level is carried out by changing the operating frequency of the processor or putting it in power efficient states when it gets idle. Processor frequency control is via a registered control. The registers are seen as a control interface to the processor, which one can read and change the its status. To ensure portability, the power manager does not directly write into the control registers. Instead, a layer of software is used to isolate the power manager from specific commands to change the processor frequency. Table 8.1 presents the functions used to control frequency scaling. This interface consists of functions to initialize internal variables that enable the frequency and voltage scaling capability, functions to get information about the frequency and voltage scaling levels available on the processor and a function to set the frequency and voltage value of the processor. Information about the frequency and voltage scaling levels include which values of voltage are appropriate to each frequency level and what time overhead is incurred by switching among different levels. The frequency and voltage values to be set at a given time are results of the power management policy being applied.

Table 8.1. Summary of Hardware and Operating System Interface

Service	Functions
Frequency/Voltage scheduling	initialize_processor_power_management() processor_get_max_frequency() processor_set_frequency(frequency_value) processor_get_frequency()
Low-Power state transitions	processor_get_states_info(processor_power_states_info) processor_set_state(state)

A similar interface is used to manipulate the processor states. For instance, the SA-1110 processor can be in one of the following three states: running, idle, sleep. The transition among these states is done by executing special sequences of instructions or writing values into special SA-1110 registers. In the Table 8.1, we show a simplified version of the functions needed to manipulate those states. We basically have a function to get information about each state and a function to set the processor into a specific state. Information about each processor state must include the overhead of switching to that state and the power consumed by it. If the processor needs to provide information to the scheduler or to the applications, this layer of software can be used to attach ISRs (interrupt services routines) to the processor interrupts (if necessary).

For other OS services such as IO or sensors management [15], for example, a similar software layer should be provided, enabling those services to become power-aware. For IO devices, for example, the device drivers should be augmented with the appropriate modifications that make the hardware/software knob available to the OS service that uses the device, enabling it to execute in a power aware fashion. For instance, a radio communication device can communicate at a faster and more power consuming or slower but energy saving data rate depending on the system needs.

2.2 Interface between the Operating System and the Power Manager

This section defines the interface between the OS and the Power Manager. This layer provides means to define power aware operating system services. This enables an application writer, for example, to define power-aware tasks and scheduling techniques.

Table 8.2 summarizes the major functions defined by our implementation regarding OS scheduling. The table lists the requirement numbers as previously stated and the respective functions implementing each requirement. Functions `create_task_type` and `create_task_instance` implement requirements 1

Table 8.2. Summary of Operating System and Power Manager Interface

Requirements	Functions
1 and 2	create_task_type(task_type_info) create_task_instance(type_handle,instance_handle,entry_point)
3	app_set_remain_time(instance_handle,time_remaining) app_started(instance_index) app_done(instance_index)
4	os_pred_remain_time(instance_handle,remain_time,time_stamp)
5	killer_thread() thread_register_as_hard(instance_handle)

and 2 respectively. The structure `task_type_ info` convey the real-time info characteristics of each task into the kernel and the result of the function call `create_task_type` is the creation of a task type, which is maintained internally within the kernel by means of tables. Function `create_task_instance` creates an instance of a task type using the real time characteristics passed in by means of `type_handle` and the code passed in as `entry_point`, resulting in a task instance accessible by the variable `instance_handle`.

Function `app_set_pred_remain_time` implements requirement 3. The remaining execution time of a task can be estimated in a per-task basis according to the data manipulated by the task. For example, depending on the size of the input being processed, different worst case execution times can be estimated. With such interface the application can do a more accurate prediction based on the current workload and the current execution mode in which it is executing.

The computation of the estimated remaining time is application specific. The functions `app_started` and `app_done` must be called by every thread. These functions delimit the body of the thread. Hence all the useful work the thread has to execute periodically is delimited by these two functions. Without calling `app_done` the scheduler cannot know whether the execution of an instance has finished or missed its deadline. An example on the use of these two functions is be presented in the section 4.

Requirement 4, related to OS feedback to the application, is implemented by function `os_pred_remain_time`. The information provided by this function is computed internally in the OS kernel and is based on the task execution time estimation used. Given this estimation and the one provided by the application, the application can adapt itself (changing the processor execution frequency and voltage) in order to make a best effort for meeting its deadline or saving power. The calculation of the estimated execution time varies depending upon the estimation and/or scheduling technique being used.

Requirement 5, regarding task termination in case of missed deadlines, implies a more elaborate implementation. Basically when a task is instantiated, its deadline is known by means of the thread type. Some external entity has to keep track of thread's deadline and has to be notified when the deadline comes and the thread's execution has not yet finished. A solution is to create a task dedicated only for this purpose. Each thread has to create an alarm that expires on its deadline. When a task finishes its execution meeting its deadline it should disable the alarm associated with it (this is done by the function `app_done`). On periodic threads, the thread is recreated (in the case of missed deadlines) so that it can be executed in the next period.

The function `thread_register_as_hard` is used to enable the alarm associated with the thread which called the function. This function is invoked from the `app_started` function if the thread type is defined as hard. The function `app_done` disables the alarm in the case the thread has not missed the deadline.

3. Implementation

To demostrate the utility of the proposed API, we have implemented it to take advantage of the frequency and voltage scaling for two platforms running *RedHat eCos* OS. The choice of *eCos* is based on its public domain availability, and similar services can be built for other OS's as well, including any event driven OS such as TinyOS [5].

The frequency and voltage scaling code has been implemented for two different platforms: StrongARM SA-1110 based [8] (*Compaq iPaq*) and Intel XScale [9] called 80200EVB [7, 10], also known as LRH. These functionalities are implemented in such a way that it is easy to insert and remove them in a configurable fashion. Policy can be changed automatically through scripting and macro preprocessing, thus enabling automatic configuration and compilation of the kernel source code. This enables the implementation of different power management policies along with different scheduling policies in a automatic and systematic way.

The scheduler strategy in which the predictive power-aware scheduler was implemented consists of scheduling the highest priority available thread at a given time. Threads are scheduled when the kernel routines are invoked. These routines may result on blocking a thread or on the creation of a new thread. In both cases if a higher priority thread becomes available or if the highest priority thread blocks, a new one has to be scheduled. The piece of *eCos* kernel source code in which the context switch occurs is the point wherein the data structures are updated and the actions such as scale the frequency and voltage down/up are taken by the power-aware scheduler.

3.1 Implementation of a Predictive Power-Aware Scheduling within *eCos*

This section presents an overview of the power-aware predictive scheduling scheme that has been implemented within *eCos* kernel. This scheme is just an example on how the power-aware API can be used in order to implement power-efficient systems. Other schemes can be implemented using the same interface or if necessary new functions can be added to the API in order to accomplish the implementation of a new scheme.

The power-aware scheduling algorithm is similar to the one devised by [13]. The algorithm maintains the execution time taken by each task and the next executions of the task are predicted based on the previous ones. One strong limitation of this algorithm is that only the last task executed in a scheduling phasing is scaled down as opposed to scale down all the possible tasks by using the slack times. Therefore the structure of the scheduling scheme implemented using the API follows the same guidelines of [20]. In general, the scheduling technique consists of a modified version of the Rate-Monotonic scheduling (RMS). The algorithm differs of the RMS because the execution time of tasks are predicted, allowing in this way to scale the processor frequency and voltage according to the predicted time. Static and dynamic slow down factors are applied while the former is calculated during the schedulability test of the task set and the latter during run-time. The code implementing the algorithm includes the following items:

tables and variables - tables of thread types and thread instances and variables needed to implement the power management policy

admission control code - code to check the schedulability of a given task set and to find out the static slow down frequency and voltage scale factors so that the task set is still schedulable

dynamic frequency scaling computation - calculate the dynamic frequency and voltage scaling down factor at run-time based on the history of previous executions, or in other words, based on the time prediction for the next execution of a task or for the remaining execution time of a given thread instance

deadline management - if a task is specified as hard and its deadline is missed, such task has to be killed since its processing is no longer useful

The source code to update the tables and perform the prediction was split in separate files such that they can be easily either inserted or removed from the implementation by means of symbols definition and preprocessing.

4. Results

As mentioned earlier, our current implementation addresses only task scheduling, and specifically frequency/voltage scaling. We have successfully tested the API on two different platforms, namely *Compaq iPaq* which is based on StrongARM-SA1110 and an Intel Evaluation board based on the XScale processor. We address the issue of how an application can be modified in order to become power aware with respect to its scheduling.

We assume that an application consists of several tasks modeled as threads. Thread creation is done by means of *eCos* system calls. The tasks then communicate with the external environment and among themselves. Without introducing any modification into the system scheduler, tasks are scheduled according to their priority and all tasks execute at the same speed, independent of their deadlines and workload.

The first modification that needs to be done concerns the operating system developer. In our case, we assume that all tasks are periodic, independent and that the period is the same as the deadline. Based on the implementation sketched in the section 3.1, the operating system developer modifies the scheduling policy so that the operating voltage/frequency of each task can be adjusted according to the workload and/or real time constraints.

The operating system scheduler is modified in such a way that the real time characteristics of each task are traded-off against energy savings. Note that this trade-off is only possible if the real-time characteristics are provided to the operating system. Once the energy saving scheduling scheme is implemented within the kernel of the operating system, the application has to be modified so that it furnishes the operating system with the necessary information.

Two kinds of modifications can be applied to the application. One, which is mandatory, consists of calling the functions `app_start` and `app_done`. This is shown in Figure 8.2-(a). The other modification, which is application specific, consists on providing the operating system with a time prediction for the remaining of the execution of the task so that the voltage/frequency can be adjusted more accurately as shown in Figure 8.2-(b).

We briefly mention some of the important implementation issues found by our experience in building the proposed API. In the *Compaq iPaq* StrongARM SA-1110 [8], the core processor clock frequency is used to derive timing properties for other platform components such as the DRAM memory access timing, LCD refresh time, serial communication port baud rate and FLASH memory timings. This means that when the core processor clock is scaled up/down, the timing properties of all these components have to be scaled up/down as well if one wants to have those components working properly. We do not have any experimental data available but these updates may generate extra overheads into the system functioning, besides making the implementation more complex.

In the XScale based board [10] the operating system tick time (which gives the OS the notion of time) is derived from processor internal timers, which in turn are implemented as multiples of the core clock frequency. This means that when the frequency is scaled up/down, the timers are also scaled up/down, spoiling in this way an accurate notion of time from the OS point of view. In

```
void thread()
{
  for (;;) {
    /* start of the original code,
       usually looping infinitely */

    ...

    /* end of the original code */
  }
}
```

(a)

```
void thread()
{
  for (;;) {
    app_start();
    /* start of the original code,
       usually looping infinitely */

    ...

    /* this is optionally called,
       depending on the application */
    app_set_pred_remain_time();

    ...

    /* end of the original code */
    app_done();
  }
}
```

(b)

Figure 8.2. (a) - Code without Any Modification; (b) - Code Modified to Be Power Aware

order to fix this, the internal processor timers also have to be updated every time the frequency is scaled. This problem, although fixed, also introduces a little bit more of complexity to the implementation.

5. Conclusions and Future Work

We have successfully implemented an API that allows the OS scheduler to become power aware thus enabling application programs to take advantage of the system-level power/performance control knobs. By means of modifications on the OS source code and applications, both work cooperatively in order to achieve power savings based on the expected workload of each task.

The current API implementation focus was on process scheduling, which is one of the services carried out by the operating system. In the future, we plan to extend an API that also makes other operating system services power-aware. For instance, IO is responsible for a big chunk of power consumed in embedded systems. IO includes communication with the external world such as radio, serial or other forms of communication, opening in this way a lot of possibilities to investigate power efficient schemes.

Often times embedded application requires large amount of data that are created and destroyed at run-time. A typical example is a protocol stack implementation in which memory regions to store packets are manipulated continuously. This also opens up possibilities to explore power efficient memory management techniques.

Another service that has not been investigated and that must be investigated with respect to power efficiency are file systems on flash memory. With the ever-increasing use of PDAs (personal digital assistants), for example, in which information is stored and retrieved in flash memory often, a power efficient way to manipulate such information could bring additional power savings.

One more service that can also be explored relates to how information is encoded in the data and address buses in the embedded system. Different applications behave in different ways and so different bus encoding techniques can be applied to the individual types of application. With some assistance of the operating system, which can be modified in such a way that it knows which "type" of application is running at a given time, and with some reconfigurability with respect to the memory and IO controllers, power aware bus encoding/decoding can be tuned according to the applications.

And finally more at the application level than at the operating system level applications can also be modified to operate on a power-aware fashion, rather than only cooperate with the scheduler, changing its operating mode according to the power needs of the system. For this the operating system can be used as an information provider, which tells the applications about the state of the whole system furnishing them with the necessary information to make decisions.

Acknowledgement

This article is based in part on research funded by the DARPA PAC/C program under AFRL contract # F30602-00-C-0154. The views and conclusions expressed herein are those of the authors and should not be interpreted as necessarily representing the official policies or endorsements, either expressed or implied, of the DARPA, Air Force Rome Laboratories, or the U.S. Government.

References

[1] Ana Azevedo, Radu Cornea, Ilya Issenin, Rajesh Gupta, Nikil Dutt, Alex Nicolau, and Alex Veidenbaum. Architectural and compiler strategies for dynamic power management in the copper project. In *IWIA 2001 International Workshop on Innovative Architecture*, Maui, Haiwai, January 2001.

[2] L. Benini, A. Bogliolo, G. Paleologo, and G. De Micheli. Policy optimization for dynamic power management. *IEEE Transactions on Computer-Aided Design of Integrated Circuits and Systems*, 18(6):813–833, June 1999.

[3] Luca Benini and Giovanni De Micheli. *Dynamic Power Management: Design Techniques and CAD Tools*. Kluwer Academic Publishers, 1997.

[4] T. Burd, T. Pering, A. Stratakos, and R. Brodersen. Dynamic voltage scaled microprocessor system. In *International Solid-State circuits Conference (ISSCC 2000)*, pages 294–295, 2000.

[5] Jason Hill. A software architecture supporting networked sensors, December 2000.

[6] Chi-Hong Hwang and Allen C.-H. Wu. A predictive system shutdown method for energy saving of event-driven computation. *Design Automation of Electronic Systems*, 5(2):226–241, 2000.

[7] Intel. Intel 80200 processor based on intel xscale microarchitecture developers manual, 2000.

[8] Intel. Intel strongarm sa-1110 microprocessor developers manual, 2000.

[9] Intel. Intel xscale core developers manual, 2000.

[10] Intel. Intel 80200 processor evaluation platform board manual (80200EVB), 2001.

[11] Intel and Microsoft. Bios interface specification, 1996.

[12] Intel, Microsoft, and Toshiba. Advanced configuration and power interface specification, 1996.

[13] Pavan Kumar and Mani Srivastava. Predictive strategies for low-power rtos scheduling. In *Proceedings of the 2000 IEEE International Conference on Computer Design: VLSI in Computers and Processors*, pages 343–348, 2000.

[14] Yung-Hsiang Lu, Luca Benini, and Giovanni De Micheli. Operating-system directed power reduction. In *Proceedings of the International Symposium on Low Power Electronics and Design*, pages 25–27, Rapallo Italy, 2000.

[15] R. Min, T. Furrer, and A. Chandrakasan. Dynamic voltage scaling techniques for distributed microsensor networks. In *IEEE VLSI Workshop*, 2000.

[16] Brian D. Noble, M. Satyanarayanan, Dushyanth Narayanan, James Eric Tilton, Jason Flinn, and Kevin R. Walker. Agile application-aware adaptation for mobility. In *Sixteen ACM Symposium on Operating Systems Principles*, pages 276–287, Saint Malo, France, 1997.

[17] J. Pouwelse, K Langendoen, and H. Sips. Dynamic voltage scaling on a low-power microprocessor. In *Mobile Computing Conference (MOBICOM)*, 2001.

[18] Johan Pouwelse, Koen Langendoen, and Henk Sips. Energy priority scheduling for variable voltage processors. In *Proceedings of the 2001 International Symposium on Low Power Electronics and Design*, pages 28–33, 2001.

[19] Gang Quan and Xiaobo (Sharon) Hu. Energy efficient fixed-priority scheduling for real-time systems on variable voltage processors. In *Pro-*

ceedings of the Design Automation Conference, pages 828–833, June 2001.

[20] Vijay Raghunathan, Paleologos Spanos, and Mani Srivastava. Adaptive power-fidelity in energy aware wireless embedded systems. In *To be presented at the IEEE Real-Time Systems Symposium*, 2001.

[21] RedHat. eCos: Embedded configurable operating system. http://www.redhat.com/products/ecos/, 1998.

[22] Y. Shin, K. Choi, and T. Sakurai. Power optimization of real-time embedded systems on variable speed processors. In *Proceeding of the International Conference on Computer-Aided Design (ICCAD)*, pages 365–368, 2000.

[23] Youngsoo Shin and Kiyoung Choi. Power conscious fixed priority scheduling for hard real-time systems. In *Proceedings of the Design Automation Conference*, pages 134–139, 1999.

[24] T. Simunic, L. Benini, and G. De Micheli. Event-driven power management of portable systems. In *International Symposium on System Synthesis*, pages 18–23, 1999.

[25] Frances Yao, Alan J. Demers, and Scott Shenker. A scheduling model for reduced CPU energy. In *IEEE Symposium on Foundations of Computer Science*, pages 374–382, 1995.

[26] T. Simunic Yung-Hsiang Lu and G. De Micheli. Software controlled power management. In *CODES, Proceedings of the IEEE Hardware / Software Co-design Workshop*, pages 157–161, May 1999.

IV

COMPILER LEVEL POWER MANAGEMENT

Chapter 9

PACT HDL: A COMPILER TARGETING ASICS AND FPGAS WITH POWER AND PERFORMANCE OPTIMIZATIONS

Alex Jones, Debabrata Bagchi, Sartajit Pal,
Prith Banerjee, Alok Choudhary
*Northwestern University**
{akjones, bagchi, satrajit, banerjee, choudhar}@ece.northwestern.edu

Abstract Recently, there has been a focus on high-level languages, C/C++ in particular, for hardware synthesis. At the same time, power dissipation is becoming an important metric in hardware design. This work presents PACT HDL, a C to HDL Compiler with facilities for power and performance optimizations. PACT HDL is an attempt to merge together automated hardware synthesis optimizations for power and performance with the capability to leverage high-level algorithms written in C and target arbitrary hardware architectures. By making the compiler modular and flexible, optimizations may be executed in any order and at different levels in the compilation process. PACT HDL generates HDL codes, such as RTL Verilog and VHDL that can be synthesized and profiled for power using commercial tools.

Keywords: compiler, FPGA, ASIC, synthesis, low-power, high-performance, FSM

1. Introduction

As chip fabrication processes continue deep into the sub-micron level and Integrated Circuits (ICs) and Field Programmable Gate Arrays (FPGAs) can support larger and larger amounts of logic, system designers require increasingly high-level tools to keep up. Recently, industry has targeted C/C++ and variants as potential long-term replacements for Hardware Description Languages (HDLs) such as VHDL and Verilog currently employed for today's

*This work was supported by DARPA's power-Aware computing and communications program under contract No. F33615-00-C-1631.

hardware design. Also, as technologies increase in density in both the fabricated and reconfigurable areas, power-consumption becomes an progressively more important problem. While some work has been done in targeting C/C++ as an HDL and considering power consumption in hardware synthesis, combining the two tasks creates a new challenge. This work presents PACT HDL, a compiler targeting the C language that produces synthesizable HDL usable for either FPGA designs or Application Specific ICs (ASICs) with a framework for both power and performance optimizations. PACT HDL supports arbitrary target architectures and allows for both power and performance optimizations at the C level or at an HDL level. The compiler uses a back-end to generate synthesizable Register Transfer Level (RTL) HDL codes, such as VHDL and Verilog, current industry standards. It can be easily extended to support new HDL codes as they are developed.

1.1 PACT

PACT HDL is part of the PACT (Power Aware Architecture and Compilation Techniques) project. The objective of PACT is to develop power-aware architectural techniques and associated compiler and CAD tool support that will take applications written in C and generate efficient code that runs on power-aware systems. Unlike many current projects targeting power savings, PACT is a comprehensive approach to creating a power-aware computing solution. Figure 9.1 shows an overview of PACT.

1.2 Power-Aware Compilation (PACT HDL)

The work in this chapter is part of Task B: Power-aware compilation techniques. Traditionally, the main goal of the compiler optimizations has been performance optimization. In the embedded computing domain, it has been to optimize resource needs and code size under performance (e.g., throughput) constraints. Compilation techniques can play an even bigger role in the area of minimizing power consumption under performance and resource constraints.

The work on Task B is multi-fold: (1) PACT ARM: the development of a compiler that targets the ARM processor and potential variants from the work done in Task A (2) PACT HDL: the development of a compiler that targets ASICs and FPGAs and interfaces with the work done in Task. On top of both of these compilers is a scheduler that determines which compiler should be targeted for a particular code unit. Only work related to the PACT HDL portion of the compiler task is described in this paper.

As shown in Figure 9.1, with the exception of the development of power-aware applications and algorithms, the power-aware compilation strategy is at the highest-level in the hierarchy. At this level there is a lot of room for potential power and performance improvement. While RTL code synthesis is

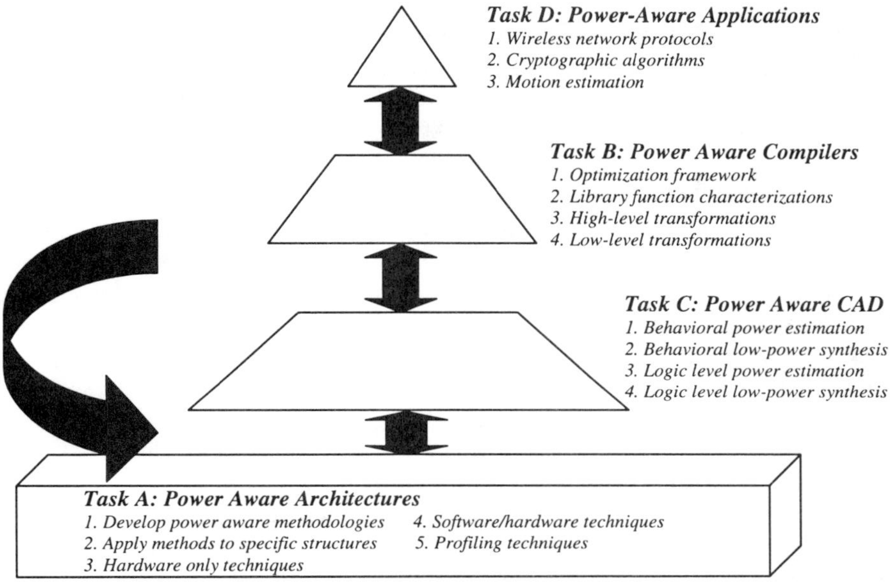

Figure 9.1. PACT Research Task Overview

a fairly well understood problem, synthesis originating from C is much less well defined. Optimizations can happen at multiple levels in the compilation process: closely related to the C form, an intermediate high level RTL form, and during the RTL code generation in the back-end.

1.3 Related Work

Recently, there has been a lot of work in the use of the C programming language and other high-level languages to generate synthesizable HDL codes or hardware implementations. David Galloway at the University of Toronto has developed the Transmogrifier C compiler. This compiler uses a subset of the C language and targets a Xilinx 4000 series FPGA[14]. Giovanni De Micheli et al. have developed HardwareC, a C like language that contains many HDL extensions. The Olympus Synthesis System synthesizes digital circuit designs written in HardwareC[10]. The Esterel-C Language (ECL) developed at Cadence Berkeley Laboratories is an HDL and compilation suite based on the C language[11]. CoWare, Inc. has developed the N2C layer for existing languages, C in particular, by adding clocking and cycle information to allow hardware modeling[2]. CynApps, Inc. has developed a suite of tools based on macro and library extensions for C/C++. These macros allow coding in C++ with an HDL style and the tools generate RTL HDL codes[8]. SystemC is another C like language developed by Synopsis to allow C-like HDL coding that is particularly popular. The suite of tools associated with it can generate hardware directly[22].

Of particular interest in the synthesizable C community is the handling of pointers and other derived constructs. At Stanford, Séméria and De Micheli have done work to handle the problem by attempting to eliminate the loads and stores generated by pointers in C through special control flow[24]. C Level Design Inc. has developed the System Compiler that claims to handle the entire C/C++ language, including pointers, arrays, and higher level constructs[27].

Some related work has been to target other high-level languages for hardware generation. Superlog is a language based on both Verilog and C by Co-Design Automation, Inc[7]. The MATCH group at Northwestern University has built a synthesizable RTL VHDL compiler from the MATLAB programming language[4]. Xilinx, Inc. targeted Java as an HDL for its Forge-J HDL compiler[9].

As the proliferation of battery powered portable electronics has increased, so has the work on power-optimized hardware. Many of these approaches are applicable toward automated power optimizing compilers. At UC-Berkeley, power optimization work has been done in conjunction with the wireless research center dealing at the CMOS level [6] and using transformations[5]. Additional work on power-optimized synthesis has been done at Polytechnic University of Catalonia[20] and princeton University[16],[17].

1.4 Outline

The remainder of this chapter is outlined as follows: Section 2 describes the PACT HDL Compiler infrastructure. All stages of the compilation are discussed from the front-end C parsing to the back-end code generation. Implementation issues and the suitability for optimizations at different phases are discussed. Section 3 describes the HDL AST. Section 4 explains the synthesis flow for both ASICs and FPGAs using the compiler's output code. Specific optimizations for power and performance at different compiler levels are discussed in Section 5. Section 6 presents some initial results as well as ongoing work. References are listed at the end.

2. PACT HDL

PACT HDL is a fully modularized three-stage C to HDL Compiler. In the first stage, the C code is parsed into a high-level C type Abstract Syntax Tree (AST). At this stage, both power and performance optimizations can be executed. In general, these optimizations will not require specific information about an HDL representation or about the target architecture, such as precision analysis, constant-propagation, or loop unrolling. During the second stage, the

C AST is converted into Finite State Machine (FSM) style HDL AST. Target architecture specific information is inserted into the AST in this phase. Again, power and performance optimizations can be executed. In general, these optimizations require specific information about clocks, cycles, or the target architecture, such as memory pipelining, and clock-gating. Finally, RTL code is generated in the back-end phase. This phase is not designed for optimization; however, if language-specific issues arise they are handled in this phase. The compiler flow is shown in Figure 9.2.

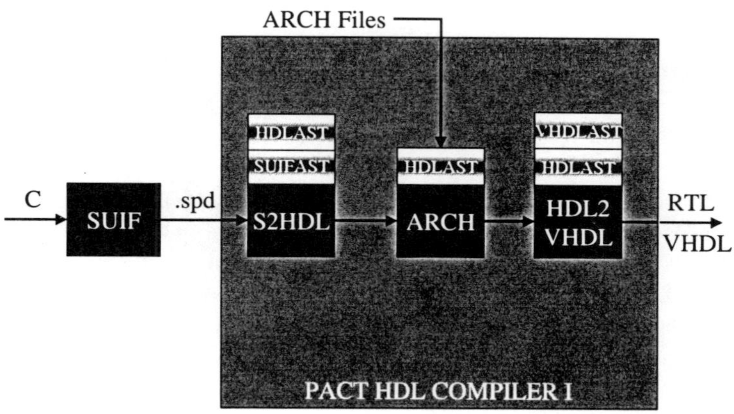

Figure 9.2. PACT HDL Compiler Flow

2.1 SUIF

PACT HDL leverages the SUIF C Compiler Version 1 from Stanford University[29] for its C front-end. SUIF itself leverages the lcc front-end developed in collaboration between AT&T Bell Laboratories and Princeton University[12]. Lcc is a capable compiler front-end that has been shown to generate reasonable code but with a much simpler implementation than some other examples like the GNU C Compiler (gcc)[13]. Lcc is not best suited to compiling the most complicated C codes such as OS kernels, however, codes that target synthesis rarely fall into this category.

SUIF (Stanford University Intermediate Format) is a widely used research compiler. It is the particularly well designed intermediate format that makes this a good choice for a front-end. Once a code is compiled into this format, it can be saved to disk through SUIF's own routines. At this point, an arbitrary set of optimizations can be run in an arbitrary order on the AST to study the effects. In addition to the set of pre-designed optimizations included with SUIF, there is a well documented and well organized implementation making it very practical to develop new optimizations at the C level in its framework. Using a combination of user designed and pre-designed optimizations poses no problem. SUIF retains both high and low-level information simultaneously in the AST. The high-level tree nodes come in just a few varieties, procedures and code blocks, if nodes, for loop nodes, and non-standard loops (e.g. while loops). The low-level nodes are in the form of MIPS like instructions grouped into categories. The most prolific is the three-operand instruction that includes arithmetic, relational, and shifting operations as well as memory loads and stores. Other instructions include constant loading, branch and jump, function call, array, and multi-branches. The three-operand, load constant, and branch and jump instructions are very similar to MIPS instructions. The array, function call, and multi-branch instructions retain more high-level information for optimizations and to target a variety of back-ends. These instructions may be dismantled into simpler ones using tools in the SUIF suite. The high-level nodes contain lists of instructions and other high-level information such as control information. Facilities for symbol tables, symbols, and types are included in SUIF. There is a global symbol table, a symbol table associated with each function level and code block node. Symbols mirror the built-in variable types in C, including pointers, arrays, structs, and enumerated types. SUIF also maintains a nice facility for annotations to AST nodes, and methods for manipulating the AST nodes.

3. HDL AST

Rather than directly translating the SUIF C like AST into a VHDL or Verilog specific AST, an interim level of abstraction was created. This approach is very different than the related MATCH project, which directly translates the high-level language, in that case MATLAB, to a VHDL specific AST. All optimizations for hardware generation are done at the VHDL AST level[4]. In contrast, the HDL AST is designed to model the RTL HDL concept and be suitable for HDL optimizations without being language-specific. The synthesizable subset of both RTL VHDL and Verilog are similar enough to make this abstraction possible and allows optimizations that would require separate implementations to be designed only once. As previously mentioned, the underlying framework of the HDL AST is a FSM structure. The top-level node is an entity node that corresponds to the entire C code hierarchy with a single target. Each entity contains one or more process nodes. The process nodes are considered to execute in parallel and correspond to C functions. Each process contains a FSM node that describes the behavior of the C function in a state machine style. The hierarchy is displayed in Figure 9.3

The FSM node contains a list of one or more state nodes. The states may contain one or more statement nodes and contain explicit control flow. State-

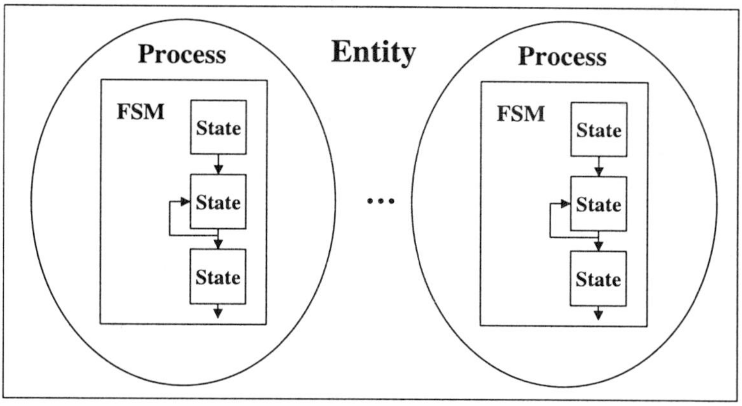

Figure 9.3. HDL AST Hierarchy

ment nodes are built out of expression nodes and operator nodes and come in several varieties: assignments statements correspond to SUIF three operand instructions, if statements for control flow, set statements correspond to SUIF set style statements (e.g. set less than), read statements correspond to SUIF loads, write statements correspond to SUIF stores, and read/write statements correspond to SUIF memory copy instructions. The state node can contain an arbitrary number of these statements with one exception, there may be a maximum of one if statement. States contain a next state exit for all states without an if statement. In the case that there is an if statement, the next state exit corresponds to the then part, and the alternate state exit corresponds to the else part, see Figure 9.4.

Like many types of ASTs, statements are built from expressions and operators. Expressions are categorized into four groups: binary expressions with two sub-expressions and an operator, unary expressions with one sub-expression and an operator, primary expressions, and symbols discussed in Section 3.1. Primary expressions also have several varieties: identifiers, numbers, binary constants, and memory variables. Operators are also categorized into groups: logical, arithmetic, relational, shifts, and unary. With this relatively simple set of AST nodes, the HDL AST can handle a very general set of codes.

3.1 HDL Symbols and Symbol Tables

The HDL AST also maintains a symbol table hierarchy. Symbols are perhaps the most complex nodes in the AST structure and come in quite a few varieties. The most common symbol is the variable symbol. Variable symbols retain certain hardware-specific information as well as higher level C type information. Variable symbols can be either hardware signals or hardware vari-

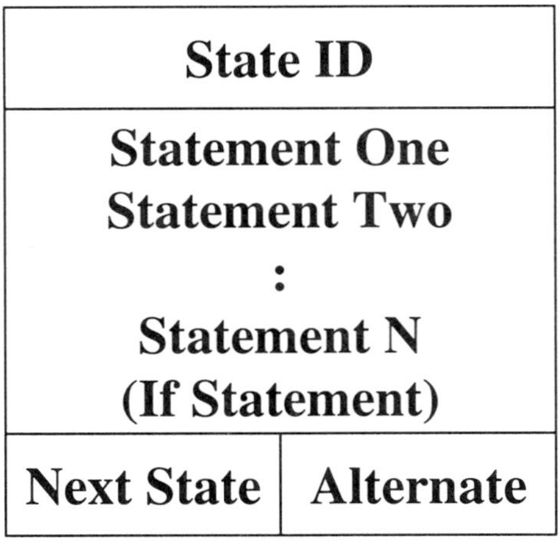

Figure 9.4. State Node Representation

ables. In this respect, the VHDL model is followed such that hardware signals are visible across processes and can be used in concurrent assignments, and hardware variables are visible only within a local process and can be used in sequential assignments. In addition, variable symbols can be used to represent C type scalars such as integers and real numbers, C type pointers and arrays, or user-defined C constructs. The other important symbol node, the label symbol, comes from low-level SUIF information. These symbols are placeholders for control-flow labels from SUIF. Several other architecture specific symbols are defined and are discussed in Section Target Architecture Independence. The entity node holds a global symbol table. This symbol table may hold only variable symbols that are hardware signals, so they are visible everywhere in the entity. Similarly, each process holds a local symbol table, which may hold either variable symbols that are hardware variables or hardware signals, or label symbols. While the hardware signals from this table would be visible to other processes, they are treated as local variables, and used only for their concurrent assignment proprieties. The hierarchy is shown in Figure 9.5 Three other architecture specific symbol tables are contained in the entity node, and will also be discussed in Section 3.3.

3.2 SUIF to HDL Translation

During the conversion from the SUIF AST to the HDL AST there are two major passes. The first pass converts instructions into states and statements; the second does a control flow pass for the explicit control flow required by the FSM. The first pass is by far the most complex. During this phase, each

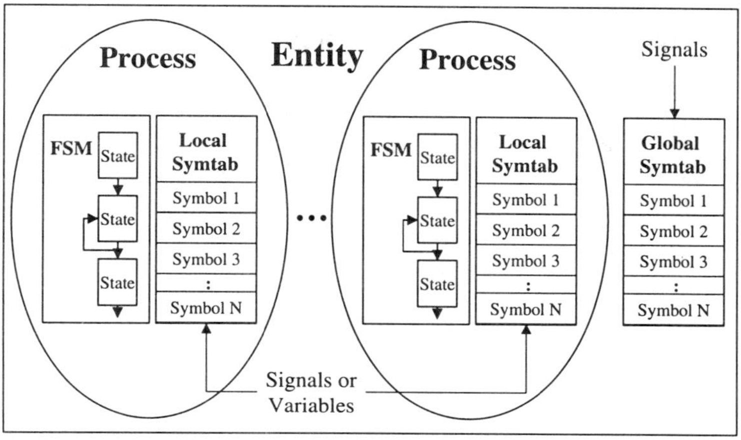

Figure 9.5. AST Hierarchy with Symbol Table

type of possible SUIF instruction must be handled, temporary variables generated for communication of data between instructions, and HDL AST nodes generated and added to the AST. The second pass connects the states together using explicit control flow. It is during this phase that the state exits described in Section 3 are set.

3.2.1 Conversion Pass. During this initial conversion phase, SUIF instructions are each converted into HDL AST statements and local SUIF symbols are converted to the HDL symbols. These newly created AST nodes and symbols are added to the HDL AST and symbol tables respectively. In order to keep things simple during the initial pass, each converted statement is placed in a unique state. Each SUIF instruction has a unique ID number, which is preserved as the State ID number in the HDL AST. Because SUIF instructions are already levelized, this guarantees very simple states. During later AST optimization passes, some states may be combined together based on data dependencies, resource allocation, and instruction scheduling. In a few cases converted instructions require multiple statements and states. This is particularly true of the address calculation of array indices with multiple dimensions. Currently, PACT HDL only supports static size arrays of an arbitrary number of dimensions. The memory is allocated by determining the array size at compile time, and each array and pointer is assigned a base address in memory. When an array is indexed, the memory address is calculated through a series of statements using a variant of Horner's Rule. For arrays with more than two dimensions, temporary variables are used to calculate offsets. The offset for dimension N is calculated by multiplying the offset value of dimension N-1 by the bound of N. Each offset value is multiplied with the appropriate index and accumulated along with the base address to generate the index address. This address is then used to access memory in a read or write statement. If a naïve approach without temporary offsets is used, the number of multiply cycles is $O(n^2)$ instead of $O(n)$ by Horner's Rule.

One other notable point is the creation of temporary variables. SUIF instructions are levelized, often requiring the result from other instructions to complete the result. The SUIF instructions are also maintained in trees such that an instruction requiring the temporary result of another instruction is that instruction's parent. The tree is traversed using a depth first search. Each time a child is explored, the information must be communicated to the parent through a temporary variable. A naïve approach was used in the initial implementation. Each time a temporary variable was required, a new unique one was created. Since on the average more than 50% of the SUIF instructions require communication through temporaries, this created temporaries on the order of the number of states. The number of temporaries can be significantly reduced through the following method. Information about the number of temporaries of each type created thus far is stored in the local symbol. Each time a new temporary of that type needs to be created, the information in the local symbol table is updated. This new temporary variable is uniquely identified through a combination of a type ID number and a unique number from the local symbol table. During each descent, tree specific temporary variable information is maintained. While exploring a child node, a new scope is created. Using the tree specific information, the next appropriate temporary variable is used. The local symbol table information is checked to see if a new temporary is required, otherwise the appropriate existing one is used. Once the deepest point is reached, the scope of the parent is restored. When a new descent occurs, the temporaries from the previous descent that are no longer needed for current instructions can be reused. Figure 9.6 shows an example of the temporary creation scheme using a simple AST. For a more complex code like the Sobel transform, the number of temporaries required was reduced by an order of magnitude from over 100 to 14.

3.2.2 Control Pass. The second pass over the SUIF AST is used for control. In general, control is not specified in the SUIF AST and it is assumed that each instruction falls through to the next one in the list. Only cases of explicit control are specified through the use of branch or jump instructions. However, this is not sufficient for the FSM style HDL AST. During traversal of the SUIF AST, each corresponding state/statement in the HDL AST is connected to the previous one. Branch and jump instructions are handled somewhat differently. The label specified by the branch or jump points to the SUIF instruction to jump or branch to. The equivalent state from the HDL AST may

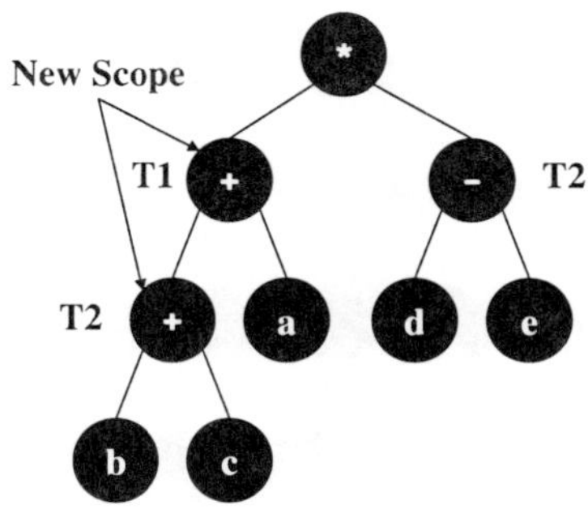

Figure 9.6. Example of Temporary Reuse Scheme for a Simple AST

be obtained through the unique ID number. In the case of a jump, the next exit is just linked to the jump target rather than the next instruction. In the case of a branch, two targets are required, the next state and the state from the label. For branch true, next becomes the label target, and alternate becomes the next state in the list. For branch false the reverse is true. Some higher-level SUIF nodes require special control flow, but supply only certain corresponding control instructions. If statements require control flow based on the result of the test portion of the statement. In the case that the then part of the if statement is executed, SUIF does not include the jump instruction to skip the else instructions. If the else art is to be executed, there is a jump instruction to the else part, but no corresponding target label. To correct this, during the conversion phase the omitted label is added to the symbol table. Then during the control phase, the missing label instructions and jump instructions are added to the SUIF AST so that the correct control may be followed. While handling SUIF for nodes, instructions related to the loop test, break and continue label instructions are omitted. These instructions are created in the conversion phase and the control flow is done at that time, since the corresponding SUIF instructions do not exist. This is necessary since there is no record in the SUIF AST for the ID numbers for these new states and thus no way to reference them for control. The SUIF documentation claims that there is a way to automatically add these missing instructions in the if and for nodes to the SUIF AST; however, it does not appear that this was ever implemented.

3.3 Target Architecture Independence

PACT HDL is an architecture independent compiler, but does assume the processor and local memory paradigm. The HDL code generated from the compiler interfaces with templates and libraries that describe the target architecture during simulation and synthesis. In order to correctly accomplish this, the compiler requires some target specific information from the templates. Certain signals and libraries and functionality will need to be described to allow the generated HDL code to work correctly. Memory access for FPGAs and ASICs are far from standardized. Things vary from the number of cycles required to accomplish the access to the address size and bandwidth. Thus, architecture description files for the target architecture are required as compiler inputs. There are four files required: a file describing libraries and packages to be included in the design, a file describing signals to be included from a template, a file describing a memory read, and a file describing a memory write. The library and signal architecture files are parsed just once and the data is stored in HDL symbol tables alluded to in Section 3.1. Library and package symbols are HDL symbols that contain the names and packages of libraries for inclusion in the design from he library symbol table stored in the entity node. A library may contain one or more packages. The signal architecture file is slightly different. The file is broken up into two groups: port signals that are directly interpreted by the compiler, and hidden signals that require additional specification. These signals are mapped to symbols and stored in a port symbol table and hidden symbol table respectively. The port signals correspond to the VHDL concept of a port, and specify signals that are inputs or outputs to a template. These signals must exactly match the template's specification, even if the compiler does not directly understand how to handle them. Thus signals specified in a bit form are handled directly by the compiler but higher-level VHDL constructs, such as records, are not. In this case, the record type name is recorded in the port section of the file, and stored in the port symbol table. The fields of the record that are required in the design must be specified in the hidden signals section and thus in the hidden symbol table. These symbols are available for use in the read and write architecture files but are not placed in the port. One signal is specified as the main design clock, and a second as the design reset. Other signals may be specified as sensitivity list signals. These signals are handled specially by the backend, described in Section 3.4. The information in the read and write architecture files is used during a third AST pass. Whenever a read or write statement is found, these files are parsed to read the architecture specific method for accomplishing the memory task. The file specifies one or more states used for accessing memory, the control required between the states, and the settings for memory specific signals. Placeholders are included where compile time specific information is required such as the memory address variable and the variable to read into or write from. These nodes are created, from the architecture file templates and inserted into the HDL AST for each memory access.

The architecture files have been designed for future extensibility. Additional information about the architecture specific to optimization passes can be included. For example, memory access delays might be included in the architecture file for optimizations like memory pipelining. Expansion is also possible for new architecture files that describe things like multiple functional unit implementations with different characteristics. These files can be annotated with power and area numbers by the backend tools described in Section 4. Details about functional units power and area consumption can then be fed back into the compiler so decisions can be made about which functional units are needed to meet both power and latency requirements.

3.4 Backend

Once code has been converted to the HDL AST and all applicable operations have run, it is necessary to use a back-end for code generation. Currently, the compiler has a RTL VHDL back-end. The code is converted from the HDL AST to a VHDL AST based on the work done by the MATCH team[4] and the VHDL grammar as specified in Ashenden[3]. The VHDL AST was re-written to better utilize some hierarchical structures of C++ and to include some additional VHDL structures that were omitted in the prior implementation. The VHDL grammar, even when limited to the synthesizable subset, is quite a bit more complicated than the HDL grammar. The AST is designed for code generation only. For example, there are no symbol tables of any kind. Variable and Signal names are just identifiers in the AST. Optimizations are done at the HDL level so that an arbitrary back-end may be targeted. During the AST conversion, libraries and packages are included from the library symbol table. The port to the language specific architecture templates is generated from the port symbol table. Signals are declared from the global symbol table, and each of the local symbol tables. During this phase, each process creates a state machine control signal, with an enumerated-type that corresponds to all the state names in the process. The processes with sensitivities to the global clock and reset signals are created next, and all of the variables from the local symbol table are declared. A conditional that checks for the global reset is created. During the reset state, all local signals and variables are reset. If these correspond to scalar values, they are initialized to zero, pointers and arrays are initialized to the base address value from the memory scheduler. Finally the FSM structure is generated with each of the states. In general, the AST conversion is relatively simple because the HDL nodes convert to similar counterparts in the VHDL AST. Some other simple conversions happen at this stage, such as conversion of integer values to bit representations, some precision checking for operations and assignments, and translation to handle some VHDL specific idiosyncrasies. Once the conversion to the VHDL AST is complete, the code is generated for synthesis.

4. Synthesis Flow

Once the PACT HDL compiler has generated RTL VHDL code it is simulated, synthesized and analyzed for both power and performance. There are two sets of tools used, one to work with the FPGA targets, and a second to target ASICs. The FPGA path uses tools from Model Technologies[19], Synplicity[26] and Xilinx[30]. The ASIC path uses tools from Synopsys[25]. The tool flow for both FPGA and ASIC designs is outlined in Figure 9.7

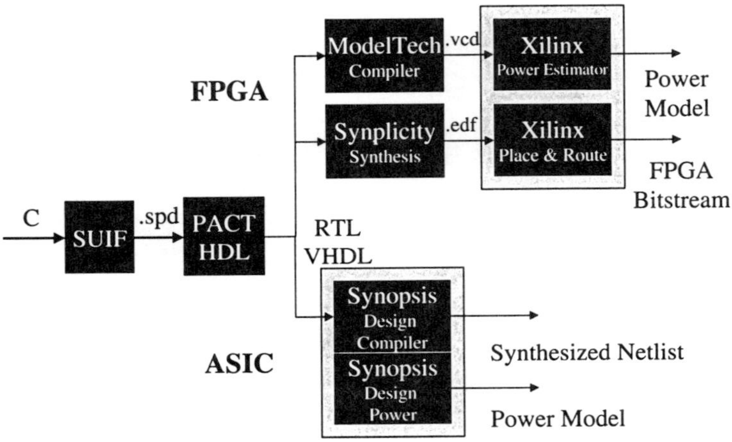

Figure 9.7. Synthesis Flow

4.1 FPGA Design Path

The FPGA designs are targeting the WILDSTAR multi-FPGA board from Annapolis Micro Systems[1] described in Figure 9.8. The design is simulated for correctness using ModelSim from Model Technologies in combination with the WILDSTAR VHDL templates supplied from Annapolis Micro Systems. The design is synthesized using Synplify from Synplicity and placed and routed using the Design Manager from Xilinx Foundation Tools, resulting in a bitstream that can be used to program the FPGA. The design can be run on the FPGA to determine the design's performance. Estimating power is more difficult. Design test-benches must be created to accurately model the typical behavior of the design. These test-benches are used with the design in ModelSim to generate a file that describes the behavior of the system with the current input. This is sent into the Xilinx power estimation tool along with the placed and routed design, to generate a power consumption estimate. These power and performance results can then be used to determine the effectiveness of the optimizations.

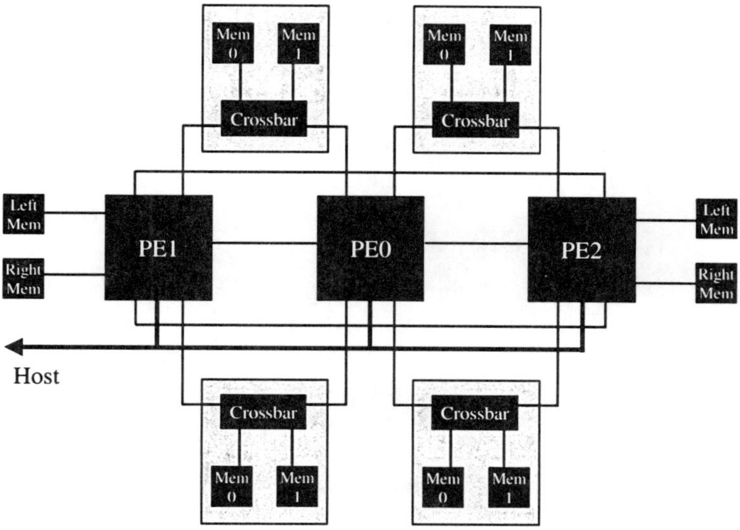

Figure 9.8. Wildstar from Annapolis Micro Systems

4.2 ASIC Design Flow

The ASIC design flow is quite a bit simpler. The ASIC VHDL code is sent into Design Compiler by Synopsys, Inc. It is through Design Compiler that the code is simulated, synthesized, and placed and routed. The Design Compiler suite of tools has a power estimator called Design Power. Thus far, the power libraries from Synopsis for academic work are a 1.5mm process from 1992, operating at 5V. These libraries are clearly too far out of date to provide anything but a very rough estimate of how much power is consumed. An effort is being made to acquire a .25mm process operating at 2.5V from Leda Systems[18]. The new ASIC libraries should give much more realistic results.

5. Optimizations for Power and Performance

The purpose of this paper is to describe the PACT HDL compiler infrastructure for generating real HDL codes from C and the facilities for optimizations at different levels in the compiler hierarchy. It is important to note that SUIF already supplies a fairly extensive toolkit of standard compiler optimizations for which the performance versus power tradeoff may be explored such as constant propagation, common sub-expression elimination, and dead-code elimination. Of course, the SUIF optimizations were generally designed for performance with a focus on parallelization techniques and contain no direct consideration of power issues. It is important to discuss some of the optimizations currently being pursued to examine the power and performance tradeoff. Some of these optimizations will be created at the SUIF AST level, and others will be done at the HDL level. However, some simple optimizations have already been included in the compiler such as the array address calculation by Horner's Rule and temporary variable reduction technique discussed in Section 3.2.1.

5.1 Memory Pipelining

When creating hardware designs, either by hand or through automated methods, memory pipelining is a common performance optimization. It is particularly effective in numerically intensive applications particularly image processing and DSP algorithm implementations. When the hardware target has sufficient logic resources, the bottleneck for these applications becomes memory bandwidth which this optimization can help alleviate. Memory pipelining is a form of functional unit pipelining, allowing the stages of a single ported RAM to occur in parallel. The result for an N cycle memory is an N cycle delay, followed by data every subsequent cycle. It is only possible to pipeline a group of reads or a group of writes. When switching from read to write or vice versa, the N cycle delay plus any switching delay is imposed. Thus, the goals of a memory pipelining algorithm is to group as many reads and writes together as possible, and execute them in a pipelined manner. The pipelining framework in PACT HDL targets loop bodies, particularly the inner most loop. This loop often contains loads and stores corresponding to array accesses. The earliest and latest possible time to execute the load or store is found. These calculations amount to a degree of freedom for reordering the instructions. Sometimes, this allows loops to be unrolled the reads and writes across loop iterations can be grouped for pipelining.

While pipelining often leads to a significant performance improvement it can have an impact on power consumption as well. When considering the power consumed by the memory controller, if the switching from read to write or vice versa is significant, pipelining can prevent this switching and save power. Conversely, pipelining can require instantiation of more registers for temporary storage across loop iterations and significant control logic requiring use of more area and power. If these registers are reused from previously instantiated registers, the switching activity can still cost power. In general, it is expected that pipelining will cost more power and improve performance. This trade-off can be tuned within the power and performance requirements.

5.2 SUIF AST Optimizations

If an optimization can be done at the SUIF AST level, it is generally best to do so. The SUIF infrastructure for creating and executing passes over the

AST is well conceived and implemented. SUIF optimizations may also take advantage of AST passes already available in the SUIF toolkit, such as forward propagation, induction variable detection, and a variety of node dismantling passes. Since the SUIF AST is a much higher-level representation than the HDL equivalent, removing nodes at the SUIF level often has a big impact at the HDL level where a similar code reduction might not be possible. Optimizations that are currently under consideration at this level are

- Bit-width analysis and reduction
- Common sub-expression elimination
- Reverse levelization
- Loop analyses

In C, very few options exist for declaring variables of custom width. Often variables are declared with little regard to the actual values it might hold. In many cases, analysis can determine the minimum required precision is much less than declared in the code. In hardware synthesis, this often leads to reduced register precision and functional unit requirements which in turn lowers power and area requirements and potentially improves circuit performance. Common sub-expression elimination finds identical occurrences of the same calculation where the operands are the same value. The calculation is done once before all the occurrences in the code and stored in a temporary register. The calculations are then replaced by previously calculated value. This can reduce the number of functional units required for the code, allow more instructions to occur in parallel, and potentially necessary functional units to be idle more often, improving area, performance, and power respectively. Reverse levelization increases the number of operations done by one instruction in a single clock cycle. This translates to longer critical paths which in hardware synthesis can result in slower clock speeds and instantiation of more functional units. The performance benefit is fitting more computation into a clock cycle. The success of this optimization will be examined through the power versus performance trade-off. It is well known that caching can have a significant effect on performance; however, its power implications can be equally significant. If a poor caching technique requires more accesses to main memory, the power required to refresh memory will impact the overall power dissipated. Several high-level loop transformations that improve locality and performance can also equally impact power. These optimizations, such as loop reordering, are easily done at the SUIF level and the information can be passed through annotation to the HDL AST for further caching work.

5.3 HDL AST Optimizations

Many of the optimizations considered for this project will map well to the HDL AST. They could potentially require concurrency information, clocking, resource allocation, target architecture specification, and other HDL specific information. The HDL AST is well suited for optimization because it has a simple but general implementation and, like SUIF, allows for arbitrary ordering of AST passes, even before or after the target architecture information has been added. Some optimizations being targeted at this level are

- Memory Caching
- Loop invariant code motion
- Resource Allocation Strategy
- Superfluous control and dead state elimination

As introduced in Section 5.2, caching has a big impact on both performance and power savings. However, unlike general-purpose processor targets, there is a facility for an arbitrary size cache. Issues of data packing, locality, and cache size will all be considered in an optimization set to yield power and performance results. SUIF has a facility for moving loop invariant outside of the loop called a landing pad. Some general invariant code optimizations can be done using this facility in SUIF. Invariant code, such as the code generated from array address calculations described in Section 3.4.1, must be handled at the HDL level. In addition to performance improvement, this optimization will save power by keeping the function units required in the invariant code idle. Bit-width analysis at the SUIF level may make resource sharing more difficult. Operations that could share functional units before the optimization may either require distinct width functional units or must share the higher precision of the two required units. In addition, different instantiations of these functional units with different power and delay numbers may be available as described in Section 3.3. A resource allocation strategy is necessary which decides which units to replicate or share and which instantiation to choose. Most likely a heuristic solution will be required to generate the best possible result in a realistic amount of time. During the SUIF to HDL conversion dead and empty states can be created, often from explicit control. These states add unnecessary complexity to the FSM control logic and can be eliminated improving power, performance, and area consumption. Empty states should be easily eliminated, but dead states will require tests to determine if the state can ever be executed.

6. Results, Conclusions, and Future Work

6.1 Results

The PACT HDL compiler has been tested with several benchmarks and has been shown to generate correct code. The benchmark set is still somewhat limited due to some compiler limitations. The complier currently only supports code without function calls, and with integer arithmetic. The benchmarks include

- Vectorsum
- Matrix Multiplication
- FIR FIlter
- Laplace Transform
- Sobel Transform

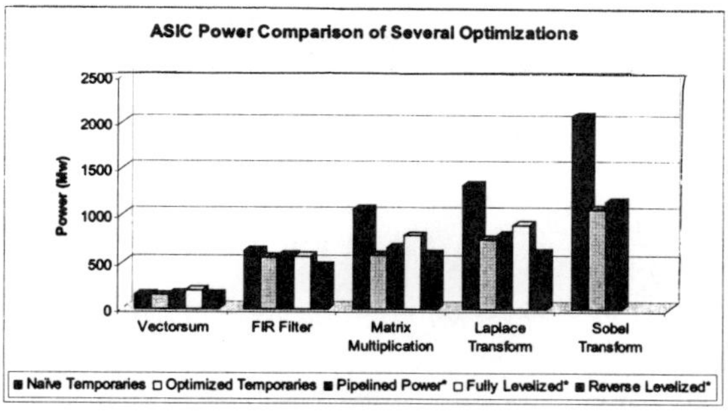

Figure 9.9. Power Comparison between Temporary Variaible Generation Schemes, Memory Pipelining, and Code Levelization optimizations

The initial results are very promising. The compiler is generating correct code for some real applications. All of the benchmarks were synthesized and profiled for power and area requirements using the commercial tools described in Section 4 targeting both the Xilinx XCV400 Virtex FPGA and a 1.5 microm process ASIC using libraries from Synopsys. Unfortunately, when these results were taken, the Synopsis libraries were all that were freely available to universities. The technology is from around 1992 and is useful only for extremely rough estimates on power consumption. There are currently negotiations to acquire 0.25 microm process digital libraries from Leda Systems[18] or more realistic results. Results for the benchmarks can be found in detail in [15].

While many of the optimizations described in Section 5 are still in development a several have already been implemented such as smart temporary variable creation described in Section 3.4.1, memory pipelining, and reverse levelization Optimizations. Figure 9.9 shows a power comparison between the first-cut approach for temporary generation, the improved temporary generation, memory pipelining optimization, and code levelization techniques. The results for memory pipelining, fully levelized and reverse levelized code are based on the optimized temporary generation during code conversion. It is important to note that the default code generated in the compiler is not completely

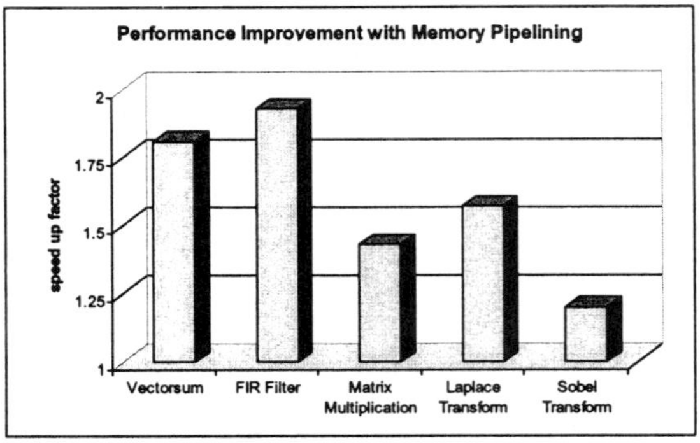

Figure 9.10. Performance Comparison of Pipelined vs. Non-Pipelined Code

levelized, some array address computations contain a multiply accumulate operation done in one clock cycle during array address calculation. The fully levelized code shown here levelizes these statements. It is clear that significant power is saved through the use of smarter temporary generation. Memory pipelining can be considered as a power versus performance trade-off. As seen in Figure 9.9, the power required by pipelined designs does increase, although not dramatically. The speedup achieved by the designs shown in Figure 9.10 can range from 1.25X to nearly 2X. Completely levelizing the code does not have a favorable power results, but reverse levelization did decrease power consumed. It is hoped that inclusion of addition optimizations targeted in Section 5 will continue to contribute to these initial power and performance results.

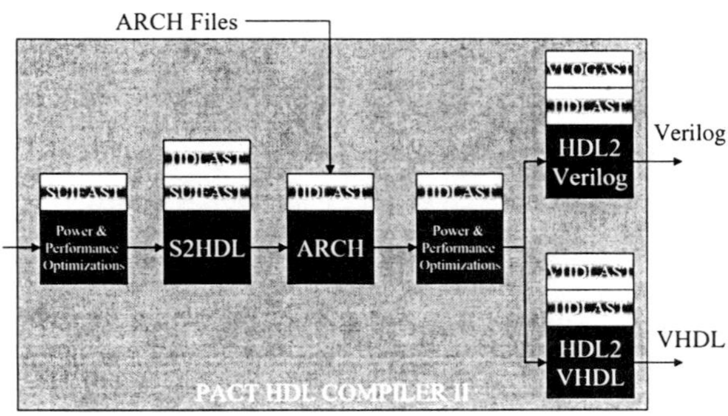

Figure 9.11. Next Generation PACT HDL Compiler

6.2 Future Work

While PACT HDL is certainly a functional tool in automated synthesis, there is still work to be done for improvement and added functionality. More optimizations for power and performance as discussed in Section 5. Support for function calls and real numbers will be added. Support for high level data-structures will be considered. A RTL-Verilog back-end will be developed. Additional architecture specific information will be included. The next generation PACT HDL Compiler is displayed in Figure 9.11

References

[1] Annapolis Micro Systems. WILDSTAR. Xilinx Virtex Based Multi-FPGA Board. www.annapmicro.com.

[2] Arnout, G. *C for System Level Design*. Design, Automation, and Test in Europe Conference and Exhibitions 1999.

[3] Ashenden, P. *The Designer's Guide to VHDL*. Morgan Kaufmann. 1995.

[4] Banerjee, P. Shenoy, N. Choudhary, A. Hauck, S. *MATCH: A Matlab Compilation Environment for Configurable Computing Systems*. Submitted to IEEE Computer 1999.

[5] Chandrakasan, A. P. Potkonjak, M. Mehra, R. Rabaey, J. Brodersen R. W. *Optimizing Power Using Transformations*. IEEE Transactions on Computer Aided Design of Integrated Circuits and Systems 1995.

[6] Chandrakasan, A. P. Sheng, S. Brodersen, R. W. *Low Power CMOS Digital Design*. IEEE Journal of Solid-State Circuits 1992.

[7] Co-Design Automation, Inc. Superlog Website. www.superlog.org.

[8] CynApps Suite. Cynthesis Applications for Higher Level Design. www.cynapps.com.

[9] Davis, D. Edwards S. Harris J. *Forge: High Performance Hardware from Java*. Xilinx Whitepaper, www.xilinx.com.

[10] De Micheli, G. Ku D. Mailhot, F. Truong T. *The Olympus Synthesis System for Digital Design*. IEEE Design & Test of Computers 1990.

[11] Esterel-C Language (ECL). Cadence website. www.cadence.com.

[12] Fraser, C. W. Hanson, D. R. *A Retargetable Compiler for ANSI C*. SIGPLAN Notices 1991.

[13] Free Software Foundation GNU C Compiler. www.gnu.org/software/gcc.

[14] Galloway G. *The Transmogrifier C Hardware Description Language and Compiler for FPGAs* IEEE Symposium on FPGAs for Custom Computing Machines (FCCM) 1995.

[15] Jones, A. Bagchi, D. Pal, S. Choudhary, A. Banerjee, P. *PACT HDL: A C Compiler Targeting ASICs and FPGAs with Power and Performance Optimizations*. Submitted to DATE 2002.

[16] Lakshminarayana, G. Jha, N. K. *Technical Report No. CE-J97-003: Synthesis of Power-Optimized Circuits from Hierarchal Behavioral Descriptions*. IEEE Design Automation Conference (DAC) 1998.

[17] Lakshminarayana, G. Raghunathan, A. Khouri, K. S. Jha, N. K. Dey, S. *Common-Case Computation: A High-Level Technique for Power and Performance Optimization*. Proceedings of the Design Automation Conference (DAC) 1999.

[18] Leda Systems. .25 um Process Standard Cell Libraries. www.ledasystems.com.

[19] Model Technology. ModelSim. HDL Simulator. www.model.com.

[20] Musoll, E. Cortadella, J. *High-level Techniques for Reducing the Activity of Functional Units*. Proceedings of the International Symposium on Low Power Design 1995.

[21] Muchnick, S. S. Compiler Design and Implementation. Morgan Kaufmann. 1997.

[22] Overview of the Open SystemC Initiative. SystemC website. www.systemc.org.

[23] Pitas, I. *Digital Image Processing Algorithms and Applications*. John Wiley & Sons. 2000.

[24] Smria, L. De Micheli, G. *SpC: Synthesis of Pointers in C: Application of Pointer Anaylsis to the Behavioral Synthesis from C*. IEEE/ACM International Conference on Computer-Aided Design (ICCAD) 1998.

[25] Synopsys. Design Compiler. Synthesis and Power Estimation Toolset. www.synposys.com

[26] Synplicity. Synplify. Synthesis Toolset. www.synplicity.com.

[27] System Compiler: Compiling ANSI C/C++ to Synthesis-ready HDL. Whitepaper. C Level Design Incorporated. www.cleveldesign.com.

[28] Thomas, D. E. Moorby, P. R. *The Verilog Hardware Description Language: Fourth Edition*. Kluwer Academic. 1998.

[29] Wilson, R. P. French, R. S. Wilson, C. S. Amarasinghe, S. P. Anderson, J. M. Tjiang, S. W. K. Liao, S. W. Tseng, C. W. Hall, M. W. Lam, M. S. Hennessy, J. L. *SUIF: An Infrastructure for Research on Parallelizing and Optimizing Compilers*. SIGPLAN Notices 1994.

[30] Xilinx. Foundation Tools. Place and route tools for Xilinx FPGAs. www.xilinx.com.

[31] Zeidman, B. Verilog Designer's Library. Prentice Hall. 1999.

Chapter 10

COMPILER OPTIMIZATIONS FOR LOW POWER SYSTEMS

Mahmut Kandemir, N. Vijaykrishnan, Mary Jane Irwin
Microsystems Design Lab
Pennsylvania State University
*University Park, PA**

Abstract Most current compiler optimizations focus on improving execution time. With the increasingly widespread use of embedded systems, however, power/energy consumption is also becoming an important issue. This is particularly true for battery-operated devices where power consumption has first class status along with performance and form factor.

This paper makes the following contributions. First, we present two low-level (back-end) compiler optimizations for energy reduction. An important conclusion drawn from evaluating the impact of these optimizations is that compiling for power/energy is different from compiling for execution cycles. Second, we evaluate widely used state-of-the-art high-level compiler optimizations from a power consumption perspective. Further, we compare the relative impact of these high-level optimizations on both energy and performance metrics in order to identify any differences in optimizing for energy and performance. Finally, we cover a set of optimizations that are designed specifically for exploiting low power features supported by the hardware. In particular, we show how loop and data transformation can be used to exploit low-power mode control mechanisms available in some memory architectures.

Keywords: Compiler Optimizations, Low Power Modes, Energy Optimization, Loop and Data Transformations

*Partial funding provided by a grant from GSRC and NSF CAREER awards 0093082 and 0093085 and NSF grant 0073419

1. Introduction

Energy has become an important design consideration, together with performance, in computer systems. While energy conscious design is obviously crucial for battery driven mobile and embedded systems, it has also become important for desktops and servers due to packaging and cooling requirements where power consumption has grown from a few watts per chip to over a 100 watts. As a result, there has been a great deal of interest recently in examining optimizations for energy reduction from both the hardware and software points of view.

While hardware optimizations has been the focus of several studies and are fairly mature, software approaches to optimizing power are relatively new. Progress in understanding the impact of traditional compiler optimizations and developing new power-aware compiler optimizations are important to overall system energy optimization. Software has a significant impact on the overall energy consumption being the main determinant for activity on the processor core, interconnect and memory system, which are, collectively, responsible for significant percentage of total power dissipation. Despite this observation, to date, most of the compiler techniques consider only delay and area as their main performance metrics. With the growing demand for power-aware software, there is an acute need for investigating energy-oriented compilation techniques and their interaction and integration with performance-oriented compiler optimizations. In this chapter, we seek answers for the following questions:

- Is the most efficient code from the performance perspective the same as that for the energy viewpoint? If not, why?
- What is the impact of current performance-oriented software optimizations (that primarily aim at maximizing data locality and enhancing parallelism [16]) on energy? How do they affect the energy consumption of different system components (memory system, datapath, etc.)?
- What are the relative gains obtained using software and hardware optimization techniques? How can one exploit the interaction between these optimizations to reduce energy further?

We believe that any progress made in answering them will pave the way for our understanding of impacts and interactions of hardware and software optimizations.

The rest of this chapter is organized as follows. In Section 2, we present two energy-aware low level compiler optimizations. Next, we evaluate the influence of performance-oriented software optimizations on energy in Section 3. In Section 4, we show how hardware and software optimizations for energy interact with each other. Finally, we conclude in Section 5.

2. Energy-Aware Low-Level Compiler Optimizations

In this section, we present two low-level compiler techniques to reduce energy consumption. First, we show how instruction scheduling can be performed to reduce energy consumption in the datapath. Second, we present a post-compilation optimization that relabels registers to minimize switching energy in the buses.

2.1 Influence of Instruction Scheduling on Energy

Traditionally, optimizing compilers use instruction scheduling or reordering instructions to improve datapath performance. There have been several efforts at tuning the instruction schedulers to optimize energy consumption. A key observation behind all these optimizations is that the major determinant of the energy consumption is the switching activity in the datapath. Thus, by ordering instructions in the schedule such as to reduce this switching activity between successive instructions, an energy-aware schedule can be created. An instruction scheduling algorithm called *cold scheduling* that prioritizes the selection of each instruction based on the energy cost of placing that instruction next into the schedule is proposed by Su et al. [12]. Tiwari et al. [13, 14, 8] and Russell and Jacome [11] used instruction-level energy characterization based on current measurements to study the impact of instruction ordering on energy consumption.

In order to capture, the essence of the energy-oriented scheduling mechanisms, we explain a pure energy-oriented version of list scheduling given in Figure 10.1. Each instruction is assumed to have a base cost, denoting its average energy consumption. A weighted edge between nodes `i` and `j` gives the energy consumed by the activity of switching from instruction `i` to instruction `j`.[1] This edge weight is also called *circuit-state effect* (*circuit-state cost* or *inter-instruction cost*) [13]. Since the type and number of instructions (a process called *instruction selection*) are fixed prior to instruction scheduling, a successful scheduling algorithm can only reduce the total circuit-state effect. The pure energy-oriented scheduling accomplishes this as follows. First, it selects one of the schedulable nodes (say `i`) and schedules it. In the next step, it attempts to select a node `j` such that the circuit state effect between `i` and `j` is minimal among all possible alternatives; in the following step, a node `k` is selected such that the circuit-state cost between `j` and `k` is minimum and so on. This approach is also greedy, in that it tries to minimize the cumulative circuit-state cost up to the current point in each step. One noticeable difference between this approach and the classical performance-oriented list scheduling is that the latter considers the maximum delay among the candidate nodes (that

```
energy_{i,j} – inter-instruction (circuit-state) cost between
        the instructions i and j
last – last instruction executed

  schld=∅
  while can-sch ≠ ∅ do
    if schld = ∅ then
      n = can-sch[1]
    else
      min-energy = + ∞
      for each j ∈ can-sch do
        if energy_{last,j} < min-energy then
          min-energy = energy_{last,j}
          n = j
    last = n
    schld ⊕ = [n]
    can-sch - = {n}
    for each i ∈ succ(n) do
      if ∀m ∈ pred(i) ∃j such that schld[j]=m then
        can-sch ∪ = {i}
  return (schld)
```

Figure 10.1. Energy-Oriented Scheduling

is, it considers the nodes that have not been scheduled yet) whereas the former considers the total circuit-state cost so far (that is, it considers the nodes that have already been scheduled). The algorithms proposed by [12] and [13] also use a similar approach.

Through extensive experimentation presented in [9], we found that a pure performance-oriented scheduling technique does not necessarily generate the most energy-efficient code. A pure energy-oriented scheduling, on the other hand, is found to be quite effective in reducing the energy consumption; additionally, it was only 6% worse than the pure performance scheduling as far as the execution cycles are concerned for studied configurations. Instead of optimizing for energy and performance separately, instruction scheduling techniques that consider energy and delay simultaneously in a unified setting can also be designed. Parikh et. al. show that such combined metrics are effective in reducing both datapath energy and performance.

One of the difficulties of instruction scheduling for energy is in estimating the impact of the data portions of the instructions. Thus, in addition to instruction scheduling, techniques such operand re-ordering (e.g., swapping operands in ALU or floating point operations) can be very effective in reducing switching activity on buses. In the next subsection, we show how to reduce switching activity in the operand fields of instructions.

2.2 Influence of Register Assignment on Bus Energy

In this section, we focus on reducing the switching activity on the Icache data bus (between the processor core and Icache) by relabeling the register fields of the compiler generated instructions. Sample traces are used to record the transition frequencies between register labels (encodings) in the instructions executed in consecutive cycles using *SimplePower* [15]. This information is then used to obtain new encodings for the registers such that the switching activity (and consequently the energy consumption) in the Icache data bus is reduced. The technique is applicable to any system where switching activity imposed by register labels has an impact on overall energy consumption. The encoding of instructions using relabeling can be considered similar to data encoding techniques investigated for data.

To illustrate the idea, let us consider two consecutive add instructions in the *SimplePower* assembly language:

```
add si,sj,sk
add sl,sm,sn
```

In this simple sequence, there are three switching activities between registers: (1) from s_i to s_l in the destination-register slot, (2) from s_j to s_m in the first source-register slot, and (3) from s_k to s_n in the second source-register slot. Depending on the encodings of the registers involved, the impact of these switching activities can be quite significant. These *register transitions* can also occur between different types of instructions.

It should be clear that for register fields that have frequent transitions, we need to use register numbers whose Hamming distance is minimum. The problem is that register assignments are done by the compiler using sophisticated algorithms and considering a number of other important issues such as minimizing register spills and maximizing register reuse. Therefore, we cannot arbitrarily change the register numbers, just to minimize power consumption. Such modifications, among other things, can also violate data dependences across instructions, thereby changing the semantics of the program being optimized. On the other hand, other alternatives, namely, determining a near-optimal register assignment considering both power and performance is very difficult.

We developed a post-pass, polynomial-time algorithm that relabels the registers *after* global register allocation performed by the compiler back-end. Relabeling registers is always legal as long as it is performed throughout the code. For example, if we decide to relabel s_i as s_j, *all* the occurrences of s_i should be changed to s_j. Also, if we are to perform relabeling for multiple pairs, this should be done simultaneously for all pairs. Informally, our post-pass algorithm takes a compiler-generated register assignment (register allocation) as input and generates an alternative assignment that reduces the power, maintaining the same performance as the original assignment.

The register relabeling optimization was incorporated in the *SimplePower* compilation framework by modifying the Simplescalar toolset. When the optimization was evaluated using several applications, we observed a 12% reduction in the total energy reduction in the Icache data bus using the register relabeling optimization [18].

3. Influence of High-Level Loop Optimizations on Energy

Software techniques aimed at memory power optimizations are more recent as the traditional emphasis has been on improving performance. The goals of these transformations are to reduce the redundancy in data transfers, and to introduce more locality in the accesses so that more data can be retained in registers local to the datapaths and the part of the memory hierarchy closer to the processor. Typically, accessing memory smaller and closer to the datapath reduces the effective capacitance that is switched, thereby reducing the overall energy consumption. Some of the optimizations may, however, result in energy consumption to increase in other system components. In order to understand these tradeoffs better, we evaluate the impact of three widely used high-level compiler optimizations on a simple matrix multiply code.

3.1 Overview of High-Level Optimizations

In this subsection, we provide an overview of the optimizations considered. They are as follows:

Linear Loop Transformations: The linear loop transformations attempt to improve cache performance, instruction scheduling, and iteration-level parallelism by modifying the traversal order of the iteration space of the loop nest. The simplest form of loop transformation, called loop interchange [16], can improve data locality (cache utilization) by changing the order of the loops. From the power consumption point of view, by applying this transformation we can expect a reduction in the total memory power due to better utilization of the cache [6]. The power consumed in other parts of the system can potentially increase as some loop transformations can result in complex loop bounds and array subscript expressions. This may occur, for example, with loop transformations used for eliminating data dependencies alone.

Loop Tiling: Another important technique used to improve cache performance is blocking, or tiling [17]. When it is used for cache locality, arrays that are too big to fit in the cache are broken up into smaller pieces (to fit in the cache). When we consider power, potential benefits depend on the changes in power dissipation induced by the optimization on different system components. We can expect a decrease in power consumed in memory, due to better data reuse [10, 2]. On the other hand, in the tiled code, we traverse the same iteration space of the original code using twice as many loops (in the most general case); this entails extra branch control operations and macro calls. These extra computations might increase the power dissipation in the components of the system. The exact increase/reduction in power consumed in this example will depend on the problem size (`n`), the tile size (`t`), and the architecture under consideration. Note that improving the energy consumption in the memory system using compiler optimizations may make the energies consumed in the core and the memory system comparable, rendering the core power consumption more significant than before.

Loop Unrolling: This optimization unrolls a given loop, thereby reducing loop overhead and increasing the amount of computation per iteration. From the power point of view, fewer computations means less power dissipation. In addition, we can also expect a reduction in the power consumed in the register file and data buses. The power consumed in the datapath will also be affected by the low-level optimizations performed by the back-end compiler.

3.2 Experimental Evaluation

We evaluated the energy consumptions for the matrix multiply code for different cache topologies (configurations) and program versions (each corresponding to different combinations of three optimizations mentioned above). The first observation we made is that all optimizations except loop unrolling increase the core power. This is due to the fact that the optimized versions generally have more complex loop structures; that, in turn, means extra branches and more complex subscript and loop bound calculations. Loop unrolling is an exception, as it reduces loop control overhead and enables better loop scheduling.

When considering the memory power, on the other hand, we made the following observations. First, with the increasing cache size and/or associativity, tiling performs better than pure linear loop transformations and unrolling. Unlike those optimizations, tiling exploits locality in all loop nest dimensions; increasing associativity helps to eliminate conflict misses between different array tiles. Second, in the original (unoptimized) code, the memory power is 5 to 47 times larger than the core power. However, after some optimizations, this picture changes. In particular, beyond a 2K, 2-way set associative cache (i.e., higher associativities or larger caches), the core and memory powers become comparable when some optimizations are applied. For example, when tiling is applied for a 2K, 4-way associative cache, the memory energy is 0.0764 J, which is smaller than the core energy, 0.0837 J. Similarly, for the most optimized version (that uses all three optimizations), the core and memory energy

Version	↓→	Miss Rates			
		1-way	2-way	4-way	8-way
original	1K	0.1117	0.1020	0.1013	0.1013
	2K	0.0918	0.0989	0.1013	0.1013
	4K	0.0737	0.0330	0.0245	0.0150
	8K	0.0680	0.0214	0.0117	0.0117
linear transformed	1K	0.0278	0.0119	0.0113	0.0104
	2K	0.0185	0.0107	0.0099	0.0099
	4K	0.0135	0.0100	0.0099	0.0099
	8K	0.0118	0.0099	0.0099	0.0099
unrolled	1K	0.0678	0.0384	0.0359	0.0359
	2K	0.0479	0.0362	0.0359	0.0359
	4K	0.0358	0.0198	0.0145	0.0173
	8K	0.0294	0.0135	0.0077	0.0077
tiled	1K	0.0180	0.0055	0.0039	0.0039
	2K	0.0105	0.0028	0.0016	0.0016
	4K	0.0046	0.0016	0.0012	0.0013
	8K	0.0027	0.0008	0.0007	0.0006

Figure 10.2. Miss Rates for the Matrix Multiply Code

consumptions are very close for a 4K, 4-way set associative cache. This shows that when we apply optimizations, we reduce the memory energy significantly making the contribution of the core energy more important. Since we expect these optimizations (in particular, loop tiling) to be applied frequently by optimizing compilers, reducing core power using additional techniques might become very important. Overall, the power optimizations should not focus only on memory, but need to consider the overall system power. In fact, the choice of best optimization for this example depends strongly on the underlying cache topology. For instance, when we consider the total energy consumed in the system, for a 4K, 2-way cache, the version that uses only loop permutation and unrolling performs best. Whereas for an 8K, 8-way cache, the most optimized version (that uses all three optimizations) outperforms the rest. In fact, given a search domain for optimizations and a target cache topology, an optimizing compiler can decide which optimizations will be most suitable.

3.3 Cache Miss Rates versus Energy Consumptions

We now investigate the correlation between cache miss rate and energy consumption. Figure 10.2 gives the miss rates for some selected cases. This subsection will make some correlations between miss rates and energy consumptions. Let us first consider the miss rates and energy consumption of the original (unoptimized) code. When we move from one cache configuration to another, we have a similar reduction rate for energy as that for miss rate. For instance, going from 1K, 1-way to 1K, 2-way reduces the miss rate by a factor of 1.10 and reduces the energy by the same factor. As another example, when we move from 1K, 1-way to 4K, 8-way, we reduce the miss rate by a factor of 7.45, and the corresponding energy reduction is a factor of 7.20. These results show that the gain in energy obtained by increasing associativity is not offset, in general, by the increasing complexity of the cache topology. As long as a larger or higher-associative cache reduces miss rates significantly (for a given code), we might prefer it, as the negative impact of the additional complexity is not excessive. However, we note that when moving from one cache configuration to another, if there is not a significant change in miss rate (as was the case in our experiments when going from 1K, 4-way to 1K, 8-way), we incur an energy increase. This can be expected as, everything else being equal, a more complex cache consumes more power (due to more complex matching logic).

Next, we investigate the impact of various optimizations for a fixed cache (and memory) topology. The following three measures are used to capture the correlation between the miss rates and energy consumption of the original and optimized versions.

$$\texttt{Improvement}_{\texttt{m}} = \frac{\text{Miss rate of the original code}}{\text{Miss rate of the optimized code}},$$

$$\texttt{Improvement}_{\texttt{e}} = \frac{\text{Memory energy consumption of the original code}}{\text{Memory energy consumption of the optimized code}},$$

$$\texttt{Improvement}_{\texttt{t}} = \frac{\text{Total energy consumption of the original code}}{\text{Total energy consumption of the optimized code}}.$$

In the following discussion, we consider four different cache configurations: 1K, 1-way; 2K, 4-way; 4K, 2-way; and 8K, 8-way. Given a cache configuration, the following table shows how these three measures vary when we move from the original (unoptimized) version to an optimized (tiled) version of the matrix multiply code.

	1K, 1-way	2K, 4-way	4K, 2-way	8K, 8-way
$\texttt{Improvement}_{\texttt{m}}$	6.21	63.31	20.63	19.50
$\texttt{Improvement}_{\texttt{e}}$	2.13	18.77	5.75	2.88
$\texttt{Improvement}_{\texttt{t}}$	1.96	9.27	3.08	1.47

We see that in spite of very large reductions in miss rates as a result of tiling, the reduction in energy consumption is not as high. Nevertheless, it still follows the miss rate. We made the same observation in different benchmark codes as well. We have found that $\texttt{Improvement}_{\texttt{e}}$ is smaller than $\texttt{Improvement}_{\texttt{m}}$ by a factor of 2 - 15. Including the core (datapath) power makes the situation worse for tiling (from the energy point of view), as this optimization increases the core energy consumption. Therefore, compiler writers for energy-aware systems can expect an overall energy reduction as a result of tiling, but not as much as the reduction in the miss rate. Thus, optimizing compilers that estimate the miss rate (before and after tiling) statically at compile time can also be used to estimate an approximate value for the energy variation. The following table gives the same improvement measures for the loop unrolled version of the matrix multiply code.

	1K, 1-way	2K, 4-way	4K, 2-way	8K, 8-way
$\texttt{Improvement}_m$	1.65	2.82	1.67	1.52
$\texttt{Improvement}_e$	2.07	3.53	2.07	1.83
$\texttt{Improvement}_t$	2.03	3.37	1.97	1.68

The overall picture here is totally different. First, $\texttt{Improvement}_e$ is larger than $\texttt{Improvement}_m$, which proves that loop unrolling is a very useful transformation from the energy point of view. Including the core power makes only a small difference, as this optimization reduces the core power as well. We should mention that our other experiments (not presented here due to lack of space) yielded similar results. We now look at the loop transformed version of the same code:

	1K, 1-way	2K, 4-way	4K, 2-way	8K, 8-way
$\texttt{Improvement}_m$	4.02	10.23	3.30	1.18
$\texttt{Improvement}_e$	3.42	8.51	2.74	0.99
$\texttt{Improvement}_t$	3.17	6.84	2.32	0.94

Here, $\texttt{Improvement}_e$ closely follows $\texttt{Improvement}_m$. Including the core energy brings the energy improvement down further, as in this example, the loop optimization results in extra operations for the core. In the experiments with other cache configurations, we observed similar trends: $\texttt{Improvement}_e$ generally follows $\texttt{Improvement}_m$; but it is slightly lower. And, $\texttt{Improvement}_t$ is smaller than $\texttt{Improvement}_e$ by a factor of 1.05 to 1.80.

We can conclude that the energy variations do not necessarily follow miss rate variations in the optimized array-dominated codes.

4. Interaction of Hardware and Software Optimizations

In this section, we focus specifically on memory system energy due to data accesses and illustrate how software and hardware optimizations affect this energy.

4.1 Hardware Optimizations

A host of hardware optimizations have been proposed to reduce the energy consumption. In this section, we focus on two cache optimizations, namely, block buffering and cache subbanking [5]. Note that none of these optimizations cause a noticeable negative impact on performance. In the block buffering scheme, the previously accessed cache line is buffered for subsequent accesses. If the data within the same cache line is accessed on the next data request, only the buffer needs to be accessed. This avoids the unnecessary and more energy consuming access to the entire cache data and tag array. Multiple block buffers can be thought of as a small sized Level 0 cache. In the cache subbanking optimization, the data array of the cache is divided into several subbanks and only the subbank where the desired data is located is accessed. This optimization reduces the per access energy consumption.

We studied the energy consumed by the matrix multiply code in the data cache with different configurations of block buffers and subbanks (the number of block buffers being either 2, 4 or 8 and the number of sub-banks varying from 1 to 4) for a 4K cache with various associativities. This result showed that increasing the number of sub-banks from one to two provides an energy saving of 45% for the data cache accesses. An additional 22% saving is obtained by increasing the number of sub-banks to 4. It must be observed that the savings are not linear as one may expect. This is because the energy cost of the tag arrays remains constant, while there being a small increase in energy due to additional sub-bank decoding. We found that for block buffering adding a single block buffer reduced the energy by up to 50%. This reduction is achieved by capturing the locality of the buffered cache line, thereby avoiding accesses to the entire data array. However, access patterns in many applications can be regular and repeating across a varied number of different cache blocks. In order to capture this effect, we varied the number of block buffers to two, four, and eight as well. We observed that, for our matrix multiply benchmark, an additional 17% (as compared to a single buffer) energy saving can be achieved using four buffers.

We also found that using a combination of eight block buffers and four subbanks, the energy consumed in 4K (16K) data cache could be reduced on an average by 88% (89%). Thus, such hardware techniques can reduce the energy consumed by processors with on-chip caches. However, if we consider the entire memory system including the off-chip memory energy consumption, the energy savings from these techniques amount to only 4% (15%) when using a 4K (16K) data cache. Thus, it may be necessary to investigate optimizations at the software level to supplement these optimizations.

4.2 Combined Optimizations for Memory Energy

It was found that when a combination of different software (loop tiling, loop unrolling, and linear loop transformations) and hardware (block buffering and subbanking) optimizations is applied, tiling performs the best among the three individual compiler optimizations applied in terms of memory system energy across different cache configurations. Since, we mentioned earlier that tiling increases the cache energy consumption, subbanking and block buffering are of particular importance here. For the tiled code, moving from a base data cache configuration to one with eight block buffers and four subbanks reduces the overall memory system energy by around 10%. Thus, it is important to use a combination of hardware and software optimizations in designing an energy-efficient system.

Further, we observed that the linear loop transformed codes exploited the block buffers better than the original code and other optimizations. For example, when using two (eight) block buffers in a 4K 2-way cache, the block buffer hit rate was 69% (82%) as compared to the 55% (72%) for the unoptimized matrix multiply code. Thus, it is also important to choose the software optimizations such that they provide the maximum benefits from the available hardware optimizations.

Overall, we observe that even performance based compiler optimizations provide a significantly higher energy savings as opposed to those gained using the pure hardware optimizations considered. However, a closer observation reveals that hardware optimization become more critical for on-chip cache energy reduction when executing optimized codes. We refer the reader to [5] for more discussion on this topic. In the next two sections, we show how we could design software optimizations to improve the effectiveness of low power hardware features.

4.3 Improving Effectiveness of Power Mode Control Mechanisms Using Code Transformations

Memory (DRAM) modules can be placed in different power modes characterized by different energy consumption per cycle and different latencies to service requests. While it is possible to exploit these power modes in optimizing the DRAM energy without modifying the original code (except for inserting instructions to set operating modes), it is also possible to obtain further energy savings by employing computation (loop nest-based) transformations [4]. In this section, we show how loop fission, a widely-used high-level optimization technique can be used to improve the effectiveness of power mode control.

Loop fission (also known as loop distribution [16]) takes a nested loop that contains multiple statements in it, and creates multiple nested loops each with a subset of the original statements. An optimizing compiler can use loop fission for a number of reasons which include improving instruction cache locality and enhancing iteration-level parallelism. This optimization first builds a statement-level dependence graph (in which the nodes denote statements and the edges correspond to data dependences between them) for the body of the nested loop. Then, if there are no cycles in the dependence graph, the optimization creates a separate loop for each statement. Otherwise, the statements in a (dependence) cycle remain in the same loop. If the nested loop contains multiple loops, loop fission is applied starting from the innermost position, and in fissioning the outer loops, the inner loops are treated as single block statements. Note that even in cases where each statement can be put into a separate loop nest an optimizing compiler may choose not to do so for some other reason such as improving data cache locality or reducing code expansion.

Loop fission helps to improve the effectiveness of array allocation by allowing a finer-granular control over the allocation of arrays. For instance, in the example shown below (assuming that the arrays are of the same size and each memory bank can hold at most two arrays), with the original nest, the two banks that contain the four arrays should be in the active mode throughout the entire execution. After the loop fission, on the other hand, only a single bank needs to be in the active mode during the execution of each loop (assuming that array allocation places a and b into one bank, and c and d into the other). The other bank can be put into a low-power mode, thereby saving energy.

```
for(i=0;i<N;i++)
{                          for(i=0;i<N;i++)
  {U[i],V[i]}        ==>     {U[i],V[i]}
  {W[i],X[i]}              for(i=0;i<N;i++)
}                            {W[i],X[i]}
```

The overall algorithm for loop fission for energy is given in Figure 10.3. The purpose of this algorithm is to try different loop fissioning strategies for a given nested loop. If there are K instructions within the loop nest, the main 'for loop' in the algorithm enumerates $K-1$ alternatives. The ith alternative is formed by breaking up the nest into two nests after the ith statement from the beginning of the nest (if it is legal to do so). We also add two more alternatives to these $K-1$ alternatives: the original nest and a code with K nested loops each with its own statement (again, if this is legal). The $K+1$ alternatives considered by the algorithm in Figure 10.3 are as follows:

```
                            for(...)
              for(...)      {               for(...)
for(...)        S1            S1            {           for(...)
{             for(...)        S2              S1          S1
  S1          {             }                 S2        for(...)
  S2            S2          for(...)          S3          S2
  S3     ,      S3     ,    {        ,....,   ...   ,   for(...)
  ...           ...           S3              SK-1        S3
  SK-1          SK-1          ...           }             ...
  SK            SK            SK-1          for(...)    for(...)
}             }               SK              SK          SK
                            }
```

For each alternative, the estimated dynamic energy consumption is calculated, and the alternative with the minimum energy consumption is selected. Although we evaluate only a limited set of alternatives, our experimentation indicated that there was no other fissioning strategy for the codes in our experimental suite that would result in more energy savings.

While it is possible to develop a loop fissioning strategy that targets the entire procedure (by taking into account the inter-nest interactions), in this study, we have just focused on the most energy consuming nest (called *dominating nest*) from each benchmark, and applied the loop fissioning algorithm only to that nest. This approach is expected to improve the overall energy consumption, possibly at the expense of an increase in energy consumption of the other nested loops, because in all the codes that we experimented with, there were only one or two dominating nests. If there are two nested loops that consume exactly the same (maximum) amount of energy, we designate the first one (in the textual program order) as the dominating nest.

Figure 10.4 shows the percentage decrease in energy (when the *entire application* is considered) brought about by each fissioning alternative over mode control plus clustering. It should be noted that we consider only the codes that can benefit from loop fission, and different most costly nests (of different codes) have different number of fissioning options (alternatives). The main reason that prevented the application of loop fission to other codes was the fact that the dominating nests in these codes contain only a single instruction; so, there was only one option, which is the original nest. We also observed that applying loop fission to the *second dominating nest* (instead of the dominating nest) in general increased the overall energy consumption as the array layouts suggested by the second most costly nest are usually not suitable for the most costly nest.

Consequently, we selected the most effective alternative for each benchmark, and applied it. The average improvement (%) over all benchmarks

INPUT: *A perfect loop nest containing K instructions*
OUTPUT:
An energy-optimized loop-fissioned code
(B contains the number that identifies the loop fissioned version with minimum energy)
Begin
Let E_{min} *be the least computed energy consumed by the loop nest*
Let B be the number of the more energy saving option.
$E_{min} \leftarrow$ *The energy consumed by the original loop nest (without fission)*
$B \leftarrow 0$
For *i = 1 to K - 1*
Create two loop nests instead of the original loop nest
• *the first containing the first i instructions and the second the remaining* $K - i$ *instructions*
Compute the energy E_i *consumed by this new program.*
If $E_{min} > E_i$
$E_{min} \leftarrow E_i$
$B \leftarrow i$
EndIf
EndFor
Create K loop nests, instead of the original loop nest
• *each one contains one instruction of the original nested loop*
Compute the energy E_K *for this new program*
If $E_{min} > E_K$
$E_{min} \leftarrow E_K$
$B \leftarrow K$
EndIf
End

Figure 10.3. An Algorithm that Improves Memory Energy Consumption using Loop Fission

(that are amenable to loop fission) was around 55.5%. Since the array allocation considers only the dominating nest, there might be a negative impact (energy-wise) on other nests if the best layout strategy for the most costly nest is not suitable for the other nests. Figure 10.5 shows the energy impact of this dominating nest-centric optimization on other nests. We observe that different benchmarks behave quite differently. For example, in `eflux`, all the other nests (in addition to the dominating nest) show energy improvement. For `vpenta`,

Benchmark	Alternative Fissioning Strategies for the Dominating Nest (%)								
	#1	#2	#3	#4	#5	#6	#7	#8	#9
adi	47.0	47.0	61.2						
dtdtz	48.5	48.5	48.5	48.5					
eflux	47.0	45.2	43.3	66.9					
matvec	49.8	33.2	16.6	58.2					
tomcatv	49.5								
vpenta	8.2	23.5	16.6	24.9	18.0	16.6	20.7	8.2	48.4

Figure 10.4. Percentage Energy Improvements due to Different Loop Fission Alternatives

Benchmark	Loop Nests (%)								
	#0	#1	#2	#3	#4	#5	#6	#7	#8
adi	61.2	0.0							
dtdtz	48.5	0.0	-33.0						
eflux	47.0	19.9	24.7	19.9	19.9	25.0			
matvec	58.2	-48.1	0.0	0.0					
tomcatv	49.5	0.0	0.0	0.0	-1.5	0.0	16.3	0.0	0.0
vpenta	-33.2	-39.8	0.0	48.4	-33.3	-33.3	-0.3	25.0	

Figure 10.5. Percentage Variations in Energy of All Nested Loops in the Benchmarks

on the other hand, the second nest experiences a 39.8% increase in energy consumption. Overall, the nests *other than the dominating ones* show only a 2.8% increase in energy. Thus, we conclude that focusing only on the dominating nest and performing an array allocation based on that works well with loop fission.

4.4 Improving Effectiveness of Power Mode Control Mechanisms Using Data Transformations

While loop transformations can be used to improve the effectiveness of low power memory mode operation as shown in previous subsection, they can be problematic to apply in certain cases. For example, unlike other languages such as C or Fortran, changing access pattern of Java codes is more problematic as the language imposes a precise exception requirement. That is, it is not safe to change the execution order of loop iterations unless one is certain that the original occurrence order of exceptions will be preserved after this modification. This is the primary obstacle before any transformational approach based on loop transformations and instigated some efforts in eliminating array bounds checking and in developing constrained loop optimizations [7]. Data transformations on the other hand are independent of access pattern and cannot modify/constrain occurrence order of exceptions. In the following two subsections, we summarize the anticipated impact of using two data optimization approaches on energy consumption. The details of the techniques can be found in [1] and are omitted here for lack of space.

4.4.1 Data Layout Transformations. The data layout transformations can be used to reduce the energy savings of partitioned memory modules with low power modes. By changing the storage order and improving the spatial locality of the arrays using the transforms, one can expect to increase the average inter-access times between two memory operations to the same module. The increased inter-access times would provide an opportunity for operating in a lower power mode for a longer time. Note that the increased inter-access time could be due to two different reasons. First, improved cache behavior can increase the time between two accesses to memory. Next, the memory accesses can be confined to a particular memory module for a period of time allowing the other modules to operate in a lower power mode for extended periods of time. For example, consider storing a large array that spans across different modules. If it is stored in row-major form and accessed in column-major form, successive references may access different modules. However, if it is stored and accessed in column-major form, successive accesses will be confined to the same module except at module boundaries.

4.4.2 Array Interleaving. In this section, we discuss a data space transformation technique, array interleaving, that transforms a number of array variables simultaneously. It achieves this by interleaving the memory allocation for the elements of a set of arrays in a common heap space. This can be used to eliminate inter-variable conflict misses, that is, the conflict misses that are due to different variables. In our partitioned memory architecture, such an optimization can also maximize opportunities for placing memory modules into a low power operating mode.

For the partitioned-memory architecture, only those memory modules containing the parts of the arrays currently being accessed need to be active. If we use the array interleaving strategy, it would co-locate portions of different arrays which are accessed at the same time. This can provide an opportunity for transitioning more modules into a lower power mode. For instance, let us assume that the values of elements from one array `U` are copied to the elements of another array `V` and that the two arrays are allocated in two different memory modules. When copying the elements of one array to another, we need to keep both modules active for the entire duration of loop execution. After interleaving, the corresponding rows of each array are co-located. Hence, half of each array spans each module. Thus, during the first half of the loop execution one of the modules can be transitioned into a lower power mode and in the second half of the loop execution the other module can operate in a lower power mode.

Through experimentation on a set of Java applications [1], it is observed that using layout transformation provides an average of 9.68% energy saving in addition to the savings provided by mode control [1]. Similarly, the array interleaving optimization provides an average of 14.96% additional energy savings.

5. Summary

In this chapter, we have shown that software optimizations play a critical role in determining the system energy. These optimizations can be applied at different stages of the compiler, either at the high-level or the low-level. Further, we demonstrated that optimizing for energy and performance are not the same. The optimizations proposed also stress the importance of increasing the synergy between the underlying low power hardware and the software executing on it in order to gain significant energy savings. While most of this chapter has focussed on dynamic power consumption, it will be vital in future to focus on leakage power management. Also, it would be important to dynamically generate code that adapts to changing energy constraints in mobile environments.

Acknowledgments

The authors wish to acknowledge the contributions of the students from the Microsystems Design Group who have worked on several projects reported in this paper.

Notes

1. Note that, even if there is no edge between two instructions, we will still have a weight between them, as these two instructions can be executed in the final schedule one after another.

References

[1] R. Athavale, N. Vijaykrishnan, M. Kandemir and M. J. Irwin. Influence of Array Allocation Mechanisms on Memory System Energy. In Proc. of *International Parallel and Distributed Processing Symposium*. April 2001.

[2] F. Catthoor, S. Wuytack, E. D. Greef, F. Balasa, L. Nachtergaele, and A. Vandecappelle. Custom memory management methodology – exploration of memory organization for embedded multimedia system design. *Kluwer Academic Publishers,* June, 1998.

[3] A. Chandrakasan et.al., Optimizing power using transformations, *IEEE Trans. on CAD*, TCAD-14(1):12-31, Jan. 1995.

[4] V. Delaluz, M. Kandemir, N. Vijaykrishnan, and M. J. Irwin. Energy-oriented compiler optimizations for partitioned memory architectures. In

Proc. of *International Conference on Compilers, Architecture, and Synthesis for Embedded Systems*, November 2000.

[5] G. Esakkimuthu, N. Vijaykrishnan, M. Kandemir, and M. J. Irwin. Memory system energy: Influence of hardware-software optimizations. In Proc. of *ACM/IEEE International Symposium on Low Power Electronics and Design*, Rapallo/Portofino Coast, Italy, July, 2000.

[6] M. Kandemir, N. Vijaykrishnan, M. J. Irwin, and W. Ye. Influence of Compiler Optimizations on System Power. In Proc. of *the 37th Design Automation Conference*, Los Angeles, CA, June 5–9, 2000.

[7] J. Moreira, S. Midkiff, M. Gupta, P. Artigas, M. Snir, and R. Lawrence. Java Programming for High-Performance Numerical Computing. *IBM Systems Journal,* 39(2), 2000.

[8] M. Lee, V. Tiwari, S. Malik, and M. Fujita. Power analysis and minimization techniques for embedded DSP software. *Fujitsu Scientific and Technical Journal,* 31(2), pp. 215–229, 1995.

[9] A. Parikh, M. Kandemir, N. Vijaykrishnan and M. J. Irwin. Instruction scheduling based on energy and performance constraints. In Proc. *The IEEE CS Annual Workshop on VLSI,* Orlando, FL, April 27-28, 2000.

[10] K. Roy and M. C. Johnson. Software design for low power. *Low Power Design in Deep Sub-micron Electronics,* Kluwer Academic Press, October 1996, ed. J. Mermet and W. Nebel, pp. 433–459.

[11] J. T. Russell and M. F. Jacome. Software power estimation and optimization for high performance, 32 bit embedded processors. In *Proc. ICCD'98,* Austin, TX, October 1998.

[12] C.-L. Su, C.-Y. Tsui, and A. M. Despain. Low power architecture design and compilation techniques for high-performance processors. In *Proc. IEEE COMCON,* 1994, pp. 489–498.

[13] V. Tiwari, S. Malik, A. Wolfe, and T.C. Lee. *Instruction Level Power Analysis and Optimization of Software*, Journal of VLSI Signal Processing Systems, Vol. 13, No. 2, August 1996.

[14] V. Tiwari, S. Malik, and A. Wolfe. Power analysis of embedded software: A first step towards software power minimization. *IEEE Transactions on VLSI Systems,* December 1994.

[15] N. Vijaykrishnan, M. Kandemir, M. J. Irwin, H. Y. Kim, and W. Ye. Energy-driven integrated hardware-software optimizations using SimplePower. In Proc. *the International Symposium on Computer Architecture,* Vancouver, British Columbia, June 2000.

[16] M. Wolfe. *High Performance Compilers for Parallel Computing*, Addison Wesley, CA, 1996.

[17] M. Wolf and M. Lam. A data locality optimizing algorithm. In *Proceedings of ACM SIGPLAN 91 Conference Programming Language Design and Implement ation*, pages 30–44, June 1991.

[18] W. Ye, N. Vijaykrishnan, M. Kandemir, and M. J. Irwin. The design and use of SimplePower: a cycle-accurate energy estimation tool. In Proc. *37th Design Automation Conference*, Los Angeles, June 2000.

Chapter 11

POWER-PERFORMANCE TRADE-OFFS IN SECOND LEVEL MEMORY USED BY AN ARM-LIKE RISC ARCHITECTURE

Kiran Puttaswamy⋆, Lakshmi Narasimhan Chakrapani⋆, Kyu-Won Choi⋆, Yuvraj Singh Dhillon⋆, Utku Diril⋆, Pinar Korkmaz⋆, Kyoung-Keun Lee⋆, Jun Cheol Park⋆, Abhijit Chatterjee⋆, Peeter Ellervee◇, Vincent John Mooney III⋆, Krishna V Palem⋆, Weng-Fai Wong*

⋆ *School of Electrical and Computer Engineering,*
Georgia Institute of Technology.
◇ *Department of Computer Engineering,*
Tallinn Technical University.
∗ *School of Computing,*
National University of Singapore.

Abstract Power consumption is an important dimension in microprocessor and digital system design. This is especially true in the embedded setting where microprocessors have to operate without the luxury of a large power supply or cooling structures. In this paper, we describe an infrastructure setup for the study of power-performance tradeoffs in microprocessor architecture and compiler optimizations. This infrastructure distinguishes itself from those already proposed in the literature in its use of power estimations based on synthesis of the architecture and the full integration of a well-established optimizing compiler framework. We present some preliminary results where we show how the circuit level and architectural techniques can be combined to save overall system power. In particular we reduce the clock frequency and supply voltage of level two memory accesses (circuit level technique) and compensate for the resulting increase in the completion time by implementing a non-blocking store buffer (architectural technique) resulting in up to 39 % less power and up to 28 % less energy on a set of candidate benchmarks.

1. Introduction

In embedded systems where battery life is often limited, the trend of increasing power consumption by modern microprocessors is an important concern.

Power Aware Computing
Edited by Graybill and Melhem, Kluwer Academic/Plenum Publishers, 2002

New devices, architectural features, and even compilation techniques are being examined to trade-off performance for power. An infrastructure to investigate and optimize for power dissipation must include the following:

* Device Technology: Power consumption is obviously dependent on the technology and design parameters used.
* Circuit Geometry: The geometry of the circuit also affects power. An example would be the length of buses; longer buses imply higher capacitance and more power dissipation.
* Architectural Parameters: Architectural parameters such as the size and associativity of the caches, etc., are known to have a significant impact on power consumption.
* Compiler Optimizations: Compiler optimizations change the actual code that is executed on the processor.

Given all the above considerations, it is necessary to have an integrated framework in order to explore the power-performance design space. In this paper, we present a fully integrated infrastructure starting from an aggressive compilation front-end to a synthesized Verilog model of an ARM-like architecture. In particular, we use a Register-Transfer Level (RTL) Verilog Hardware Description Model (HDL) model of an ARM-like RISC processor developed by the University of Michigan [13]. The power estimates are obtained using the synthesized models. The rest of this paper will outline the infrastructure and describe the results.

2. Previous Work

Power dissipation in a system can be divided into dynamic power dissipation and static power dissipation. Dynamic power dissipation consists of switching power and short-circuit power dissipation, whereas static power dissipation mainly consists of the leakage power dissipation [15]. Since switching activity depends on the high level features such as the algorithm and microarchitecture, much of the research on reducing dynamic power revolves around system level innovations, algorithmic transformations and architectural innovations. We summarize some of this research in the following paragraphs.

At the system level, there have been studies on power-performance tradeoffs using profiling hardware to identify tightly coupled regions of code and optimizing the configuration of the microprocessor (eg. need based resource allocation and aggressive clock gating styles) dynamically so as to save energy with minimum performance penalty [24]. Also, there has been work on transforming the algorithms in such a way as to be able to dynamically adjust the speeds by either lowering the frequency or shutting down the unused modules, whichever gives the best results [25]. At the algorithmic level, various transformations perform the most significant computations first and throw away the less significant computations based on a quality metric, thus reducing the load of computation and achieving reduction in the total energy and hence reducing the power[18]. Also there has been work on reducing the transition count on the main memory address bus by power-optimally mapping the behavioral array references to physical memory addresses which improves the cache locality, reducing the cache miss rate and thus impacting power [16]. For power reduction within the core, some techniques reduce the hamming distance between branch addresses so as to reduce the switching and hence dynamic power dissipation [17]. At the technology level, which has a direct bearing on the power consumption, various techniques have been researched such as gating the supply voltage to cache memories[19] . A similar framework to our infrastructure is introduced in Simplescalar ARM [14]. It is a framework for architectural-level power analysis and optimizations, with the main components being the power models for various hardware structures and the architectural simulator SimpleScalar. The best feature of this framework is the simulation speed and the ease of configuring the architecture. On the other hand, our work can more accurately model the specific RISC architecture as the results are based on a more detailed analysis. One similarity between the two works is that we use the same simulation models. An additional aspect of our work is the inclusion of the external memory and the printed circuit board bus power models. A survey of the previous work related to reducing power dissipation shows that the results so far have tended to focus on either the compiler or the micro-architecture or the circuit level techniques. We, on the other hand, attempt to achieve a tight integrated flow between all three levels so that the interdependencies are more closely tracked and the interplay amongst the various parameters is better analyzed across a broad spectrum. It enables us to investigate the trade-offs between compiler, micro-architecture and some circuit level power optimizations.

3. Experimental Methodology

The experimental setup consists of three main parts, a C compiler, a Verilog simulation environment for performance and power estimation, and analytical models for off-chip memory and Printed Circuit Board (PCB) bus [figure 11.1] between processor and memory.

- Compiler: The C compiler infrastructure we use is an open-source optimizing compiler infrastructure called Trimaran[9]. The compiler backend has been modified to generate code targeted towards the ARM-like RISC architecture.
- Simulator: The simulator consists of a cycle accurate Verilog model of the ARM-like RISC architecture, obtained from the University of Michi-

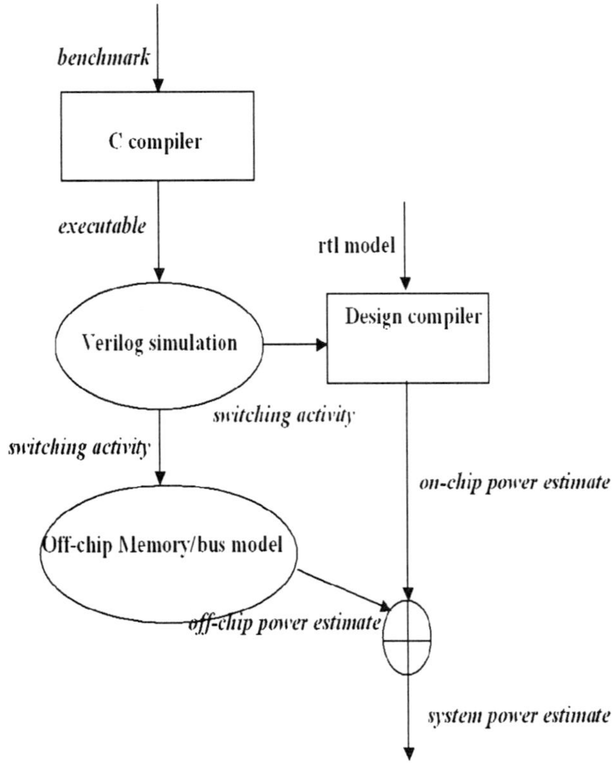

Figure 11.1. Experimental Infrastructure

gan[13]. We interfaced the ARM-like RISC model to an off-chip memory model from Denalisoft[20]. The external memory bus switching activity and the memory access patterns are collected from the simulation environment and fed to the external bus and memory models. We synthesized the RTL core using Synopsys Design Compiler [10] targeted towards a 0.25-micron TSMC library from LEDA Systems [11]. (Note that the main difference between the simulation and synthesis models is the memory modules. The RTL model has synthesizable cache memory components instantiated using Synopsys Designware library and mapped to the TSMC technology library. At the heart of the above infrastructure is a power estimation and analysis tool called Power Compiler from Synopsys, Inc. This tool has two interfaces, one to the simulation environ-

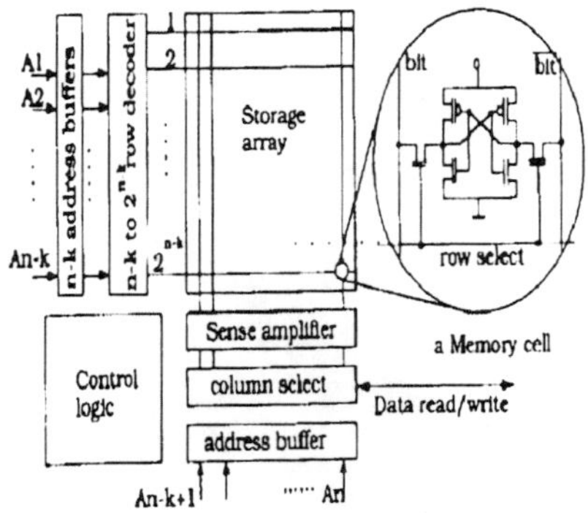

Figure 11.2. Memory Circuitry

ment and the other to the synthesis environment. It obtains the switching activity of each net (wire) of the various functional modules based on an execution of the benchmark in the simulation environment. Then, it annotates this switching activity onto the synthesis environment and provides estimates of the dynamic and static power dissipation for the particular technology chosen (in our case, 0.25 micron TSMC CMOS Technology).

- Memory: The memory circuit [figure 11.2] is a basic SRAM design using six transistors per memory cell. The components modeled are memory cells, input buffer, output buffer, row decoders, column decoders, sense amplifiers and read/write control logic. A typical memory is divided into five parts viz , memory cell, row decoder, row driving, column selection, sense amplifier and load. The total power consumed is equal to the sum of power consumed by each of these components.

For our experiments, we assume the off-chip memory to be a single eight Megabyte memory chip. We use the analytical model based on [23].The analytical model was combined with the parameters for the TSMC 0.25 μ library and the various parameter values were obtained. The variations of power and read/write access delay for the memory with supply voltage V_{dd} and switching activity were found.

The graph in figure 11.3 describes the dependence of the power dissipation of the memory with the power supply voltage V_{dd} and the switching activity period. (Note that switching activity period is the reciprocal of effective switching frequency. Effective switching frequency is how often a signal actually switches values. Clearly, effective switching frequency of a signal is data- or profile-dependent, i.e., dependent on a particular benchmark and data.) As the

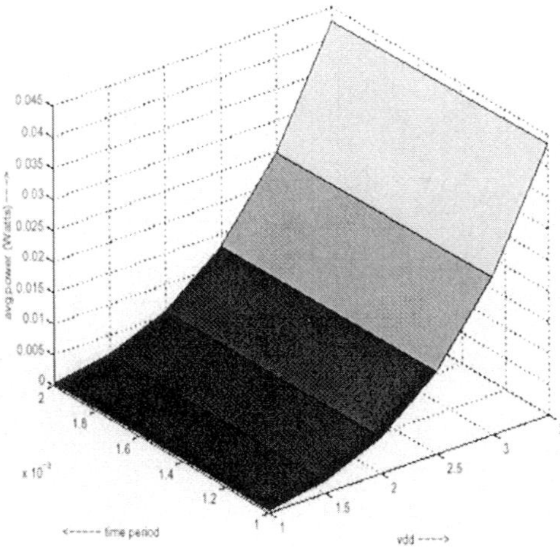

Figure 11.3. Memory Power Dissipation Pattern

supply voltage is decreased, the average power dissipation is found to decrease quadratically. Also the memory delay was found to double as we reduced the voltage from 3.3 Volts to 2 Volts, reducing the maximum clocking frequency possible from 100 MHz at 3.3 Volts to 50 MHz at 2 Volts.

- PCB Bus Power Model Any implementation of a processor and the second level off-chip memory will typically require at least two chips connected by buses on a Printed Circuit Board (PCB). We are currently using a PCB built with an Intel StrongARM processor (SA-110). The PCB was designed at Compaq and is called "Skiff" [21]. The "Skiff" board was used as the prototype for obtaining the external bus parameters such as length of PCB buses and capacitance values for individual PCB bus wires. The stack-up information of the board, which consists of the number of layers, number and types of planes (power and/or ground) used, ordering or sequence of the layers, and spacing between the layers is used to obtain the capacitances of the external buses. With the given stack-up information, the following micro strip equations are used to calculate the characteristic impedance (Z_0) of buses [12]:

$$Z_0 = \frac{60}{\sqrt{\varepsilon e}} * ln(\frac{8d}{W} + \frac{W}{4d})$$

where ,

ε_e, *the effective dielectric constant is given by,*

$$\varepsilon_e = \frac{\varepsilon_r + 1}{2} + \frac{\varepsilon_r - 1}{2} \frac{1}{\sqrt{1 + 12d/W}}$$

where,
ε_r *is the relative dielectric constant.*
d is the thickness of the dielectric substrate.
W is the conductor width of the micro strip line.
The capacitance per unit length, C is found using the following equation:

$$C = \frac{\sqrt{\varepsilon_e}}{cZ_0}$$

where,
c is speed of light in free space.

Example 1: Consider a bus of length 12 cm which passes through only one conductor layer of the Printed Circuit Board.
Assume the thickness of substrate, d to be 6 mil[1] and conductor width, W to be 4 mil.
The relative dielectric constant of the dielectric substrate, ε_r is 5.
Using the above equations for calculating ε_e

ε_e= 3.45.

Now, characteristic impedance Z_0 is calculated and is found to be 80.7 Ω .
Therefore, capacitance per unit length is given to be 76.7 pF/m

Finally, using the measured length of the bus to be 12 cm, total Capacitance C_{bus} is 9.2 pF.

Since most of the capacitance values were clustered around 10 pF, we considered an average capacitance value of 10 pF for each bus line. Using the measured capacitance values, the total memory bus power, P is calculated using the following equation:

$$P = \frac{V_{dd}^2 * C_{bus} * f}{2}$$

where, V_{dd} is the voltage supply for the bus
C_{bus} is the capacitance of bus
f is the effective switching frequency on the bus[2]

The graph in figure 11.4 describes the dependence of the power dissipation on the switching activity of the memory bus and also the power supply voltage V_{dd}. The power dissipation is found to decrease as the switching activity and the supply voltage is decreased.

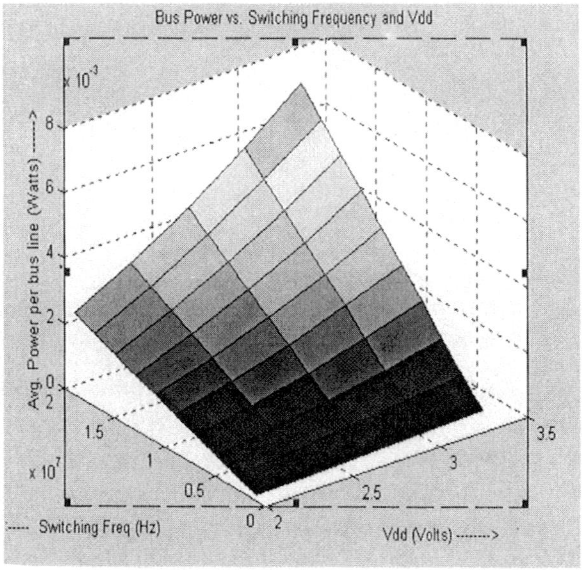

Figure 11.4. Off-Chip Bus Power Dissipation Pattern

4. Methodology

Each benchmark is compiled as follows. First, the C benchmark code is compiled to relocatable ARM assembly code using the Trimaran code generator backend called "Triceps". Then the GNU cross-assembler is used to generate a binary executable that can run on our Verilog model in the Synopsys VCS simulator. Using the switching activity information collected by the Verilog simulation, the Synopsys Power Compiler gives an estimate of dynamic and static power dissipation in the ARM-like RISC core. Furthermore, the Verilog simulation for each benchmark gives the switching activity (toggle count) for each output pin to external memory. This activity is fed to the power models for external bus and memory. The power estimates from external memory and buses are added to the power estimates of the core to get system level power estimates. In particular, we analyze the case where we reduce the voltage of the off-chip memory and off-chip memory buses by half. The resulting increase in memory delay (it doubles) is taken care of by cutting down the off-chip frequency (we cut the frequency by half). Of course, reducing off-chip bus and memory frequency increases program execution time. We help offset this effect by adding a write buffer to the CPU model. Thus, our infrastructure allows us to explore the design space consisting of the compiler, architecture, and the technology parameters. The algorithms, architectural parameters and tech-

Table 11.1. List of Benchmarks

benchmark	explanation
sort_int	Integer Array Sort
fib	Fibonacci
fir	Fir Filter
matmul	Matrix Multiply

nology parameters can now be modified and analyzed for power-performance trade-offs.We consider the following three metrics to measure the effects of various trade-offs:

- Workload Completion Rate (WCR): WCR is defined as the workload units completed per unit time and is given by the following formula: WCR = (Mobile Workload Completed) / (Time to Complete Workload)
- Workload Completion Efficiency (WCE): WCE is defined as the workload units completed per unit energy and is given by the following formula: WCE = (Mobile Workload Completed) / (Energy Consumed to Complete Workload)
- Workload Completion Rate Efficiency (WCRE): WCRE is defined as the workload units completed per unit time per unit energy and is given by the following formula: WCRE = (Workload Completion Rate) / (Energy Consumed to Complete Workload)

5. Results

We analyze the effect of scaling the off-chip memory bus frequency and supply voltage. Although our calculations do not include the extra overhead of multiple supply voltage generation, we assume that the board already has the multiple supply voltages needed. For example, the "Skiff" board from Compaq has 2 Volts, 3.3 Volts and 5 Volts power supplies. Therefore, we think it is reasonable to assume that our system will have several supply voltages readily available.

The benchmarks [table 11.1] considered are chosen to be computationally intensive. These benchmarks constitute the kernel of many signal-processing algorithms. These Benchmarks are available at the Trimaran website [22].

First, we present the results related to the core in table 11.2. All the benchmarks were found to dissipate similar amounts of average power. A sample representative breakup of the average power dissipation in various functional modules is presented in figure 11.5. The memory power was calculated using memory models proposed by D. Liu[23]. The results are presented in table 11.3. The switching activity is fed into the bus power model and the power numbers are generated for the off-chip memory data bus as shown in table

Table 11.2. Core Power Dissipation

benchmark	core power(Watts) without write buffer	core power(Watts) with write buffer
sort_int	1.01	1.31
fib	1.00	1.50
fir	0.72	0.82
matmul	1.01	1.31

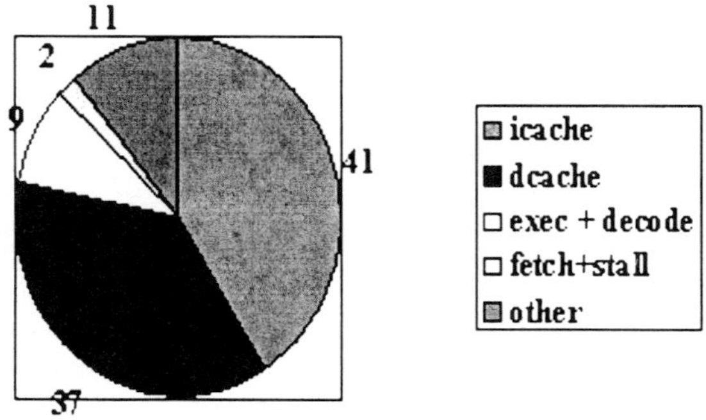

Figure 11.5. Core Power Dissipation Percentage Break-Down

Table 11.3. Memory Power Dissipation

benchmark	memory (100)	memory (50)	% savings
sort_int	232.52	84.65	64
fib	580.66	57.29	90
fir	816.763	103.27	87
matmul	732.802	27.105	96

11.4. The total execution times for the 100 MHz and 50 MHz are tabulated in table 11.5. The system level power estimate is obtained by adding up the core power, bus power and the memory power as shown in table 11.6.

Table 11.4. Memory Bus Power Dissipation

benchmark	bus power(100)	bus power (50)	% savings
sort_int	4.46	2.19	50
fib	21.50	1.38	93
fir	24.86	1.83	93
matmul	28.96	13.36	54

Table 11.5. Program Completion Time

benchmark	execution time (100)	execution time (50)	% increase
sort_int	11.758	11.767	0.07
fib	8.982	8.985	0.03
fir	0.043	0.052	20.9
matmul	1.250	1.280	2.4

benchmark	system power (100)	system power(50)	% savings
sort_int	1.287	1.417	-10
fib	1.601	1.517	5.2
fir	1.560	0.94	39
matmul	1.770	1.35	24

Table 11.6. System Level Energy Estimates

benchmark	system energy (100)	system energy (50)	% savings
sort_int	15.13	16.67	-10
fib	14.38	13.63	5
fir	0.07	0.05	28
matmul	2.21	1.73	22

The choice of off-chip bus frequency is currently assumed to be made statically by a mechanical switch choosing between 2 V and 3.3 V. The trade-off shown, for example the matmul benchmark, shows a performance penalty of 2.4 % in return for a power reduction of 24 %.The technique used also modifies the energy consumption in a favorable manner.Table 11.6 presents the energy statistics. Finally, we also present the statistics about the metrics WCR, WCE and WCRE. The first three columns in Table 11.7 are for the case of the system running at 3.3 Volts and 100 Mhz and the middle three are for the case where Off-Chip memory components are running at a reduced speed of 50 Mhz and at a reduced voltage of 2 Volts. The last three columns of Table 11.7 show the differences in each of the metrics. As can be seen from Table 11.7, the trade-off considered marginally affects the workload completion rate which suggests that the benchmarks take slightly longer time to complete. But the workload completion efficiency is greatly enhanced suggesting that the energy is more efficiently utilized. The metric WCRE, which provides the combined effect suggests that the overall effect is positive in most of the cases.

6. Conclusion

In conclusion, we have shown how a simple trick of cutting off-chip memory access frequency in half, together with the enabled (due to extra latency avail-

Table 11.7. Metrics for Various Benchmarks

	100 Mhz, 3.3 V			50 Mhz, 2 V			% Decrease		
benchmark	WCR	WCE	WCRE	WCR	WCE	WCRE	WCR	WCE	WCRE
sort_int	.085	.066	.006	.085	.060	.005	-.07	-9.23	-9.3
fib	.111	.070	.008	.111	.075	.008	-.03	7.63	7.59
fir	23.26	14.29	332.2	19.23	20.00	384.6	-17.3	40.00	15.77
matmul	.800	.452	.361	.781	.578	.452	-2.34	27.74	24.75

able) reduction in supply voltage of off-chip bus and memory, reduces both power and energy. The basic point is that both the compiler and the programmer can take advantage of smart architectural and memory hierarchy features, which allow the reduction of power with some corresponding tradeoff in reduction of performance. For example, the choice of 100 MHz versus 50 MHz for off-chip bus frequency could be made programmable dynamically. Writing to a special on-chip register or a memory-mapped location could do this. In this case, then, code could be written which operates at high performance and high power during critical times, but scales down to lower performance and low power during non-critical time periods. This paper lays the groundwork for such a system where off chip buses and memories and possibly more peripherals have their power consumption modulated either by the user or by the compiler.

7. Future Work

We will look at further hiding the load-instruction memory access latency caused by slowing down the off-chip buses and off-chip memory by using compiler optimization techniques. We will look at making some of the circuit improvements like clock gating visible to the compiler by extending the instruction set.

Acknowledgements

This research was funded by DARPA under contract number F30602-00-2-0564 . We also acknowledge donations received from Sun, Intel, Synopsys, Cadence, Mentor Graphics, Hewlett-Packard, Denali and LEDA systems.

Notes

1. 1000 mils = 1 inch
2. $f = \frac{togglecount}{2*exectime} * systemclock$

References

[1] J. Montanaro, R.T. Witek, K. Anne, , A. J. Black, E. M. Cooper, D. W. Dobberpuhl, P. M. Donahue et. al., "A 160 Mhz 32 B 0.5 W CMOS RISC Microprocessor," IEEE journal of Solid-State Circuits, Vol. 31, No. 11, pp. 1703-1714, November 1996.

[2] D. Seal, ARM Architecture Reference Manual, Second Edition, Addison Wesley Publications, England, 2001.D. Seal et. al., *ARM Architecture Reference Manual*, Addison Wesley Publications, 2001.

[3] S. Furber , ARM System Architecture, First Edition, Addison Wesley Publications, England, 1996.

[4] K. Ghose et. al., "Exploiting Bit-Slice Inactivities for Reducing Energy Requirements of Superscalar Processors," Kool Chips workshop in association with MICRO33, Monterey, California, December 2000.

[5] V. Tiwari and M. Tien-Chien Lee, "Power Analysis of a 32-bit Embedded Microcontroller," Proceedings of the Asia and South Pacific Design Automation Conference, pp. 141-148, August 1995.

[6] G. C. Cardarilli, M. Salmeri, A. Salsano and O.Simonelli, "Bus Architecture for Low Power VLSI Digital Circuits," IEEE International Symposium on Circuit and Systems, pp. 21-24, May 1996.

[7] B. Prince, Semiconductor Memories, Second Edition, John Wiley and Sons, England, 1996.

[8] R. J. Evans and P. D. Franzion, "Energy Consumption Modeling and Optimization for SRAMs," IEEE Journal of Solid State Circuits, Vol. 30, No. 5, pp. 571-579, May 1995.

[9] Trimaran Consortium, *http://www.trimaran.org*

[10] Synopsys, Inc., *http://www.synopsys.com*

[11] LEDA Systems, Inc., *http://www.ledasys.com*

[12] D. M. Pozar, Microwave Engineering, Second Edition, John Wiley & Sons Inc., USA, 1997.

[13] *http://www.eecs.umich.edu/~jringenb/power/*

[14] *http://www.eecs.umich.edu/~taustin/simplescalar*

[15] A. Chandrakasan and R. Brodersen, Low Power CMOS Design, John Wiley & Sons Inc., USA, 1998.

[16] P. R. Panda and N. Dutt, "Low Power Memory Mapping through Reducing Address Bus Activity," IEEE Transactions on VLSI Systems, Vol. 7, No. 3, pp. 309-320, September 1999.

[17] N. Chang, K. Kim and H. G. Lee, "Cycle Accurate Energy Consumption Measurement and Analysis: Case Study of ARM7TDMI," Proceedings of the International Symposium on Low Power Electronics and Design, pp. 185-190, July 2000.

[18] A. Sinha, A. Wang and A. P. Chandrakasan, "Algorithmic Transforms for Efficient Energy Scalable Computation," Proceedings of the International Symposium on Low Power Electronics and Design, pp. 31-36, July 2000.

[19] M. Powell, S. Yang, B. Falsafi, K. Roy and T. N. Vijaykumar, "Gated V_{dd}: A Circuit Technique to Reduce Leakage in Deep-Submicron Cache Memories," Proceedings of the International Symposium on Low Power Electronics and Design, pp. 90-95, July 2000.

[20] Denalisoft, Inc., *http://www.denalisoft.com.*

[21] Compaq, Inc., *http://www.compaq.com.*

[22] *http://www.trimaran.org/status.html.*

[23] D. Liu and C. Svensson, "Power Consumption Estimation in CMOS VLSI Chips," *IEEE Journal of Solid-State Circuits*, Vol. 29, No. 6, pp. 663-670, June 1994.

[24] A. Iyer and D. Marculescu, "Power Aware Microarchitecture Resource Scaling," Proceedings of Design Automation and Test in Europe, pp. 190-196, March 2001.

[25] A. Acquaviva, L. Benini and B. Ricco, "An Adaptive Algorithm for Low-Power Streaming Multimedia Processing," Proceedings of Design Automation and Test in Europe, pp. 273-279, March 2001.

V

APPLICATION LEVEL POWER MANAGEMENT

Chapter 12

APPLICATION-LEVEL POWER AWARENESS

Jeffrey A. Barnett
Northrop Grumman Corporation
Automation Sciences Laboratory
El Segundo, California
jbarnett@nrtc.northrop.com

Abstract Optimizing resource allocation to best meet system goals is the essence of good engineering. It is the unifying principal in the design of resource-constrained systems such as aircraft, farming complexes, and even military forces. Well-designed systems accomplish tasks using minimal resources. They also dynamically adapt their methods and goals to best use available resources. This is as evident in artificial systems as it is in natural systems. Below, two examples–one a design time problem and the other a dynamic allocation problem–are developed to elucidate the sort of engineering that must be applied to resource management. The first example is a design tradeoff between the efficient use of electric power and the performance of an aircraft. The second example is dynamic allocation of a limited energy budget between cooperating sensors to maximize the quality of the combined measurement.

Keywords: Power aware design, dynamic energy management

1. Introduction

The most successful of nature's systems are those that have adaptations to optimally use available resources. The form factors of birds and fish, for example, are both optimized for efficient motion. Since drag, density, and buoyancy in the air and in water are different, their shapes are quite different. Thus, we observe that design optimizations are specific to environments.

Dynamic adaptation of behavior is another aspect of energy optimization. Ordinarily, animals select prey that maximize the expected return over their investment to catch it. However, in times of great need, minimal margins are readily accepted and in times of plenty, indifferent choices are made. Plants adapt their processes to such factors as the availability of sunlight, moisture, and nutrients. Even the reproductive decisions of some species are affected by environmental cues and the state of the individual at the time. Nature has indeed developed energy-aware systems that in both design and behavior exhibit resource awareness.

Current events highlight the need for energy-aware artificial systems too. Our petroleum resources are rapidly dissipating and utility companies cannot produce enough electricity to meet all of our demands all of the time. If these were the only drivers, conservation remedies, new energy sources, and normal engineering progress would hold the problems in check.

However technology, particularly the modern computer, promises new types of systems for applications that will operate in severely energy-constrained environments. Extended duration space exploration is one well-known example. Another is the construction and deployment of unattended sensor networks comprising 1000's of miniaturized elements. Each element must be put in place with sufficient power, e.g., from a battery, to last its entire lifetime and support sensing, computing, and communications. Elements, in addition, may scavenge energy from sunlight, natural vibrations, or even chemical interactions with the environment. Energy management and optimization clearly will be of overriding importance in the design of these systems as will the need to dynamically adapt their behavior by considering the urgency of goals and the available resources.

Below, two examples of power-aware applications are described. The first is aircraft design where energy conservation contributes to overall performance by reducing battery and generator weight to enable longer flight times. Section 1 develops the mathematical relationship linking power efficiency to performance. That relationship is used to examine tradeoffs in terms of two different aircraft. Section 2 considers an application where multiple sensors develop joint measurements while sharing an energy budget. The goal is to minimize the variance of the resulting measurements while using the least energy possible and still meet system objectives. Thus, the second example is dynamic decision-making in the management of resources.

Section 3 summarizes the results and relates them to similar observations in other fields. The goal of this paper is not to present the development of specific new technology. Rather, it is too provide examples of how engineers can reason about and construct application-specific systems that will operate in energy-constrained environments.

1. A Design-Time Optimization for Aircraft

The value of power awareness and energy reduction technologies are determined by the specifics of an application domain and the milieu in which its systems will operate. Intelligent power management is one of the key technology enablers to successful deployment of deep space probes, networks comprised of miniature battery-powered sensors, and the Land Warrior. Below, a simple relation between energy efficiency and maximum flight time (endurance) for aircraft is derived. That relation is used to model the value of efficient design as measured by increased endurance: Sometimes it is very valuable and sometimes it is not. The point of this exercise is not to solve aircraft problems per se. Rather, it is to introduce the sort of domain-specific analysis that is necessary to make design-time tradeoffs for an application.

1.1 Domain Considerations

Weight is the enemy of air vehicle performance. Heaviness directly decreases maneuverability, handling, range, and endurance, where endurance is maximum flight time. Since savings in the use of electrical energy will reduce the weight of the generator, the battery, or both, power awareness will increase the missions that a given aircraft can undertake and complete. Another benefit of weight reduction is that the weight budget for the fuel used to power the air vehicle can be increased when the battery or generator weight is decreased.

Reducing electric power consumption also decreases heat production. Therefore, more equipment can be packed into the same or smaller form factor. Since decreasing size increases maneuverability, range, and endurance, we have an additional argument for power awareness. Further, obvious benefits accrue when power-aware avionics can compensate for unusual situations such as generator loss. This example makes the point that power awareness is more than energy minimization: it's about doing the best you can with the resources that are actually available.

Concerns about the effects of electric energy consumption on aircraft performance are not new. For example, space and heat considerations are discussed in [2] along with their impacts on operational costs. Power awareness is even more important in the modern world. One example is the Helios, a product of AeroVironment Corporation, with the goal of indefinite sustained flight powered by scavenging solar energy. Other extreme examples are smart munitions that either have no propulsion system or only use propulsion in burst modes. Since there is no engine-powered generator, all electric energy must be supplied by batteries to support sensors, computation, communications, and control surface actuators. Batteries are a limiting resource that produce unwanted heat, consume precious space, and add weight.

The standard relation approximating the endurance achievable by an aircraft as a function of its weight and the amount of fuel it carries is derived next. The following discussion extends that relation to include the effects of electric power utilization efficiency, hence battery weight, on endurance. Finally, that relation is used to analyze the effects of power efficiency on endurance for two example aircraft.

1.2 Endurance Is a Function of Weight and Fuel

The relation between endurance and an air vehicle's weight and the amount of fuel it carries is derived here. A reasonable first-order approximation is used to simplify the mathematics: instantaneous fuel consumption rate is linearly related to total vehicle weight for a constant velocity and altitude. The next section extends the relation to consider the effects of electric energy conservation on weight, hence, its relation to endurance. See [7] for justification of the assumptions made here as well as derivations of more detailed relationships when other variables are considered.

Let $f = f(t)$ be the amount of fuel remaining at flight time t and let $f_0 = f(0)$ be the amount of fuel loaded on an aircraft that weights W_d dry (fuel weight is zero). The assumption is that the amount of fuel, measured by weight, necessary per second of flight time per unit of aircraft weight is a constant. Thus,

$$f' = -c(W_d + f)$$

where $f' = df/dt$ and $c > 0$ is the fuel consumption constant. The solution to this simple ode, subject to the constraint that $f(0) = f_0$, is

$$f(t) = W_0 e^{-ct} - W_d,$$

where $W_0 = W_d + f_0$ is the initial weight of the aircraft at $t = 0$.

The endurance, E, is defined to be the maximum flight time. Clearly, $f(E) = 0$ is the condition to find that unique value of E. Therefore,

$$E = \frac{1}{c} \times \log\left(\frac{W_0}{W_d}\right). \tag{12.1}$$

An explanation, in part, for this, perhaps unexpected result, is that in order to increase endurance, more fuel is obviously needed. But, additional fuel is necessary to carry the original extra fuel, etc. Thus, weight has a nonlinear effect on endurance. Note, this derivation is a reasonable approximation for many land as well as airborne vehicles.

1.3 Endurance Is a Function of Energy Conservation

In order to extend the above derivation to account for a battery and, hence, to account for the efficiency with which that battery is used, a few details must be added to our simple model. Let W_S be the structural weight of the aircraft and W_B be the weight of its battery. Thus,

$$\begin{aligned} W_d &= W_S + W_B \\ W_0 &= W_S + W_B + f_0. \end{aligned}$$

In order to focus on the relevant tradeoff for this discussion—battery weight versus fuel—define $L = W_B + f_0$ as the total weight budget for fuel plus battery and β as the battery weight needed for one unit of mission time. Thus, $W_B = \beta E$. β is interpreted as the efficiency of the *energy consumers* not of the *battery*, though the analyses are practically identical in either case. Now,

$$\begin{aligned} W_d &= W_S + \beta E \\ W_0 &= W_S + L, \end{aligned}$$

where the first equation assumes that there is just enough battery to last for the whole mission. In other words, L is split between fuel and battery weight so that both resources expire simultaneously. There is one important difference between fuel and battery that makes these equations non-symmetric: when fuel is consumed, its weight decreases but no such reduction will be observed for batteries. Substituting the formulas into (12.1) yields

$$E = \frac{1}{c} \times \log\left(\frac{W_0}{W_S + \beta E}\right).$$

What we desire is an expression for E as a function of β so that the payoff of efficient energy utilization can be measured in terms of endurance enhancement. A quick inspection shows that an elementary expression isn't available. However, it is straightforward to express β as a function of E:

$$\beta = \frac{W_0 - W_S \exp(cE)}{E \exp(cE)}. \qquad (12.2)$$

Clearly, the value of E must be positive and is bounded from above by the case where $f_0 = L$ and, hence, $\beta = 0$. Thus,

$$0 < E < \frac{\log(W_0) - \log(W_s)}{c}. \qquad (12.3)$$

Equation 12.2 and the bounds on E are used below to numerically generate and graph (β, E) pairs for some examples.

1.4 Aircraft Examples and Analysis

Two simple examples of vehicles designed for quite different mission profiles are analyzed in this section. The first, called Explorer, is a simple unmanned sensor craft that uses a generator to support most applications. It flies

Table 12.1. Parameters Defining Two Aircraft Examples

Example	c	W_S	L	W_0	$E_{\max}$
Explorer	.00155/min	6,900 lb	3,100 lb	10,000 lb	240 min
Missile	.05108/min	15 lb	10 lb	25 lb	10 min

missions of a few hours duration and its battery is primarily for power during emergencies, e.g., for flight recorders or engine restarts after flame outs. The second example, called Missile, is a smart munition that uses batteries to power all onboard electronics. Fuel is only used in burst mode to regain altitude so as to increase target seek time. Its mission times are of the order of a few minutes. Table 12.1 summarizes the relevant parameters for these two hypothetical examples and $E_{\max} = \log(W_0/W_s)/c$ is the maximum endurance from (12.3).

E versus β pairs have been generated using (12.2) and are displayed in Figures 12.1a and 12.1b. Endurance, E, clearly increases as electric power utilization, as measured by β, decreases. There is no way to tell, at this point in the

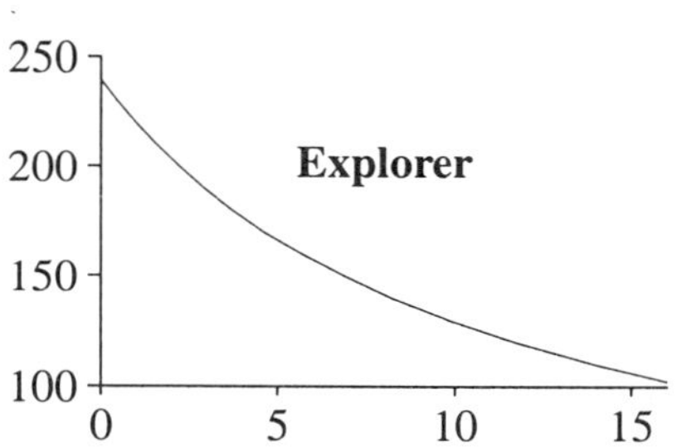

Figure 12.1a. E versus β relation for the Explorer example.

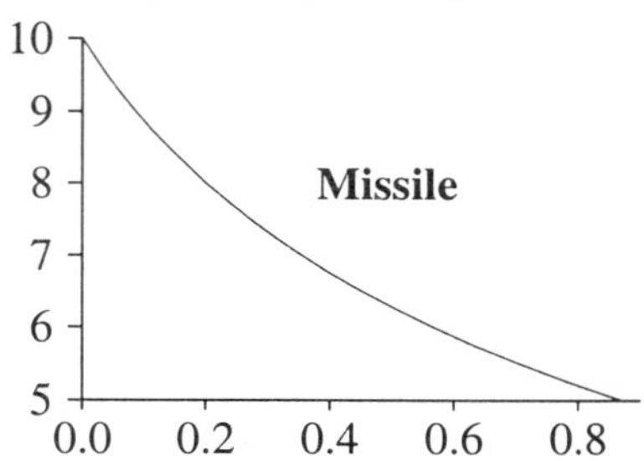

Figure 12.1b. E versus β relation for the Missile example.

development of the two examples, whether power-saving measures would be worthwhile or not. We need to know the current β, the merit of a ΔE measured as an increased fraction of successful missions, and the cost of developing and inserting power-aware technology that can provide the ΔE in order to make that determination.

Such details are beyond the scope of this paper though a simple heuristic analysis is presented. Figures 12.2a and 12.2b suggest the shapes of typical tradeoff curves: $p = p(E)$ is the probability of a successful mission if endurance is E and $C = C(\Delta E, E)$ is the cost to achieve an incremental endurance enhancement given the current value of E. Design tradeoffs are then formed in terms of this and other relevant information.

The Explorer example posits that batteries are not the main source of electric power and it is fair to assume that its designers made various decisions to

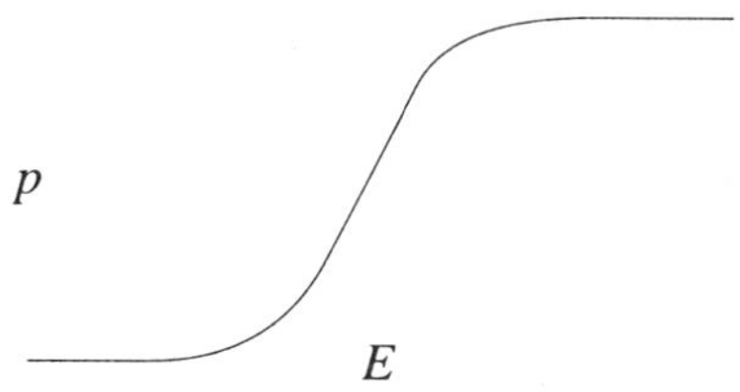

Figure 12.2a. Probability of mission success as a function of endurance.

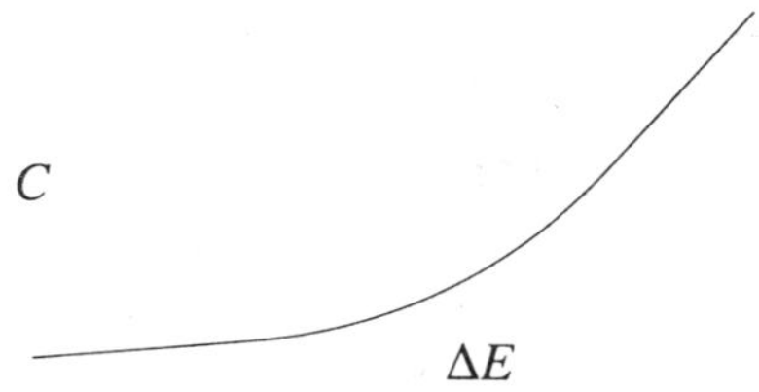

Figure 12.2b. Cost to increase endurance as a function of the desired increment.

maximize endurance. In other words, it is unlikely that the total weight of the generator and battery are a significant fraction of the total vehicle weight. Further, $f_0 \approx L$. Therefore, power-awareness, even if it could reduce the generator weight cannot increase endurance by more than a few minutes.

The Missile example is completely different. Battery weight is a significant fraction of L and L is a significant fraction of W_0. Once again, conservation can only increase endurance by a few minutes but a few minutes, compared to a current endurance of less than 10 minutes, could make all the difference in whether or not missions are successful.

1.5 Beyond Design-Time Optimization

The results and examples developed above are meant to show how considerations of power utilization can impact overall system performance. In particular, the effects of energy conservation on maximum flight time are demonstrated. Energy conservation, however, is only one aspect of power awareness. The more general problem is to dynamically determine how to optimize mission goal achievement as a function of the resources actually available. This is a much more difficult problem and is even more of an application-dependent issue than is straightforward energy minimization.

The Missile example suggest a plethora of interesting power management problems. For example, given a projected time to target impact maximize the probability of a successful strike. The problem is how best to use the remaining battery energy during the known flight time. Should sensor resource allocations be increased or should communications and computation be used to receive and fuse an ally's sensor data with onboard information? The answer depends on the resources available as well as the relative quality and costs of the two solutions. In other words, real power awareness depends on the ability to make dynamic, situation-dependent as well as application-dependent decisions. The sort of design-time tradeoffs analyzed herein are important too, but they are only a first step toward realization of fully power-aware systems.

2. Dynamic Energy Allocation for Cooperating Sensors

The problem considered in this section is how to split a limited energy budget between several sensors whose measurements will be fused. The objective is to minimize the variance of that joint measurement. A simple fusion model is defined, measurement variance is related to energy expenditures, and a criterion for an optimal allocation is derived. An extended example of a dynamic allocation problem and its solution is presented in three parts: (1) a specific model of sensor variance as a function of energy allocation is introduced, (2) the optimality problem for two sensors described by that particular model is solved, and (3) a numerical example is used to illustrate an optimal policy and the sort of decision-making that it engenders. This note is meant as an example of how one might formulate and solve dynamic energy management problems and not as a definitive practical result.

2.1 Fusing Sensor Measurements

If sensors $1, \ldots, n$ make measurements $m_1, \ldots, m_n$ that are statistically independent, the joint estimate m is taken to be

$$m = \frac{w_1 m_1 + \cdots + w_n m_n}{w_1 + \cdots + w_n}, \tag{12.4}$$

where $w_i = v_1 \cdots v_n / v_i$ and v_i is the variance of measurement m_i. The variance, v, of the joint estimate is

$$v = \frac{1}{\frac{1}{v_1} + \cdots + \frac{1}{v_n}}. \tag{12.5}$$

The function, $v = v(v_1, \ldots, v_n)$, has the following intuitive properties:

Less is better: $\frac{\partial v}{\partial v_i} \geq 0$.

Progress: $v(v_1, \ldots v_n) \leq v_i$.

As good as it gets: $v(v_1, \ldots, v_n) = 0$ if any $v_i = 0$, a corollary of progress.

No pain, no gain: $v(v_1, \ldots, v_n) = v(v_1, \ldots, v_{i-1}, v_{i+1}, \ldots, v_n)$ if $v_i = \infty$.

Unit independence: $v(rv_1, \ldots, rv_n) = rv(v_1, \ldots, v_n) = 0$.

Some justification is provided in [6] for the formulas assumed here as well as an introduction to the general topic of fusion using Kalman filters. Note, fusion formulas such as (12.4) and (12.5) are more complicated if the sensor measurements are statistically dependent or if multiple features are measured as is often the case.

2.2 Energy Allocation Determines Sensor Performance

Modern sensors offer many controls that trade energy for performance. Examples are raw power, pulse width, number of pulses per measurement, frequency of measurement, and even spectrum choices for hyperspectral sensors. Some passive sensors can vary the number of pixels or the parts of the spectrum that are processed. The more energy used, the higher the quality of the measurement. Below, quality is grossly summarized by the resulting variance of the measurement. Specific issues such as precision, etc., are assumed to be sufficiently captured by the variance estimate.

Let $v_s(e)$ represent the variance expected when sensor s makes a measurement using energy $e \geq 0$. Such a function should satisfy these criteria:

Positivity: $v_s(e) > 0$.

More is better: $\dfrac{dv_s}{de} \leq 0$.

No free lunch: $v_s(0) = \infty$.

The astute reader may be troubled by the *no free lunch* property. Isn't there always some *a priori* information available so that the initial variance would be finite? For example, an angle measurement is always in the interval $[0, 2\pi)$. The way to analyze the situation is to assume that the variance of the *a priori* estimate is v_0 and to combine it with $v_1, \ldots, v_n$ to get the combined variance, $v = (v_0^{-1} + v_1^{-1} + \cdots + v_n^{-1})^{-1}$. Since the optimality criterion derived in the next section is the same whether v_0 is accounted for or not, the math is simplified by ignoring the presence of *a priori* estimates.

2.3 Minimizing Variance Through Energy Allocation

Given a total energy budget, e, to be used to obtain independent measurements $m_1, \ldots, m_n$, the obvious objective is to divide it so that the resulting variance is minimized. Assume that $e_i \geq 0$ is used for the measurement made by sensor i. The problem is to minimize (12.5) subject to the constraint that $e_1 + \cdots + e_n = e$. Simply use Lagrangian multipliers as follows:

$$v(e_1, \ldots, e_n, \lambda) = \frac{1}{\dfrac{1}{v_1} + \cdots + \dfrac{1}{v_n}} + \lambda\left(e - \sum e_i\right).$$

Necessary minimization criteria follow from $dv/de_i = 0$:

$$\begin{aligned} 0 &= \frac{dv}{de_i} \\ &= \frac{-v'_i/v_i^2}{\left(\frac{1}{v_1}+\cdots+\frac{1}{v_n}\right)^2} - \lambda \\ \frac{v'_i}{v_i^2} &= -\lambda\left(\frac{1}{v_1}+\cdots+\frac{1}{v_n}\right)^2, \end{aligned}$$

where $v'_i = dv_i/de_i$. Since λ does not depend on i, the entire right hand side is independent of i. Therefore,

$$\frac{v'_i}{v_i^2} = \frac{v'_j}{v_j^2}, \tag{12.6}$$

for all $1 \le i, j \le n$. Standard dynamic programming techniques can be used to develop optimal allocations when there are more than two sensors.

2.4 Applying the Theory

This section develops an example of variance minimization through energy allocation. The problem of achieving a given result quality with minimum resource investment is addressed in the following section. Both problems are examples of dynamic energy management. The discussion proceeds in three parts: (1) a specific, parameterized sensor model that defines variance as a function of energy is introduced, (2) the two-sensor minimization problem is solved, and (3) the optimal allocation policy and the variance achieved are analyzed for particular sensor instances.

2.4.1 A Parameterized Sensor Model.

The form of v_i assumed in the optimization example developed in the rest of this section is

$$v_i(e) = a_i + b_i/e, \tag{12.7}$$

where $a_i, b_i > 0$. Such forms obviously possess the properties required above for variance functions. For future reference, note that

$$\frac{dv_i}{de} = \frac{-b_i}{e^2}. \tag{12.8}$$

The parameters a_i and b_i have the following interpretations in this formulation: a_i is the limit of the sensor's operation, i.e., $a_i = v_i(\infty)$, and b_i is the difference between the sensor's variance using one unit of energy and its best possible performance, i.e., $b_i = v_i(1) - v_i(\infty)$.

2.4.2 Solving the Two-Sensor Problem. Assume two independent sensors whose behaviors are defined by (12.7) and a total energy budget of e. Let energy $0 \leq x_1 \leq e$ be allocated to the first sensor and $x_2 = e - x_1$ be allocated to the second. Then from (12.7) and (12.8),

$$\begin{aligned} v_1(x_1) &= a_1 + b_1/x_1 \\ v_2(x_2) &= a_2 + b_2/(e - x_1) \\ v_1'(x_1) &= -b_1/x_1^2 \\ v_2'(x_2) &= -b_2/(e - x_1)^2. \end{aligned}$$

Now use criterion (12.6) to find the optimal allocation as follows:

$$\begin{aligned} \frac{v_1'}{v_1^2} &= \frac{v_2'}{v_2^2} \\ \frac{-b_1/x_1^2}{(a_1 + b_1/x_1)^2} &= \frac{-b_2/(e - x_1)^2}{(a_2 + b_2/(e - x_1))^2} \\ \frac{b_1}{(a_1 x_1 + b_1)^2} &= \frac{b_2}{(a_2(e - x_1) + b_2)^2} \\ \sqrt{b_1}\,(a_2(e - x_1) + b_2) &= \sqrt{b_2}\,(a_1 x_1 + b_1). \end{aligned}$$

The solutions to this linear equation for the optimal x_1^* and $x_2^* = e - x_1^*$ are

$$x_1^*(e) = \frac{a_2\sqrt{b_1}e + b_2\sqrt{b_1} - b_1\sqrt{b_2}}{a_1\sqrt{b_2} + a_2\sqrt{b_1}} \tag{12.9a}$$

$$x_2^*(e) = \frac{a_1\sqrt{b_2}e + b_1\sqrt{b_2} - b_2\sqrt{b_1}}{a_1\sqrt{b_2} + a_2\sqrt{b_1}}. \tag{12.9b}$$

Since x_1^* and x_2^* must be nonnegative, the numerators must be nonnegative. Thus, there are two validity constraints entailed by (12.9a) and (12.9b):

$$\begin{aligned} e &> \frac{\sqrt{b_1 b_2} - b_2}{a_2} \\ e &> \frac{\sqrt{b_1 b_2} - b_1}{a_1} \end{aligned}$$

If the first constraint is violated, the optimal allocation is $x_1^* = 0$ and $x_2^* = e$. If the second is violated, the optimal allocation is $x_1^* = e$ and $x_2^* = 0$. In the first case the resultant minimum fused variance, v^*, is $v^*(e) = a_2 + b_2/e$ and in the second the minimum variance is $v^*(e) = a_1 + b_1/e$. When there is an "interior" solution, v_1^* and v_2^* are calculated by substituting (12.9a) and (12.9b) into (12.7) to get

$$v_1^*(x_1) = \frac{a_1a_2e + a_1b_2 + a_2b_1}{a_2e + b_2 - \sqrt{b_1b_2}}$$

$$v_2^*(x_2) = \frac{a_1a_2e + a_1b_2 + a_2b_1}{a_1e + b_1 - \sqrt{b_1b_2}}.$$

These values are substituted into (12.5) and simplified to calculate the resultant minimum total variance as

$$v^*(e) = \frac{a_1a_2e + a_1b_2 + a_2b_1}{(a_1+a_2)e + (\sqrt{b_1} - \sqrt{b_2})^2}.$$

Table 12.2 summarizes these results for the optimal allocation policy: $x_1^*(e)$ is

Table 12.2. Formulas for Optimal Energy Allocations and Resulting Variances

	$0 \le e < \frac{\sqrt{b_1b_2} - b_2}{a_2}$	$0 \le e < \frac{\sqrt{b_1b_2} - b_1}{a_1}$	*Otherwise*
$x_1^*(e) =$	0	e	$\frac{a_2\sqrt{b_1}e + b_2\sqrt{b_1} - b_1\sqrt{b_2}}{a_1\sqrt{b_2} + a_2\sqrt{b_1}}$
$v_1^*(e) =$	∞	$a_1 + b_1/e$	$\frac{a_1a_2e + a_1b_2 + a_2b_1}{a_2e + b_2 - \sqrt{b_1b_2}}$
$x_2^*(e) =$	e	0	$\frac{a_1\sqrt{b_2}e + b_1\sqrt{b_2} - b_2\sqrt{b_1}}{a_1\sqrt{b_2} + a_2\sqrt{b_1}}$
$v_2^*(e) =$	$a_2 + b_2/e$	∞	$\frac{a_1a_2e + a_1b_2 + a_2b_1}{a_1e + b_1 - \sqrt{b_1b_2}}$
$v^*(e) =$	$v_2^*(e)$	$v_1^*(e)$	$\frac{a_1a_2e + a_1b_2 + a_2b_1}{(a_1+a_2)e + (\sqrt{b_1} - \sqrt{b_2})^2}$

the optimal allocation of total energy, e, to the first sensor and $v_1^*(e) = v_1(x_1^*(e))$ is the variance it achieves. The functions x_2^* and v_2^* are similarly defined for the second sensor. Finally, $v^*(e) = v(v_1^*(e), v_2^*(e))$ is the minimized variance of the fused measures given total energy budget e.

2.4.3 Applying the Results. This section develops a simple numerical example with two sensors. The resulting allocation policy and variance as a function of energy curves exhibit some interesting behaviors. Formulas for the optimal allocation policy and the resulting variances, derived from Table 12.2 using the constants $a_1 = 4$, $b_1 = 8$, $a_2 = 3$, and $b_2 = 18$, are shown in Figure 12.3a.

The optimal allocation of energy to the sensors is depicted in Figure 12.3b. Initially when $0 \le e \le 1 = (\sqrt{b_1b_2} - b_1)/a_1$, all of the available energy is

$$x_1^*(e) = \begin{cases} e & e \le 1 \\ \frac{e+2}{3} & e > 1 \end{cases} \qquad v_1^*(e) = \begin{cases} 4+8/e & e \le 1 \\ \frac{4e+32}{e+2} & e > 1 \end{cases}$$

$$x_2^*(e) = \begin{cases} 0 & e \le 1 \\ \frac{2e-2}{3} & e > 1 \end{cases} \qquad v_2^*(e) = \begin{cases} \infty & e \le 1 \\ \frac{3e+24}{e-1} & e > 1 \end{cases}$$

$$v^*(e) = \begin{cases} 4+8/e & e \le 1 \\ \frac{12e+96}{7e+2} & e > 1 \end{cases}$$

Figure 12.3a. Optimal allocation and resulting variances for the example where $a_1 = 4$, $b_1 = 8$, $a_2 = 3$, and $b_2 = 18$.

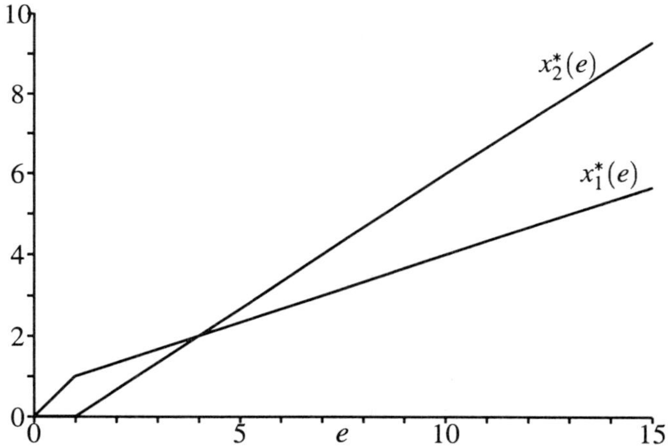

Figure 12.3b. Optimal resource allocation for the example.

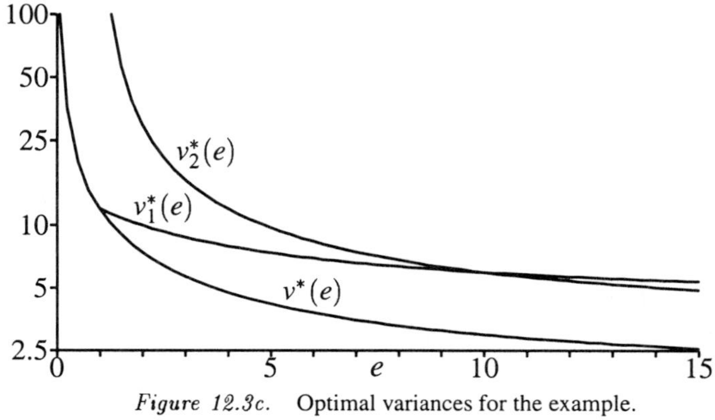

Figure 12.3c. Optimal variances for the example.

allocated to the first sensor; when $e > 1$, it is split between the two sensors. The first sensor continues to receives a larger allocation, i.e, $x_1^*(e) > x_2^*(e)$, while $0 < e < 4$ but there is a role reversal when $e > 4$ where $x_1^*(e) < x_2^*(e)$. When $e = 4$, $x_1^*(e) = x_2^*(e) = 2$.

The variances achieved using the optimal allocation policy are shown in Figure 12.3c where a log-valued vertical distance scale has been employed to improve visualization. The first sensor gives better results (has a lower variance) when a relatively small amount of energy is available but its limiting behavior is not as good as the second sensor. The variance of the fused measure, v^*, coincides with v_1^* when $0 \le e \le 1$. From then on it is lower. The contribution from the first sensor has lower variance than the second, i.e., $v_1^*(e) < v_2^*(e)$, while $0 < e < 10$. However, $v_1^*(10) = v_2^*(10) = 6$ and the second sensor's contribution finally dominates when $e > 10$.

This example was deliberately contrived to exhibit crossovers in both the optimal resource allocation policy and the quality of the results as shown in Figures 12.3b and 12.3c. The purpose of this ploy was two fold: (1) to indicate that even simple models can lead to complex or unexpected behavior and (2) to make the point that straightforward analytic methods can usually derive or approximate the actual optimal policy to be applied dynamically.

2.5 Related Problems

The area of dynamic sensor management and scheduling comprises many related problems. For example, the problem may be to develop an estimate that is good enough, i.e., to use the least energy to achieve a specific variance. This problem can be solved using formulas such as those in Table 12.2. The necessary minimum energy, e^*, is found from the inverse of v^*. This is straightforward because v^* is a necessarily monotonic function. The appropriate allocations are then calculated as the values of $x_1^*(e^*)$ and $x_2^*(e^*)$.

In a more realistic scenario, the analyst would need to account for the fact that measurements are not statistically independent, as assumed above, consider additional tradeoffs, and deal with measurement sources whose analytic descriptions are different than one another, e.g., it might be more economical to use communications to obtain an ally's measurement and fuse it than to use some or all of the local sensors.

Problems related to those above are discussed in [4] and [5] . Since some sensors can be time multiplexed amongst several applications, more applications can be serviced if the fraction of the sensor resources needed for each is reduced. A combination of filtering (fusion) and modelling is used. The models, formed from filter output parameters, predict target behavior between sensor measurements. The models also predict the variance increase as time of model use increases. That information is used dynamically to minimize the product of measurement duration and update rate for each application while maintaining track quality. Such techniques could be adapted by power-aware sensor schedulers. A benchmark facility to compare algorithms of this sort is avaible in [1].

Dynamic reasoning about the use of scarce resources is becoming more important as time goes by. It is an area that will pose many interesting science and engineering challenges in the future.

3. Afterword

Much of our understanding of the world around us is gained through hypotheses about how systems manage energy. Physicists assume least energy principals to form laws about the universe in the large and interactions of subatomic particles in the small. Biologists explain life, in part, by showing how organisms scavenge energy from the environment and store it for their own use at later times.

So concerns about power-aware systems are certainly not new. What is different today is that engineers are proposing and building systems whose functioning is all about energy management. Deep-space exploration and large unattended miniature sensor networks are two examples mentioned above. In these and other systems of the same ilk, power awareness is a primary design and operational principal, not just another support technology.

It may surprise the reader to know that Sigmund Freud proposed an energy minimization principal, as part of a cognitive economy model, to partially explain the workings of the human mind. He described dreaming and humor processes through a minimization of a quantity called *psychical energy*. I will close with a quote [3] that seems particularly germane to our current topic:

> *"I may perhaps venture on a comparison between psychical economy and a business enterprise. So long as the turnover in the business is very small, the important thing is that outlay in general shall be kept low and administrative costs restricted to the minimum. Economy is concerned with the absolute height of expenditure. Later, when the business has expanded, the importance of the administrative cost diminishes; the height reached by the amount of expenditure is no longer significant provided that the turnover and profits can be sufficiently increased. It would be niggling, and indeed positively detrimental, to be conservative over expenditures on the administration of business. Nevertheless, it would be wrong to assume that when expenditure was absolutely great there would be no room left for the tendency to economy."*

Acknowledgments

This effort is sponsored by Defense Advanced Research Projects Agency (DARPA) through the Air Force Research Laboratory, USAF, under agreement number F30602-00-1-0511. The opinions expressed are those of the author and do not necessarily reflect the opinions or conclusions of any other individual or agency.

References

[1] W. D. Blair, G. A. Watson, T. Kirubarajan, and Y. Bar-Shalom, Benchmark for radar allocation and tracking in ECM, *IEEE Transactions on Aerospace and Electronic Systems*, 34 (4) (1998).

[2] W. W. Davis, *Cargo Aircraft*, Pitman Publishing, 1946.

[3] S. Freud, *Jokes and their Relation to the Unconscious*, Leipzig and Vienna: Deuticke 1905. J. Strachey, translator, Norton Library, 1963.

[4] S. M. Hong and Y. H. Jung, Optimal scheduling of track updates in phased array radars, *IEEE Transactions on Aerospace and Electronic Systems*, 34 (3) (1998).

[5] T. Kirubarajan, Y. Bar-Shalom, W. D. Blair, and G. A. Watson, Management and tracking benchmark with ECM, *IEEE Transactions on Aerospace and Electronic Systems*, 34 (4) (1998).

[6] R. M. du Plessis, *Poor Man's Explanation of Kalman Filtering or How I Stopped Worrying and Learned to Love Matrix Inversion*, Rockwell International, 1967. (Reprinted by Taygeta Scientific Incorporated, 1996; can be purchased at `http://www.taygeta.com`.)

[7] D. P. Raymer, *Aircraft Design: A Conceptual Approach*, AIAA Educational Series, (Ed.) J. S. Przemieniecki, AIAA Press, 1992.

Chapter 13

A POWER-AWARE, SATELLITE-BASED PARALLEL SIGNAL PROCESSING SCHEME

Patrick M. Shriver*, Maya B. Gokhale*, Scott D. Briles*, Dong-In Kang◇, Michael Cai*, Kevin McCabe*, Stephen P. Crago◇, Jinwoo Suh◇

★ *Los Alamos Nat'l. Lab.*

{pshriver, maya, briles, dmc, kmccabe}@lanl.gov

◇ *Information Sciences Institute*

{dkang, crago, jsuh}@isi.edu

Abstract Satellite subsystem power budgets typically have strict margin allocations that limit the on-board processing capability of the spacecraft. Subsystems are assigned a fixed, maximum power allocation and are managed in an on/off manner according to available power and operations schedule. For a remote-sensing satellite, this limitation can result in poorer detection performance of interesting signal events as well as static instrument or data collection settings. Power-aware computation techniques can be utilized to increase the capability of on-board processing of science data and give the remote-sensing system a greater degree of flexibility.

We investigate a power-aware, signal processing scheme used to study signals from lightning events in the Earth's atmosphere. Detection and analysis of these lightning signals is complicated by the frequency dispersion experienced by the signal in the ionosphere as well as the interfering anthropogenic signals. We outline a method using multiprocessor architecture to run processing algorithms which have varying rates of power consumption. A 6 order magnitude spectrum of energy usage for these algorithms is obtained from experiment results.

Keywords: PAMA, satellite power management, power-aware remote sensing, ionospheric-dispersion signal, FORTÉ

1. Introduction

Conventional solutions to satellite power management consist of maintaining strict power budget margins during design and coarse on/off power switching of subsystems during operation. Power allocations are generally static and it is commonplace to assign a maximum power requirement to each subsystem. These methods can increase the time of the iterative design process (and thereby all associative costs of labor, overhead, etc.), be wasteful of the power resources, or require careful ground support planning of the science-instrument observation schedule, such as turning certain instruments off so that others may be turned on given the limited power available.

In this paper, we explore the capabilities of a power-aware, satellite-based computing system for on-board signal processing, to detect radio frequency (RF) events caused by natural events. The detection performance of the satellite-based remote sensing system can be improved on a moment-by-moment basis through the use of power-aware computing principles. Detection performance is directly related to how well and how often post-detection numerical computations can be executed to reduce false alarms. A greater capacity to reduce false alarms allows for a greater probability of detection. This capacity to reduce false alarms comes from the availability of computational resources, which, in turn, are dictated by power availability. This paper addresses an "intelligent" power-management technique that can be utilized in Department of Energy (DoE)/Department of Defense (DoD) as well as civil-satellite, remote-sensing applications, using the University of Southern California (USC)/Information Sciences Institute (ISI) Power Aware Multiprocessor Architecture (PAMA).

With the PAMA multiprocessors, power can be controlled in a gradient manner, in contrast to conventional techniques. This allows for more flexibility in the power budget margins and a higher degree of contingency options for the systems engineer. In our remote-sensing application, the multi-processor hardware, consisting of programmable processors and interconnect, is used to manage the data processing algorithms. The computational processing is adjusted to conserve or drain power according to the amount of power available vs. the rate of triggering events.

One remote-sensing application utilizing power-aware management techniques is in the processing of RF signals, e.g., lightning, in the Earth's atmosphere, similar to the mission of the DoE-funded Fast On-Orbit Recording of Transient Events (FORTÉ) satellite. FORTÉ was built by Los Alamos National Laboratory (LANL) and Sandia National Laboratory (SNL); flight operations have also been shared as a joint venture between LANL and SNL. The principal goal of FORTÉ is to develop a comprehensive understanding of the correlation between the optical flash and very high frequency emissions from lightning. The signal processing techniques used in analyzing RF signals is the focus for this power-aware application.

The PAMA system can be used in a more sophisticated approach as part of an "intelligent" power-management scheme. Power resources are utilized and managed in a gradient manner, as opposed to the binary on/off operations or wasted as heat. In a remote-sensing application, the computational power of processing data can be adjusted to either conserve or exploit excess power as needed. This has the advantage of more on-board processing as power is available, resulting in a quicker ground-based analysis of the science data. Additionally, in a "smart" data processing scheme, it would also be possible to reorient or re-calibrate instruments based on the incoming data while in-orbit, without the delays associated with ground communications and analysis. Processing decisions are made based upon the available power, the health status of the satellite, and the rate of "interesting" science events. During intervals of low activity, or "off-peak," there can be low rates of data collection and processing; likewise, during "peak" periods, there can be high rates of data collection and processing. Thus, with these aspects, application-oriented power management can be a valuable tool for the spacecraft designer, allowing for greater flexibility in the payload power budget margins, more contingencies for handling power fluctuations during on-orbit operations, and a better management of science data collection and analysis.

It should be noted that our approach is a management technique, not a method to minimize subsystem power, i.e., power *management* instead of power *efficiency*. Over the traditional satellite power-management methods, this approach is a "smarter" algorithmic approach to power management. The advantage is that satellites which are aware of power usage and the overall satellite state can distribute power throughout subsystems to make the best use of the available power.

2. Remote-Sensing Application

For this work, we have focused on power-aware processing for a remote-sensing application similar in nature to the mission of FORTÉ. The FORTÉ satellite was launched in August of 1997 and carries a suite of instruments used for studying the optical and RF signals from lightning in the Earth's atmosphere. The results from FORTÉ have led to a better understanding of the relationship between optical and RF lightning events, and future satellite missions can even use this knowledge to help provide global lightning and severe-storm monitoring [7]. The processing algorithm for the RF lightning signals has been chosen for the multi-processor-based, power-aware application study.

2.1 Ionospheric-Dispersed Signals

A RF lightning event in the Earth's atmosphere generates a dispersed signal, i.e., low frequencies of the signal are delayed, as it propagates through the ionosphere. This is known as a "chirped" signal. A simulated chirped signal is shown by the graphs in Figure 13.1. The top graph is an illustration of

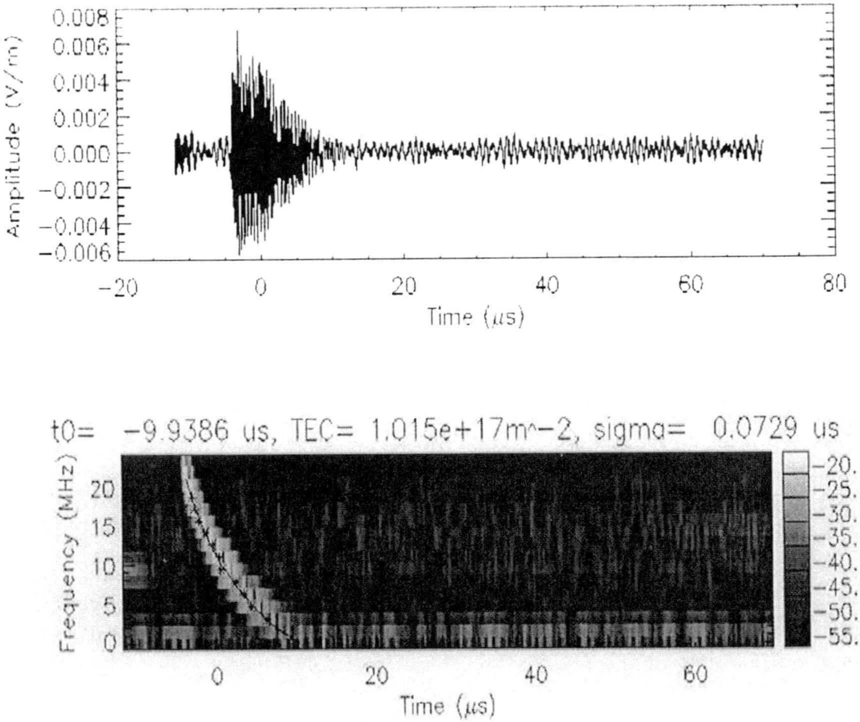

Figure 13.1. Ionic-Dispersed RF Signal

the time-domain signal, and the bottom graph is a plot of the dispersed signal frequencies vs. the corresponding times.

The time taken for a given frequency of the chirped signal to arrive at the on-orbit receiver is related to the total electron content of the ionosphere along the direction of the signal travel, the given frequency, and the signal time-of-arrival if ionospheric dispersion did not exist [1]. This relationship is illustrated in the bottom graph of Figure 13.1 and can be determined from Eq. 13.1.

$$T_f = \frac{5.3x10^{-6} N_e}{4\pi^2}\frac{1}{f^2} + T_{oa} \qquad (13.1)$$

where: T_f = frequency time-of-arrival
N_e = total electron content along the signal path
f = frequency
T_{oa} = signal time-of-arrival, neglecting ionospheric dispersion

The N_e, or total electron content (TEC), represents the number of electrons in a unit-area cross-section of an ionospheric column along the signal path. This atmospheric property is related to the propagation of radio signals through the ionosphere which can distort or bend the signals over the horizon. TEC is also related to the surface temperature of the Earth, and thus, could be viewed as an indicator for storm severity [4].

The T_{oa} is the time the signal would have arrived at the on-orbit receiver if the ionosphere did not distort the signal; on the other hand, the first term on the right-hand side of Eq. 13.1 is the additional time taken due to the frequency dispersion. Notice that for higher frequencies, the time taken approaches that of T_{oa}, and for lower frequencies the time delay is greater. The T_{oa} parameter is primarily useful for geolocation, i.e., determining the geographic origin of the signal.

2.2 FORTÉ RF Hardware

FORTÉ receives RF signals either from two orthogonal monopoles mounted at the satellite's base or by passive moderate-gain antennas mounted on a 35 foot nadir-directed boom. There are two types of receivers tunable in a 30-300 MHz band which consist of a mixer, bandpass filter, and a second mixer stage. The first mixer up-converts the antenna signal to a higher frequency then passes the signal through the bandpass filter. The second mixer then converts the band-limited signal to baseband. Depending on the type of receiver, either a 12-bit high-speed digitizer or a 8-bit digitizer is used. The digitizers are in constant operation. An analog trigger box processes the output from the second-stage mixer and determines whether or not the digitized data is to be recorded in payload memory. The recorded data can then be downlinked to the SNL or University of Alaska Fairbanks groundstations. Data analysis is carried out as part of the ground operation at LANL and SNL.

The analog signal is passed into separate channels through a set of bandpass filters in the trigger box. The triggering signal, which determines the recording of data, is generated by predetermined threshold levels in each of these channels. Setting these threshold levels causes a trade-off relationship between the probabilities of true signal detection and probability of false alarms as illustrated in Figure 13.2. As the threshold levels are increased, fewer trigger signals are seen in each channel and the probability of false alarms decreases. However, this also decreases the number of detections. Thus, for better detection performance, more false alarms must be accepted. Once these threshold levels are set, the probability of detection cannot be improved, but post-detection techniques can shift the operating points of Figure 13.2 to the left, decreasing the rate of false alarms for a given rate of detection. Therefore, it is desirable to maintain the optimum value of detections vs. false alarms, a task

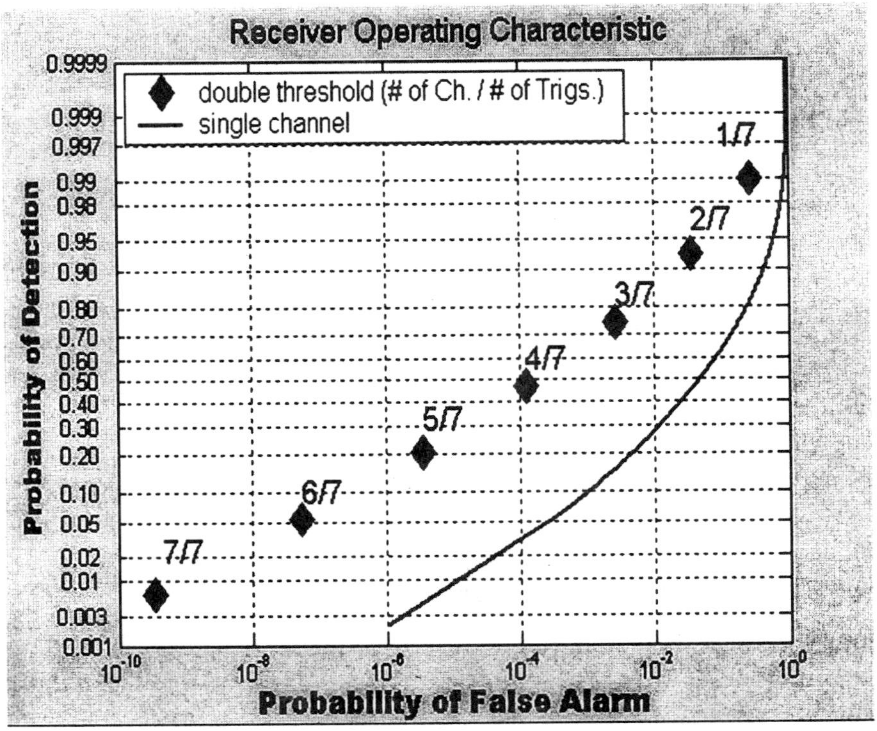

Figure 13.2. Signal Detection and False Alarm Rate Operating Curve

normally accomplished by ground staff. The RF environment is very dynamic due to anthropogenic signals. In using a method which allows more on-board processing, the remote-sensing system can adjust more quickly to incoming signals than the ground-based approach.

3. Signal Filters for Parameter Estimation

The power-management activities in this remote-sensing application study are threefold:

- Increase the amount of on-board signal processing to reduce the probability of false alarms without affecting detection performance.
- Adjust the signal sample rate or signal sample capability.
- Control the algorithm power usage by varying the clock frequency, number of active processors, and active software modules.

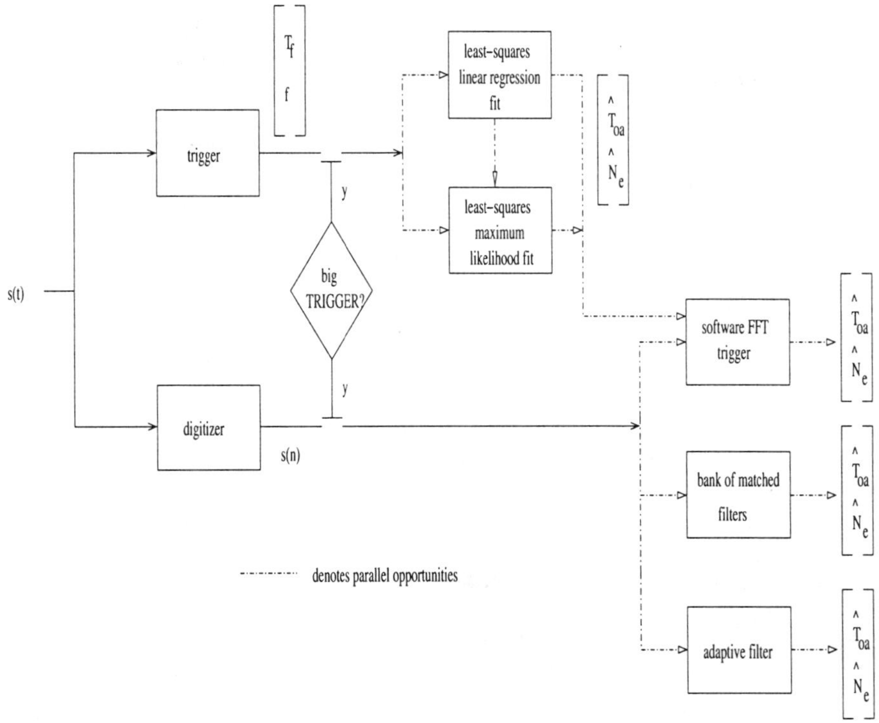

Figure 13.3. Parameter-Estimation, Algorithm Flowchart

These activities are constrained by the power state of the satellite and by the rate of incoming events.

The objective of the signal processing algorithms (see [5], [6], [2] for further background on signal processing algorithms) is to reduce the number of false alarms and estimate the N_e and T_{oa} parameters, as described by Eq. 13.1, from the chirped signals received in orbit. The parameter-estimation flowchart is depicted in Figure 13.3. Note that for FORTÉ, the trigger and digitizer output signals are normally downlinked directly to the ground for analysis; there is no on-board processing of this data.

3.1 Trigger and Digitizer Output Signals

The time-domain, analog signal, $s(t)$, is to be passed through a hardware trigger box and a digitizer. The output from the trigger box and the digitizer provides two input data sets to the signal processing algorithm software. In hardware, the processing software is not active until a signal exceeds a given number of predetermined threshold levels, i.e., an event is not cataloged until "N" many of "M" channels pass threshold, known as a "big TRIGGER."

The output from the trigger box consists of the f and T_f values obtained from filtering the signal through a set of bandpass filters, square law, and low-pass filters. Hence, f and T_f are the signal *center* frequency and *center* frequency time-of-arrival, respectively, as obtained by the filters. This data set will be passed into two routines that will perform a least-squares linear regression fit and a least-squares maximum likelihood fit to the data. These two routines will determine the initial N_e and T_{oa} estimates, $\hat{N}_e$ and $\hat{T}_{oa}$. The maximum likelihood fit is used to help reduce the number of false-alarm outliers. Both routines are deterministic code, i.e., the routines execute in a known finite amount of time.

3.1.1 Signal Filtering. The digitized signal data set $s(n)$ can be passed through the following set of filters to perform more refined estimates of $\hat{N}_e$ and $\hat{T}_{oa}$:

- **i. software FFT trigger** This routine performs a Fast-Fourier Transform (FFT) of the signal and requires the initial estimates of $\hat{N}_e$ and $\hat{T}_{oa}$ from the least-squares modules.

- **ii. bank of matched filters** The number of matched filters can be variable dependent upon the time allowed to process the current event before another event arrives, and the available computing power. This routine will take more time to execute than either the least-squares or the software FFT trigger routines.

- **iii. adaptive filter** This filter routine will follow the adaptive least-mean-squares filter algorithm and use either a fixed or variable step size. The execution time for this routine is indeterminate but depends upon the quality of the signal and initial starting conditions. This filter should yield the most accurate estimates however.

Ideally, in a *conventional* computational satellite system, given enough computing power and time to process the signal, the routines of Figure 13.3 would be executed sequentially; however, given a more realistic scenario in which there is only a finite amount of time before the next event arrives and a finite amount of computing power in which to process the signal, these activities are currently reserved for ground-based analyses. Power-aware management principles can be utilized to process the data through these routines on-board the spacecraft. These ideas are realized through the PAMA computer system by managing the number of active processors working in parallel and customizing the interconnect to the desired communications paths.

4. Adaptive Power-Aware Parallel Processing

Our goal is to determine as accurately as possible an estimate for total electron content and event time of arrival given a varying power budget and varying inter-event duration. As discussed previously in Section 3.1.1, there is a suite of successively more powerful filters that can be applied. We have available multiple nodes of the PAMA parallel processor to apply in parallel to the signal processing tasks.

4.1 PAMA System Architecture

PAMA consists of a 4-node multi-processor connected by a programmable interconnect. On the PAMA-2 board, each node is a Hitachi SH-4 processor, with 32-bit integer and floating point hardware. Systems software includes the Linux operating system, MPI-like communications between processes, and a power-aware software library that allows the application to query power levels and set processor mode, clock frequency, and voltage.

4.2 Application Partitioning

Our application designates a distinguished node 0 as the application controller. Node 0 is responsible for

- obtaining the trigger data and data samples
- determining available power
- estimating the event rate
- distributing work to the three other nodes 1–3.

Worker nodes 1–3 all run the same program, consisting of the algorithms illustrated in Figure 13.3. Each worker receives trigger data and data samples over the interconnection network from Node 0. In addition, each worker receives a control vector from Node 0 telling it which of the filters to apply to the data. Although the workers run the same program, they usually take different paths through the program. Thus we refer to the worker nodes as operating in multiple program rather than single program mode.

We distinguish two types of operation. In the first type, multiple program, multiple data-stream processing, each worker node is given a set of filters to apply to a unique data stream. This mode is used when the event rate is high. The available power determines which filters a worker node will apply. In a low-power, high-event-rate scenario, the worker nodes will just perform the least means square fit (LMS) and possibly the maximum likelihood fit (ML). These routines are very quick (see Section 5 for quantitative results) and only require the trigger box data rather than the entire data sample stream. When more power is available, the controller may choose to have one or more of

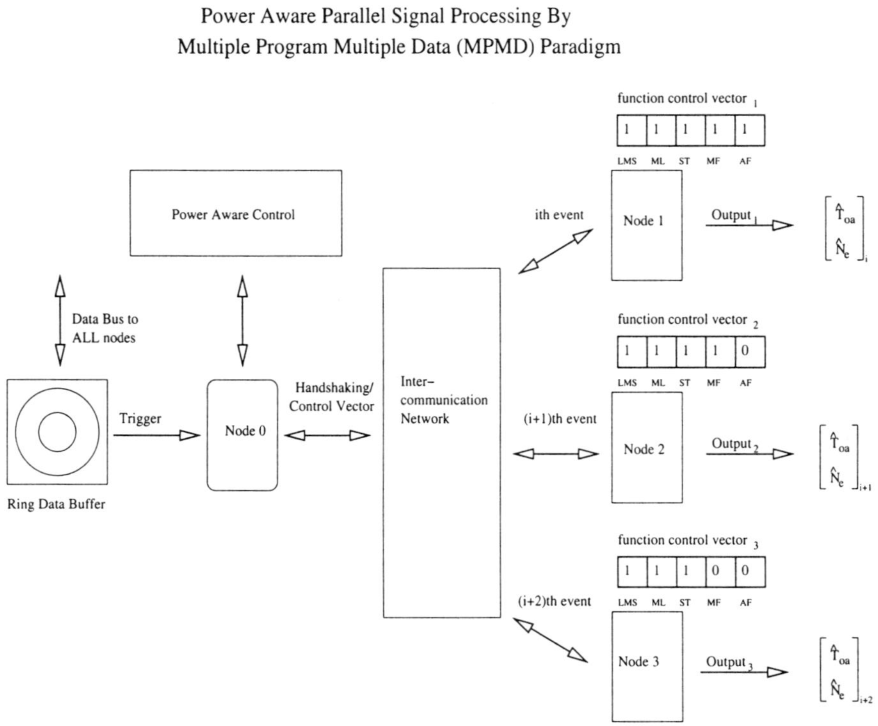

Figure 13.4. Multiple Program, Multiple Data Stream Parallel Processing

the worker nodes also compute a more time consuming filter on the data such as software trigger (ST), matched filter (MF), or adaptive filter (AF). Figure 13.4 illustrates a scenario in which Node 1 is asked to perform the complete parameter estimation suite, Node 2 runs the first four modules only, and Node 3 runs the first three modules only.

The second type of parallel processing is multiple program, single data stream mode. This mode is used in a high-power availability, low-event-rate scenario. The controller broadcasts the trigger and sample data to all workers, and, via the control vector, directs each worker to perform a different filter. In the case of another event occurring before all the workers are done, the controller may take the most accurate result computed to date and then reset the worker nodes to a new event. In the example of Figure 13.5, Node 1 performs the LMS, ML, and ST routines. Node 2 performs the LMS, ML, and MF; and Node 3 performs LMS, ML, and AF.

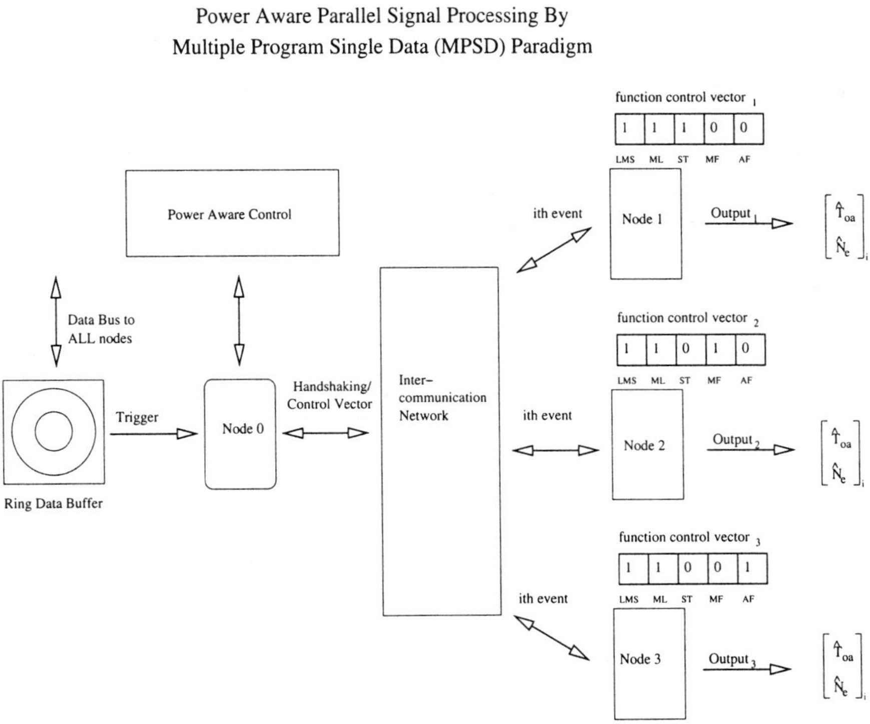

Figure 13.5. Multiple Program, Single Data Stream Parallel Processing

5. Power Availability and Usage

The control node must decide what subset of the signal filters each worker node should perform on what data. As mentioned above, this decision is based on power availability and the amount of power required by each of the signal filters.

5.1 Power Availability

The amount of power available for signal processing computation is a complex function of orbit and housekeeping workload. Typically for solar array/battery satellites, the solar arrays will both recharge the batteries and provide power during "light" times. When in the "dark," i.e., eclipse, power is taken from the batteries. FORTÉ has the ability to draw from the solar arrays or battery or a combination of both. This is accomplished automatically in the hardware depending on the load. There are a few issues to consider for estimating the amount of future power available. These include the following:

- orbit eclipse times. When the satellite is in eclipse, the arrays are not charging the batteries, and no power is drawn directly from the solar arrays, hence, the batteries are the only source of power. There is also a limitation on how much power can be drawn from the batteries, typically known as the Depth of Discharge; beyond this limit, the batteries can no longer be recharged.

  Methods for estimating the amount of time spent in eclipse consist of using orbit propagator tools. For FORTÉ, eclipse times are predicted manually by using the Satellite Tool Kit program, commercially available from Analytical Graphics, Inc. The future load on the batteries and required recharging rates is then estimated based on these predictions. A similar method can be used for PAMA by uplinking the expected eclipse times. These times can then be loaded into the power-control algorithm.

- solar array degradation due to radiation. Typically a solar array will produce more power at the beginning-of-life than at end-of-life. This is dependent on the solar array properties and the radiation exposure for the given orbit. Degradation tends to be a long-term effect and can thus be monitored from the satellite health power-state values.

- sun incidence angle. The solar array will produce max power when the array is normal to the sun vector. Attitude data relative to the sun is needed for this information. For FORTÉ, the spacecraft body is covered with solar panels to provide a constant average power generated by the panels.

- rate of future events. The event rate can vary from several events a second to an event every fifteen minutes. For our science application, the event rate is determined principally by the satellite's orbit and time of year. In general, we expect more thunderstorm activity over land and in the summer than over water during colder seasons, thus leading to a higher event rate for the former. We have developed a simple model for the event rate that factors in these considerations. In operation, we will load the event rate as a table to the on-board processor.

For further information on satellite power system requirements, design, and estimation see [3].

5.2 Power Usage

We have obtained power usage for four of the signal filters by running the filters on a variety of microprocessors. These include the following: a 733-MHz Pentium III processor running the Linux operating system and compiled with the GNU C compiler using optimization option O; a 200 MHz Hitachi SH-4 processor running the Linux operating system and compiled with the GNU C compiler using optimization option O; a Texas Instruments' TMS320C6711 with a clock frequency of 150 MHz, using no operating system and compiled using Code Composer Ver. 2.0 with the o3 optimization options; and a 266 MHz Power PC 750 running the VxWorks operating system. Time-to-execute values for each processor and for each signal-processing operation were determined. In addition, both time-to-execute values and power usage estimates (RMS and peak current) were determined for the Power PC 750.

Time-to-execute values that are presented in this paper are average values. A test set comprised of 21 test events was used for this benchmarking exercise. Each test event has hardware-trigger-box data and digitized waveform data associated with it. The data of the test set was synthetically generated using a MATLAB program that simulated a pulse event being received by a space-base receiver system containing a hardware trigger box and a waveform digitizer. Several (20 to 100) executions of the complete data set were performed to yield averaged time-to-execute values.

Time-to-execute values were determined by embedding compiler-specific timing functions into the C-language code. The timing functions involved the starting, stopping and/or reading of timers or clock-cycle counters. The method of timing was determined by what compiler-specific functions were available and familiar to the experimenter. Whatever the functions used, the placement was done in a manner as to only take into account the data manipulation only operations and not those operations associated with allocation of memory, movement of data or other operations that maybe dealt with differently in an application-specific, embedded-processors system.

Table 13.1. Timing of Signal Filters

	LMS	ML	ST	MF
Pentium/733	0.71μs	24.1μs	1.29ms	43.4ms
TI-C6711/150	4.6μs	112μs	14ms	1000ms
SH-4/200	7.14μs	152μs	11.6ms	516ms
PPC750/266	3.4μs	183μs	8.34ms	470ms

None of the development systems used to benchmark the code were developed to be application-specific system for the application presented in this paper. This being the case, the values presented in the Table 13.1 cannot be considered as inflexible values that can be used to determine design limits. However, the values do give a good presentation of the relative comparison of how the processors perform when tasked with the specific application of this paper. When examining the table, the clock frequency of the processor should

Table 13.2. Power Measurements for the PPC750

	LMS	ML	ST	MF
Current (amps-peak)	2.06	2.06	2.18	2.04
Power (Watts rms)	5.5	5.596	5.67	5.0
Execution Time	3.4μs	183μs	8.34ms	470ms
Energy (Joules)	18.7e-6	1.02e-3	47.3e-3	2.35

be taken into account. Clock frequency has a directed impact on power usage, thus Tables 13.1 and 13.2 just begin to visit some of the trade space in the design of an application-specific system.

Power usage for the Power PC 750 executing the benchmarking code is presented in Table 13.2. The Jet Propulsion Laboratory (JPL) power-aware testbed consists of a Wind River PPC750 266MHz processor board that is running VxWorks 5.4.2. The processor operates at a constant 2.67V and current consumption is measured with a Tektronix TDS 7104 Digital Phosphor Oscilloscope. Current is sampled with the Tektronix TCP202 probe that is wired to the board. Software compilation is done with the VxWorks Tornado 2.0.2 programming tools which uses the GNU C compiler.

The software is compiled and downloaded to the testbed manually with the Tornado target server shell. The programs are run until an "average" current signal snapshot is taken with the oscilloscope. The "average" signal is determined manually by watching the current response during several program runs. The snapshot is taken when the current response produces a fairly consistent signal and consistent measured values. For reference, a "background" snapshot, i.e., when there is no user-program load, is also taken for each program test. Timing information is obtained using the "tickGet()" function in the software code. Since the ticks have a low resolution (in comparison to the program execution time) of 60 ticks/sec, an average value was calculated by summing the ticks over a significant number of program runs. Representative oscilloscope measurements for the matched filter are shown in Figures 13.6 and 13.7. Figure 13.6 shows the background measurement, and Figure 13.7 shows measurement while the matched filter routine is running.

Both peak and average current were measured for a 2.67-volt power supply, supplying the processor. Table 13.2 provides the peak current, power (RMS), time-to-execute and energy-expended values for the four signal-processing operations. Since the application of this paper is satellite-based, it is believed

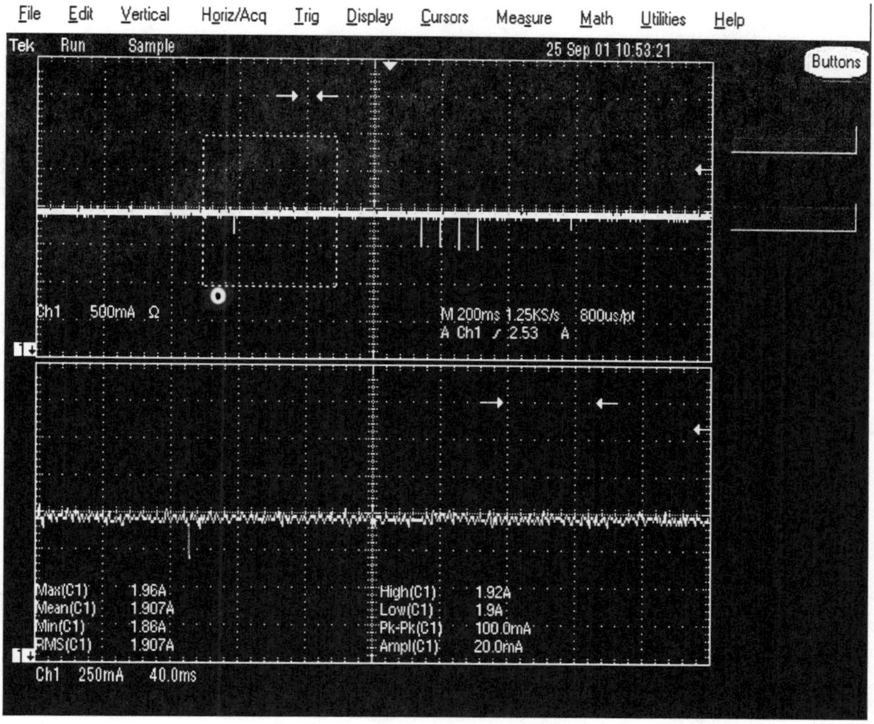

Figure 13.6. Matched Filter Background Measurements (oscilloscope snapshot taken with JPL testbench equipment)

that these values associated with power measurements would be the ones most desirable to a designer.

The test results of Tables 13.1 and 13.2 show a diverse spread of magnitudes from running the four signal processing routines LMS to MF. As expected, the LMS and ML routines execute in the least amount of time with the least amount of energy. These two routines are ideal for obtaining a quick, initial estimate of the parameters, $\hat{N}_e$ and $\hat{T}_{oa}$. The ST routine provides a more refined estimate of these parameters but also has an associated increase in the timing and energy costs. In terms of these performance metrics, the most expensive routine is the MF; it is approximately 3 orders of magnitude more costly in time than the LMS and ML routines, 6 orders of magnitude more costly than LMS in energy, and 3 orders of magnitude greater than ML in expended energy. Thus, these results illustrate the trade-off relationship between timing/energy costs and increasing parameter estimation accuracy: for higher-confidence estimates, there are more associative costs.

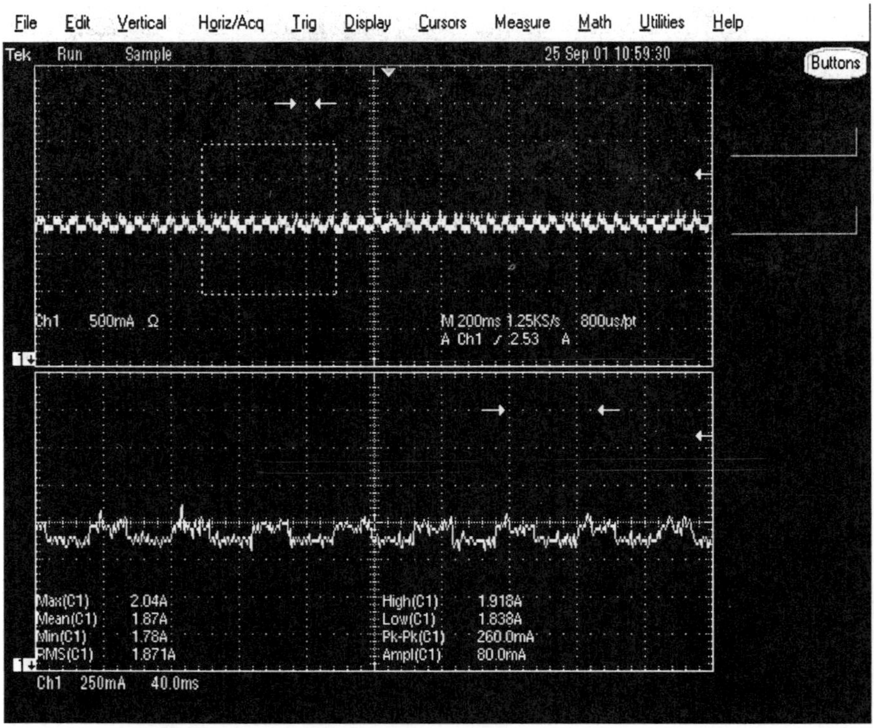

Figure 13.7. Matched Filter Run-time Measurements (oscilloscope snapshot taken with JPL testbench equipment)

This parameter estimation trade space defines the operations of the power control node. It is desirable to obtain the best estimate possible given the constraints on both current and future available power, expected rates of incoming trigger events, and the timing/energy costs associated with routine execution.

Of worth noting is our exclusion of the StrongARM processor in our presentation of data. The reason for this is the belief that the processor would not be suitable for intense floating-point operation. This belief was supported by preliminary testing. As an example, a 206 MHz StrongARM processor required 443μs to perform the LMS fit and 30.39s to perform the MF operation. Although the StrongARM processor has many desirable features in terms of power-usage control, it was clear that it is unsuitable for this application.

6. Conclusions

In this paper we discuss how power-aware management techniques can be used in satellite remote-sensing applications as a "smarter" approach to satellite power management. A remote-sensing application, similar in nature to the DoE-funded FORTÈ satellite mission, and the signal processing algorithm used is described. Using a power-aware multi-processor, we have the ability of on-board post-processing of science data to reduce the rate of false alarms and give the remote-sensing system a greater degree of flexibility. We have described how the signal processing modules are mapped onto PAMA, exploiting multiple processors for more accurate detection when power is available, while scaling back to uni-processor and lesser clock-frequency mode when power is scarce. Experimental tests on four of the signal processing algorithms have yielded a 6 order of magnitude spectrum in energy consumption between the routines. As differing algorithms produce higher-confidence parameter estimates, significant increases in energy costs are experienced. This project is currently on-going and is a collaboration between LANL and USC/ISI.

Acknowledgments

This effort is sponsored by Defense Advanced Research Projects Agency (DARPA) through the Air Force Research Laboratory, USAF, under agreement number F30602-00-2-0548. We are indebted to JPL for the use of their Power PC testbench to measure algorithm power usage, and thanks specifically to Jeffrey Namkung of JPL for his invaluable assistance with the testing. Thanks are also due to the LANL FORTÉ team for supplying details of FORTÉ operation.

References

[1] Donald C. Enemark and Martin E. Shipley. The FORTÉ receiver and subband triggering unit. In *8th Annual AIAA/Utah State University Conference on Small Satellites*, 1994.

[2] Steven M. Kay. *Fundamentals of Statistical Signal Processing: Estimation Theory*. Prentice Hall, 1993.

[3] Wiley J. Larson and James R. Wertz. Space Mission Analysis and Design, *2nd edition*. Kluwer Academic Press, 1992.

[4] NOAA. http://www.ngs.noaa.gov/GRD/GPS/ Projects/TEC, 1999.

[5] A. Oppenheim and R. Schafer. Discrete-Time Signal Processing. Prentice Hall, 1989.

[6] W. Press et al. Numerical Recipes in C, *2nd edition*. Cambridge University Press, 1992.

[7] Diane Roussel-Dupré, Phillip Klingner, et al. Four years of operations and results with FORTÉ, AIAA 2001-4627.

Chapter 14

THE CASE FOR POWER MANAGEMENT IN WEB SERVERS

Pat Bohrer, Elmootazbellah N. Elnozahy, Tom Keller, Michael Kistler, Charles Lefurgy, Chandler McDowell, Ram Rajamony
IBM Research, Austin TX 78758, USA.
http://www.research.ibm.com/arl

Abstract Power management has traditionally focused on portable and handheld devices. This paper breaks with tradition and presents a case for managing power consumption in web servers. Web servers experience large periods of low utilization, presenting an opportunity for using power management to reduce energy consumption with minimal performance impact. We measured the energy consumption of a "typical" web server under a variety of workloads derived from access logs of real websites, including the 1998 Winter Olympics web site. Our measurements show that the CPU is the largest consumer of power for typical web servers today.

We have also created a power simulator for web serving workloads that estimates CPU energy consumption with less than 5.7% error for our workloads. The simulator is fast, processing over 75,000 requests/second on a 866MHz uniprocessor machine. Using the simulator, we quantify the potential benefits of dynamically scaling the processor voltage and frequency, a power management technique that is traditionally found only in handheld devices. We find that dynamic voltage and frequency scaling is highly effective for saving energy with moderately intense web workloads, saving from 23% to 36% of the CPU energy while keeping server responsiveness within reasonable limits.

1. Introduction

There is a growing industry trend to "outsource" computing services to large data centers accessible through the World Wide Web (WWW). These data cen-

[0]This research has been supported in part by the Defense Advanced Research Projects Agency under contract F33615-00-C-1736.

ters use economies of scale to amortize the cost of ownership and system management over a large number of machines. A typical data center thus deploys hundreds or thousands of computers, densely packed to maximize floor space utilization. This form of server consolidation provides the customer with a more cost-effective approach than the alternative of operating the same services in-house.

Server consolidation, however, has its own problems as large-scale deployment pushes the limits of power supply and cooling systems in a data center. Anecdotal evidence from data center operators already points to the large contribution of power consumption and cooling to operation cost [24], and the intermittent failures of computing nodes in densely packed systems due to insufficient cooling. Furthermore, in many data centers, power supply to the server racks is a key inhibitor to increasing server density. One estimate, based on data from several modern Internet hosting facilities, puts the practical limit in power supply to a rack at around 7KW. This power is often insufficient to allow the rack to be fully populated with "1U" servers. Thus, power consumption increasingly carries an opportunity cost in lost revenue of additional systems. The problem is likely to worsen as newer server-class processors are projected to offer higher levels of performance at the expense of higher power consumption levels. This technology trend is a natural response to the growing needs for more performance. We therefore believe that these technology and market trends will make power management within servers a necessity.

Research in power management has primarily focused on portable and handheld systems. Such systems experience workloads that include substantial interactions with a single user, providing opportunities to turn parts of the system off when the user is "thinking" before issuing the next command. Typical energy saving techniques include turning off the system's display and disk, and slowing or halting the processor [18, 14].

This paper addresses power management in server systems. We focus on web server workloads because of the importance and widespread use of these servers in data centers. We first demonstrate that power management can provide significant benefits by examining workload characteristics from several real web sites, and then support this examination with measurements from an instrumented system in our lab running a workload derived from production Internet web sites. Then, we characterize the power consumption of the web server for these workloads, and use the measurements to validate a simulation model. We then use the simulation model to examine the effect of voltage and frequency scaling on the energy consumption of web servers.

This paper makes three contributions:

- Makes a case for power management in web servers and identifies the processor as the main system component where power management could yield energy savings.

- Introduces a power simulator for web serving workloads that is accurate to within 5.7% over the range of workloads we use in this paper.
- Evaluates the effectiveness of dynamic voltage scaling, an energy-saving technique typically found in handheld devices, using our simulation model.

There are several limitations to a study of this kind, including the inherent inaccuracies in measuring and modeling power consumption. In addition, web servers can be run on a wide variety of machines with vastly different power consumption characteristics. We have chosen to study a "commodity server" (often referred to as a "white box server"), since we believe this is the most common type of web server machine employed in commercial data centers. However, even commodity servers have differing power consumption characteristics, so while we believe our results are generally valid for this space of systems, there are certainly some exceptions.

The remainder of this paper is organized as follows. In Section 2 we present the case for power management in web servers. Section 3 explains our environment for data collection. Section 4 presents a characterization of power consumption in web servers. In Section 5 we describe a simulation model for evaluating various power management policies and our validation of this model. We present the results of simulating a power management policy that dynamically varies the processor voltage and frequency in Section 6. In Section 7 we discuss the implications of energy management on traditional web server performance metrics. A comparison to related work is presented in Section 8. Finally, Section 9 concludes the chapter.

2. The Case for Power Management

Web site designers typically plan web server capacity to provide acceptable service even during periods of peak demand. Additional capacity may also be planned in clustered environments for high availability, where it may be necessary to redistribute the load of a failed server to the functioning ones. As a result, a web server may not be operating at its maximum capacity for substantial durations. We show by an analysis of several real internet web servers that this is true in practice. This model of operation has implications on the power consumption of web servers. We then study the efficiency of a web server as it operates under different workloads, and establish the case for power management in such servers.

2.1 Web Server Loads

Consider the design of the web site used for the 1998 Winter Olympics in Nagano [5]. The site consists of a three-tiered structure that is a common approach to constructing high–volume web sites. The first tier consists of edge servers that interface with clients. These servers provide firewalls,

Table 14.1. Analysis of Real Internet Web Server Workloads

	Avg req / sec	Max req / sec	Ratio
Olympics98	459	1840	25%
Wimbledon	778	7166	11%
Finance	16	46	35%
Shopping	25	55	45%
Proxy	15	30	50%

caching, workload distribution, and the HTTP interface for client interaction. The second tier consists of application servers, which perform functions such as content generation and personalization. The last tier is made up of traditional database servers that provide a reliable data repository with sophisticated processing capabilities. The site was replicated over four geographic areas to cover the globe and provide high availability.

The total traffic experienced by the site during the 16 days of the Olympics was roughly 634.7 million hits, and the site successfully handled a peak of over 110,000 hits for a one minute period during the women's freestyle figure skating event [5]. No failures were detected during the 16 days of the Olympics. Thus, while the peak workload was 1840 hits/sec, the average workload over the entire event was only 459 hits/sec. It follows that on average, the site was operating at about 25% of the observed peak capacity. Note that this is not a case of careless over-engineering. The distribution of the workload was not uniform, and depended on the obvious pattern of accesses from around the globe. It was therefore necessary to design the system to meet the peak demand. Well–engineered web sites follow the same design principles. This is particularly important in commercially oriented web sites, where customer dissatisfaction with the service's responsiveness may seriously affect the profitability of the enterprise.

The Winter Olympics example supports the notion that web servers are designed to handle peak loads. A further example is the 1999 Wimbledon web site which experienced an average of 778 hits/per second over two weeks, and a peak of 7166 hits/sec in a one–minute period, implying that the site was operating at only 11% of its observed peak capacity [5].

We have also analyzed the logs of web servers employed in a variety of real internet web sites and found very similar trends in the workload of these servers. These results and the examples given above are summarized in Table 14.1.

Several studies also provide evidence that web workloads are bursty in nature, and support the intuitive notion that web servers tend to be busiest during some peak hours during the day and almost idle in others [4]. This model of operation has implications with respect to power consumption that we address next.

2.2 Energy Efficiency of Web Servers

Several power management techniques have been developed for portable and desktop computers. One general approach is to reduce the power consumed by components not currently in use. Examples of this kind of technique include placing the CPU in a "halted" state when there are no "runnable" tasks, and turning off the hard drive motor after some period of inactivity. Another category of techniques reduces power consumption of components that are in use, allowing responsiveness or quality to degrade within some acceptable range. Examples here include scaling the CPU frequency and reducing the display brightness.

In contrast, server-class systems currently do very little in the way of power management. In part, this is because most server-class systems do not incorporate the mechanisms required for many power management policies. We believe this is mainly due to the focus on performance as the only metric of relevance for server systems.

We believe that there is ample opportunity to reduce the power consumption of web server systems. As illustrated in the previous section, web server systems are typically designed to handle a peak workload that is much higher than their average workload. As a result, the system will have significant periods of low utilization. During low utilization, some components may be completely idle, and thus could be "spun down", and other components could be operated at reduced power. In web servers, such techniques would most likely be targeted at the CPU, disk, or memory. During periods of peak usage, a web server could trade off quality or responsiveness in order to reduce its power consumption. For example, the server could serve lower resolution images or throttle requests for dynamic content during periods of high activity.

To demonstrate the potential benefits of power management mechanisms, we use simulation studies to evaluate the benefits of scaling the CPU voltage and frequency in response to the server workload. Our results indicate that voltage scaling can result in significant energy savings, without substantially altering the responsiveness of the system.

3. Methodology

We begin our study by characterizing the energy consumption of a web server by measuring the energy used by its components as it executes web server workloads. This section describes the environment for collecting data and the web server workloads.

3.1 Environment

Our instrumented "white box" web server system is made up of the following components:

- Pentium III 2.0 volt 600MHz processor with 512KB off-chip L2 cache and a 100MHz memory bus in a Slot 1 package.
- Tyan S1854 motherboard with 512MB IBM 3.3volt SDRAM (2 DIMMs) – 100MHz, 32Mx72.
- IBM 60GB Deskstar ATA-100, 7200 rpm, IDE disk drive
- Deer Computer Corp DR-250 250 watt ATX power supply with a combined peak output rating of 125 watts.
- Matrox 64/MILA/16/1B2 graphics card.
- Realtek RTL-8029 10/100Mbit Ethernet card.
- Software: Red Hat Linux Release 7.0 with 2.4.3 kernel (no parameters changed) and Apache 1.3.12 web server

The default Apache configuration assumes a modest size web server. Since our goal was to test the limits of the server, we changed the `MaxClients` setting from 100 to 1,500 and increased `MaxRequestsPerChild` from 100 to 10,000. Except for these changes, all other setting were left as in the default case. We sent requests to the server from a client that is more powerful than the server, thereby ensuring that server performance was not artificially limited by client resources. All machines were on a single subnet connected through an ExtremeNetworks Fast Ethernet (100 Mbits/sec) switch.

3.2 Measurement System

We instrumented five power supply leads:

- +3.3V supply to the motherboard, powering the 512MB of RAM, video card, and Ethernet card
- +5V supply to the motherboard, powering the Pentium III processor
- +12V supply to the motherboard, powering the processor cooling fan
- +5V supply to the single disk drive, powering the controller and other components
- +12V supply to the single disk drive, powering the motor and head actuator mechanisms

We determined the energy consumed by each sub-system by measuring the drawn voltage and current. While we could directly measure voltages, we used a sense resistor in series with each sub-system to measure the current. Signals from the sense resistors were filtered with a 10kHz low pass filter, gained to be within ±10V using custom circuitry, and then passed to a PCI-6071E A-to-D board from National Instruments. Each channel was sampled at 1,000 times per second, and custom software was written to gather and process the data. The accuracy of the measurement system is within 5%, with sense resistor accuracy and amplifier voltage offset being the dominant source of errors.

3.3 Workloads

Since our goal is to study the power consumption of web servers under real workloads, we constructed workloads using web server logs obtained from several production internet servers. Each access log represents 24 hours of activity. The first workload is derived from the web server logs of the 1998 Winter Olympics web site. These logs contain all of the requests processed over the course of one day at one of the four geographically distributed sites used for the Olympics. The second workload is derived from the server access log of one day's activity on the web site of a financial services company. The third workload is derived from a log of a proxy server operated by the Information Resource Caching (IRCache) Project [16], which is affiliated with the National Laboratory for Applied Network Research (NLANR). The IRCache Project operates a number of caching proxy servers on the Internet and publishes the logs from these servers as a way to promote research into web caching. All IRCache servers run Squid, which is also supported by the IRCache project. Strictly speaking, a proxy server is not a web server in that it does not have its own content, but stores frequently referenced content of other servers for more rapid access by clients.

For each workload, we construct a stream of HTTP requests for static files based on the server access log. We exclude all requests that provide parameters as part of the URL, since these are obviously for dynamic content. Such requests account for 2.5% or less of the Olympics98 and Finance logs, and none for the Proxy server log. We treat all remaining requests to be for static files. However, to account for any remaining URLs that may generate dynamic content, all file names are augmented with the size of the file returned, so that each dynamic response (of a unique size) is treated as a separate file. We have used this approach to deal with dynamic content because we do not have access to the dynamic scripts used at any of these sites.

Persistent connections, a feature of the HTTP/1.1 protocol [26], allow a client to make multiple requests to the web server over a single TCP connection, thus amortizing the cost of connection establishment and teardown. While the access logs do indicate whether the client used the HTTP/1.0 or HTTP/1.1 protocol, there is no explicit indication of which requests were grouped onto a single connection. To approximate the effect of persistent connections, we use the following rule to group requests into connections: any request received from the same client within 15 seconds of a prior request is grouped into a connection with that prior request. Since the timestamp in the access log has only a one second resolution, we randomly distribute the arrival time of all HTTP/1.0 requests and new HTTP/1.1 connections within that second. If only one follow-up request in a persistent connection is logged in a second, we assume that it arrived at a random time within that second. We assume that multiple follow-up requests within the same second arrived at the server as a burst at a random time within that second.

It is worth noting that we have used the timestamp in the web server access log to indicate the time the request arrived at the server. In general, this is not accurate, since the timestamp in the access log generally indicates the time that the response was sent, not the time the request was received. For servers that are significantly overloaded, or for very large files, the time of the response might not be a good indicator of the time of the request. We believe neither of these factors were present to a significant degree on the Olympics98 and Finance servers on the days the logs were collected, and therefore we believe the timestamp in the access log is a good indicator of request time for our workloads. For the Proxy workload, we can determine the request time exactly since Squid logs the time required to process the request in addition to the time of the response.

Table 14.2. Characteristics of Three Web Server Workloads[a].

Workload	Olympics98	Finance	Proxy
Avg requests / sec	97	16	15
Peak requests / sec	171	46	30
Avg requests / conn	12	8.5	31
Files	61,807	16,872	698,232
Total file size	705 MB	171 MB	6,205 MB
Requests	8,370,093	1,360,886	1,290,196
Total response size	49,871 MB	2,811 MB	10,172 MB
97%/98%/99% (MB)	24.8 / 50.9 / 141	3.74 / 6.46 / 13.9	2,498 / 2,860 / 3,382

[a]Average requests per second is the average request rate over the entire 24 hour period, and peak requests per second is the highest observed rate for a one minute period. The average requests per connection is based on our technique of grouping requests into connections. Files is the number of unique files requested, and Total File Size is the total size of these unique files. Requests is the number of distinct HTTP requests, and Total Response Size is the total size of response data sent for these requests (excluding HTTP headers). 97%/98%/99% data is the amount of memory needed to hold the unique data for 97%/98%/99% of all requests.

The characteristics of our three workloads are summarized in Table 14.2. Particularly important for our study is the 97%/98%/99% data size figures, which indicate the amount of memory needed to hold the unique data for

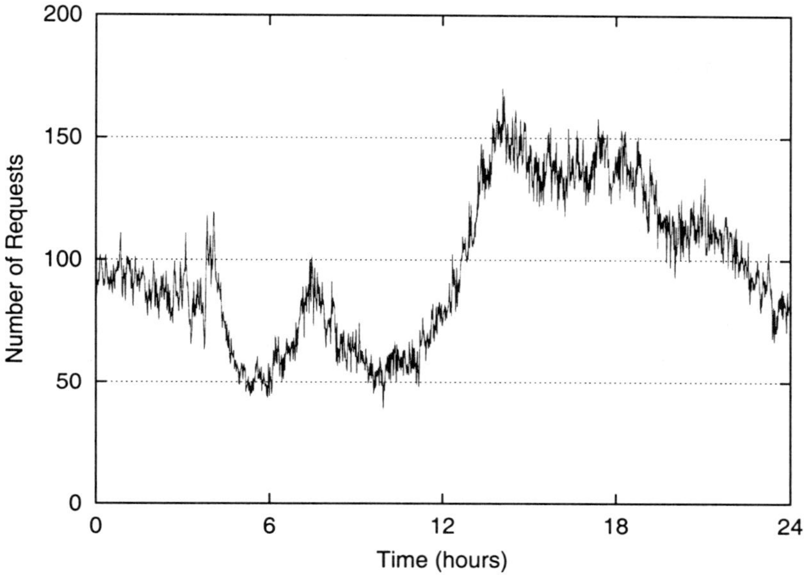

Figure 14.1. Request Rate Seen at One of the Olympics Web Sites on February 19, 1998

97%/98%/99% of all requests. For example, in the Olympics98 workload, 99% of requests could be served from memory using a cache of only 141 MB. Since our web server has 512MB of memory, caching files in memory could significantly reduce the disk activity required to serve the workload. We should note that it is difficult to make a direct correlation between the 97%/98%/99% data size figures and the cache space required by the web server, since these figures were determined with complete knowledge of the request stream. On the other hand, the figures are for a "static" cache, and could be reduced even further if cache contents could be dynamically managed. On balance, these figures should simply be viewed as indicators of what could be achieved with a reasonably effective caching strategy.

Since caching should significantly reduce disk activity for both the Olympics98 and Finance workloads, we wanted to balance them with a workload that has larger amounts of disk I/O. The Proxy workload has this feature. The total response size of Proxy workload, at 10.2 GB, is less than twice as large the total file size, indicating that first-reference requests alone will generate a significant level of disk activity.

Web serving workloads are notorious for having high variability [4] and these three workloads are true to form. As indicated in Table 14.2, the peak request rate measured in one minute intervals is anywhere from 1.5 to 3 times

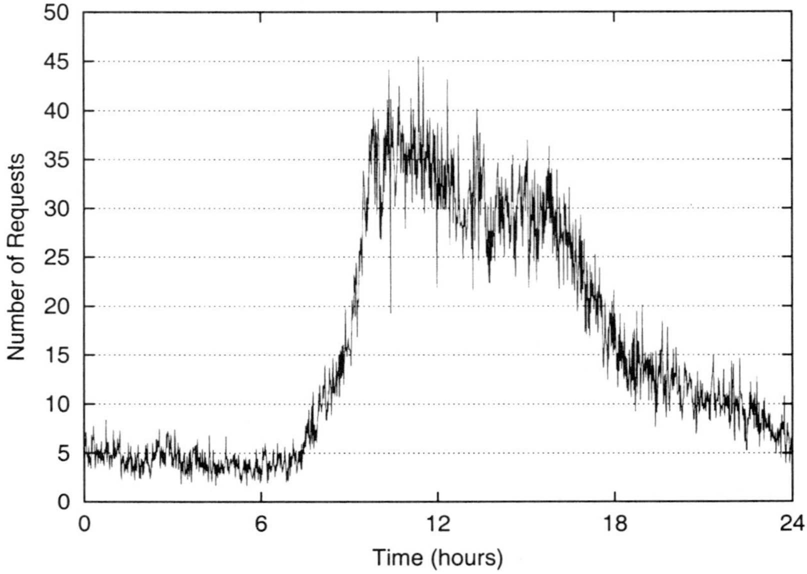

Figure 14.2. Request Rate Seen at a Financial Services Company Web Site on Oct. 19, 1999

larger than the average request rate.[1] Each workload shows its own distinct pattern of request arrivals, which are illustrated in Figures 14.1, 14.2, and 14.3. This characteristic reflects the challenges in configuring web server systems that have sufficient capacity to handle peak load and are power-efficient when serving the average load.

3.4 Replay Program

We use a modified version of the httperf workload generator [20] to generate an HTTP load on the web server. httperf can simulate a large number of http clients from a single machine by using an event-driven design that minimizes the resources required to maintain each connection. The process described in the previous section converts an access log into a trace file that contains a sequence of connections, each consisting of a sequence of HTTP requests. The trace file specifies the time interval between each connection initiation and the time interval between the initiation of each HTTP request within a connection. We modified httperf to effectively replay a server access log by

[1] The request rate data for Olympics98 does not match that given in Table 14.1 because the latter is computed over all 16 days of the Winter Olympics.

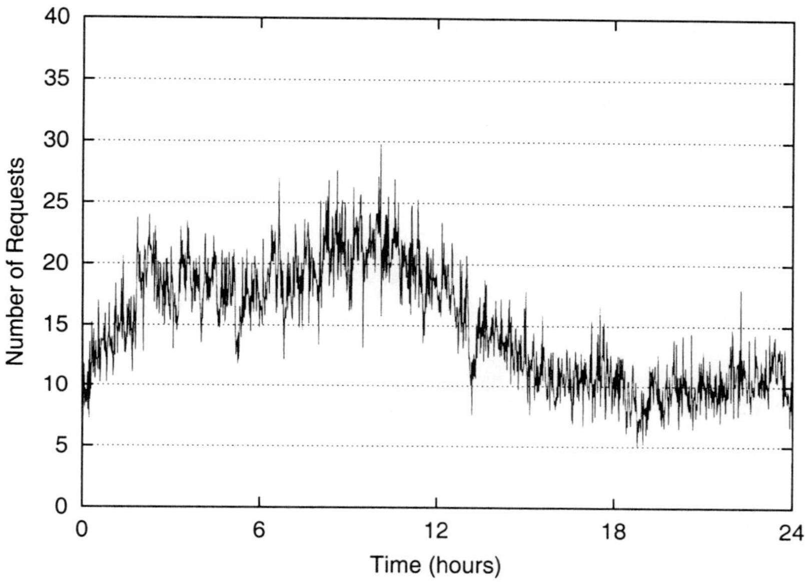

Figure 14.3. Request Rate Seen at IRCache Project Silicon Valley Proxy Server on May 2, 2001

initiating connections at the appropriate times, generating requests for specific URLs at specified time intervals, and closing the connections after all requests were complete.

In addition, we added the capability to scale the inter-arrival time of connections (but not requests within a connection) by a user–specified amount. A scalefactor of "2×" corresponds to reducing the inter-arrival time of connections by 50%. Since each connection roughly corresponds to a client, reducing the connection inter-arrival time effectively generates a heavier client load, but with the same basic pattern of connection arrivals. Inter-arrival time of requests within a connection are not scaled since these essentially represent user think time or network and client overhead involved in retrieving multiple components of a web page. We use this scaling mechanism to evaluate server performance for a range of client load intensities. Thus we are able to scale the intensity of any workload to "1×", "2×", "2.5×", etc.

4. Power Consumption in Web Servers

In this Section, we characterize the power consumption of a web server using the three workloads described in Section 3.3.

4.1 Overview of System Power Consumption

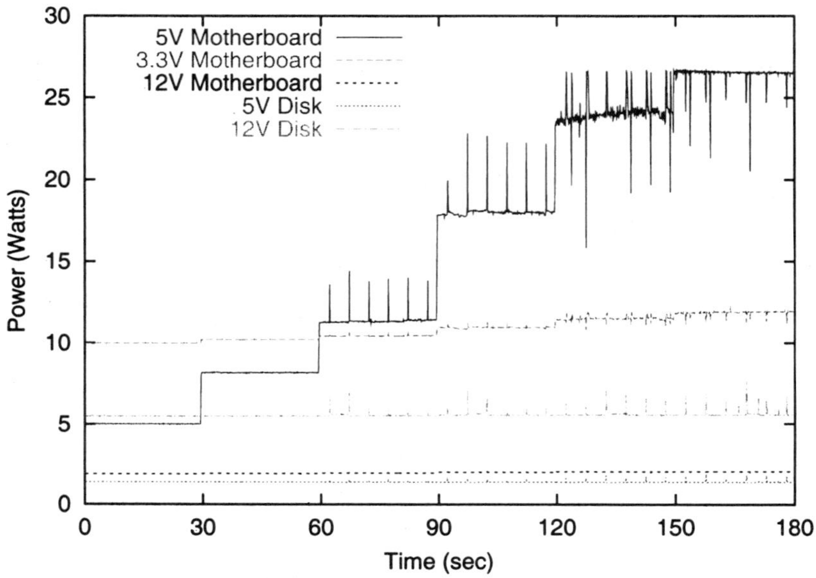

Figure 14.4. Web server power consumption for a steady request stream measured on each power supply line over time. Notice that the 5V motherboard power changes significantly as the request rate to the web server increases. The other power supply lines are mostly constant.

Our first measurement is performed using a steady request stream for a single URL at various request rates. The purpose of this measurement is to study the relationship between the load placed on the server, as measured by request rate, and power consumed by the various components of the web server. While a workload of a single URL is clearly not realistic, it nevertheless provides valuable insights into the effect of load on web server power consumption. Figure 14.4 displays the power consumed by our 600MHz web server system over time for the "idle" case and request rates of 100, 200, 400, 600, and 800 req/sec, each executed for 30 seconds. The graph shows the average power consumed from each of the five supply lines for each 100 millisecond interval.

The 5V motherboard supply, which provides power to the 600MHz Pentium III processor as well as other components on the motherboard, changes significantly during the course of the workload. When the system is not servicing requests, the processor spends nearly all the time in a halted state, consuming approximately 5.0 Watts. At 800 requests/sec, the CPU appears saturated,

Table 14.3. CPU Energy Consumed (Per Second) for Serving Requests of Different Sizes

Request Rate (req/sec)	Idle	100	200	400	600	800
CPU Energy (Joules)/sec	5.0	8.3	11.6	18.2	24.3	26.5
Response time (ms)	-	1.0	1.1	1.2	4.8	908

consuming approximately 26.5 Watts. By increasing the sampling rate of our energy measurement equipment, we determined that the CPU typically operated at one of two extremes. At one extreme, the system was idle (waiting for client requests), and the processor was in the halt state consuming only about 5 Watts. At the other extreme, the CPU was active and thus consuming nearly its peak power of 26.9 Watts. The result of this bimodal operation is that CPU energy is simply proportional to the number of cycles spent processing requests. For our simple workload that serves all requests from memory, the number of cycles spent processing requests is essentially linear in the request rate for light to moderate load, and thus energy consumed is essentially a linear function of request rate up to around 600 requests/second. We also measured average response time at each request rate, and have summarized this data along with the average CPU energy consumption in Table 14.3. The two orders of magnitude increase in response time from 600 to 800 reqs/sec confirms that the system is saturated at this level, with the CPU the apparent bottleneck resource.

The CPU and 12V disk power show pronounced spikes every five seconds, which are due to the periodic activity required to flush the web server log to disk. The 5V disk and 12V motherboard energies have almost no variation over the course of the run and are also quite small in comparison to the other three components. The 12V disk power also shows relatively little variation outside of the periodic spikes already mentioned. We note a trend in the 3.3V motherboard supply, where it remains perfectly constant at about 10 Watts while idle, but increases slightly as the load increases. We attribute this to the increased use of memory and to the network interface. Through a separate measurement, we ascertained that the cooling fan draws a constant 1.5 watts from the 12V supply, and the video card, a nearly constant 6.09 Watts from the 3.3 V motherboard supply.

Next we measured the power consumed by our web server for the Olympics98 workload when run at a scalefactor of 4.0, which results in a request rate that stresses the server but does not overload it. These results are shown in Figure 14.5. The figure shows the average energy consumed (in Joules) on each of the five instrumented power leads, averaged over intervals of 60 seconds. The scale for power is the same as in Figure 14.4. As in the single request workload above, the largest and most variable component of power consumption in this workload is CPU power. This result is not surprising,

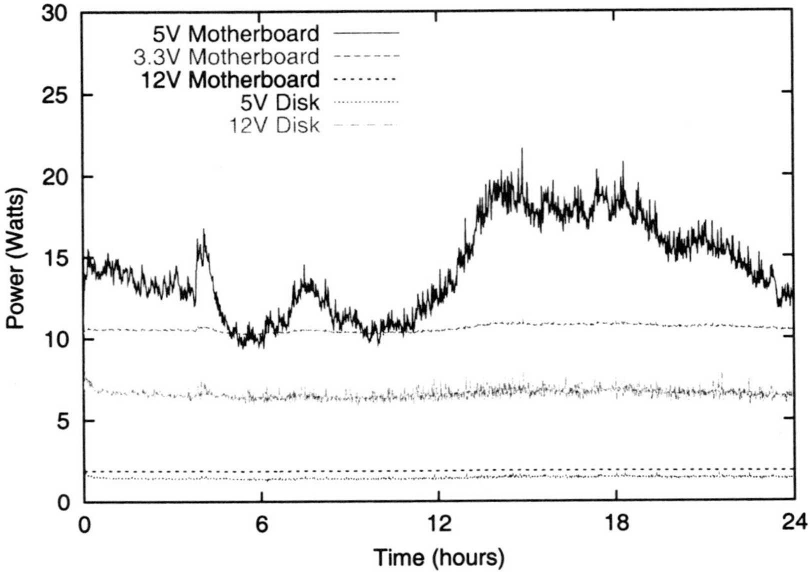

Figure 14.5. Power Consumption of the Olympics98 Workload (at 6X)

since a high percentage of requests for the Olympics98 workload can be served from the RAM cache. The total system power consumption of this workload is mostly between the 200 req/sec and 400 req/sec workloads in Figure 14.4. We also measured the response time of each HTTP request over the course of the workload, and found that the 95th percentile of the response time was less than 25 milliseconds, indicating that the server is not overloaded for any significant portion of the run. The same workload executed at a scalefactor of 6.0 resulted in 95th percentile response time of over 130 milliseconds, which indicates significant periods of overload.

We scaled the Finance and Proxy workloads to insure that the server was moderately heavily loaded, in a manner similar to that of the Olympics98 workload. Figure 14.6 shows the power measurements we obtained using the Finance workload at a scalefactor of 20.0. We also measured the power consumption of our web server for the Proxy workload at a scalefactor of 2.0. These results are shown in Figure 14.7.

The Proxy workload differs from the Olympics98 and Finance workloads in that a large fraction of requests result in actual disk I/O. This difference is clearly visible in the power measurements, where the CPU consumes much less power with little variation. On the other hand, the 12V Disk supply (disk mechanics) shows significantly higher and more variable power con-

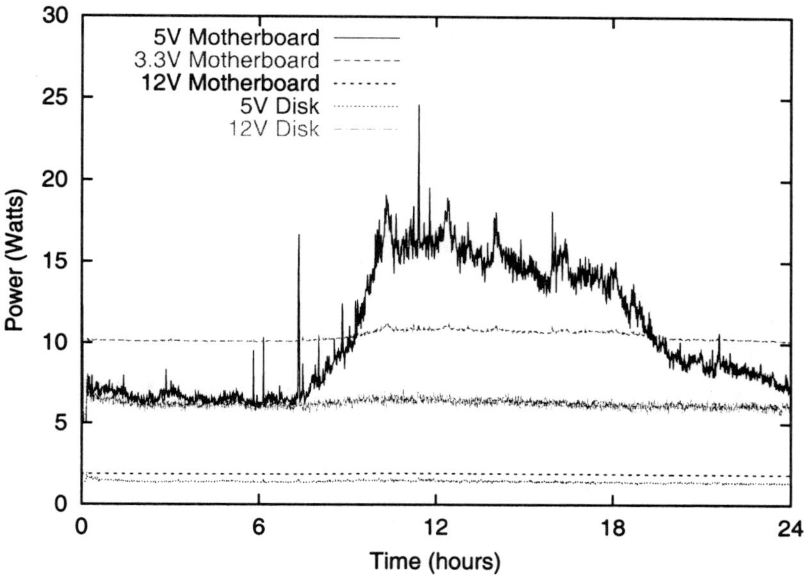

Figure 14.6. Power Consumption of the Finance Workload (at 20×)

Table 14.4. Total Energy Consumed (in Joules) of Each of the Five Power Supply Leads for Each of Our Three Workloads

	Olympics98	Finance	Proxy
5V Motherboard	1,232,710	711,415	627,977
3.3V Motherboard	914,116	882,117	869,015
12V Motherboard	164,795	161,997	160,732
5V Disk	127,007	119,610	157,679
12V Disk	569,395	522,065	724,130

sumption. Total CPU energy consumption is also significantly smaller than for the Olympics98 workload, with the 3.3V motherboard (memory) being the most dominant single component. Table 14.4 presents the total energy consumption for each of the workloads over the course of a 24 hour run.

4.2 Opportunities for Power Management

Our web server system is made up of several components that consume power. As we observe in the previous section, the CPU (5V Motherboard) is the dominant consumer of power when executing the Olympics98 and Finance

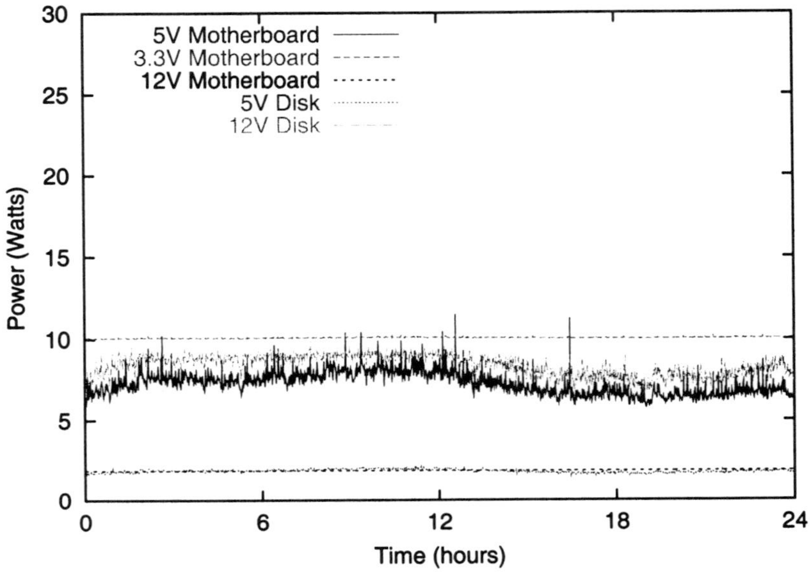

Figure 14.7. Power Consumption of the Proxy Workload (at 2×).

workloads. The CPU power consumption with these workloads also exhibits a large variation, presenting an excellent opportunity for power management. However, for the Proxy workload, it is the 12V disk and 3.3V motherboard (memory) that consume more power and show a large variation, raising the question of whether these components should be power managed.

The 3.3V motherboard supply powers the chipset, 512MB of memory, and associated components. To determine if it might be worthwhile to employ some form of power management for memory, we performed additional measurements on the 3.3V Motherboard supply. Two 256MB DIMMs make up the memory in our system. We began with two measurements: first, the power consumed (by the 3.3V Motherboard supply) when the system is idle, and next, the power consumed after removing one 256MB DIMM. The reduction in power was measured to be 0.9 Watts, implying that the 512MB of memory in our system consumes 1.8 Watts when idle. We then executed a microbenchmark on the 512MB system which saturated the memory with read and write requests. The difference between the power consumed under the memory test and the idle 512MB measurement was 2.1 watts. This difference represents an upper bound on additional power used by the 512MB of memory when active (that is, we have attributed all the additional power to the memory, even though some might actually be consumed by other components powered by the 3.3V motherboard supply). Thus, the total power consumed by our 512MB of memory (idle + active) is no more than 3.9 watts. Even if a large fraction of this power could be eliminated without increasing power use in other components, the savings would be small in comparison to the potential savings from CPU power management.

Comparing the power consumed by the 12V disk supply when the disk is spinning idly, with the power consumed when it is busy results in a similar conclusion. The power consumption of the 12V disk supply is 5.6 Watts when idle, rising to 9.9 Watts when serving a random stream of reads. As in the case of the memory subsystem, even if active power management were to reduce this power the maximum possible reduction would only be 4.3 Watts. Although power management based on inactivity may be possible, we expect its effectiveness to be markedly less than that in handheld systems because of the continuous nature of the web serving workload and the steady stream of disk writes (server logs). More complicated schemes that modify the disk's workload, perhaps by delaying or reordering activities, may be possible but are beyond the scope of this paper.

In summary, for web server systems that are built from industry standard, "white box" servers, the greatest opportunity for energy savings will be from CPU power management. Power management of the remaining components are unlikely to provide significant energy savings. We explore one approach for managing CPU power consumption in web servers in Section 6.

5. A Power Simulator for Web Server Workloads

Most web server systems today are not configured for power management, precluding direct measurement of any power management policies on a real system. We therefore constructed a simulator in order to evaluate the potential benefits of power management in this environment. Using this simulator and our three web server workloads, we evaluate dynamic voltage and frequency scaling, which has shown significant benefits in battery powered systems.

Our simulator is based on a queuing model for a server, and uses the CSIM execution engine [15]. In addition to simulating the residence time (and thus the response time) for each request, we also simulate the energy expenditure of the CPU during web serving. The input to the simulator is a stream of time–stamped requests in essentially the same format as used by the httperf tool. Each request is characterized by its arrival time, and its cost as measured by the size of the response. The simulator determines the CPU time and energy consumed to service a request from a model based on measurements of the actual energy consumed by our 600MHz web server system.

To construct the CPU time and energy model, we measured the energy consumed by the CPU when servicing requests with a variety of response sizes.

Table 14.5. Energy Consumed when Serving Requests of Different Sizes on the 600MHz System

Resp. size (bytes)	100	1000	5000	10,000	50,000	100,000	500,000
$E_{service}$ (Joules)	0.0171	0.0177	.0206	0.0294	0.112	0.277	0.761
CPU cycles (est.)	469K	485K	563K	803K	3.05M	7.57M	20.82M

Part (a). Responses Served from Memory.

Resp. size (bytes)	100	1000	5000	10,000	50,000	100,000	500,000
$E_{service}$ (Joules)	0.0484	0.0865	0.139	0.190	0.382	0.589	2.426
CPU cycles (est.)	1.32M	2.37M	3.81M	5.21M	10.45M	16.11M	66.38M

Part (b). Responses Served from Disk.

The calibration was accomplished by injecting a stream of requests with a fixed response size at a known rate, and measuring the energy consumed from the 5V Motherboard power supply over a fixed time interval. From this value, we subtract the CPU idle energy, which is the energy that would have been consumed by the CPU had it been idle for the duration of the interval. The result is the additional CPU energy required to serve requests over the interval. Dividing this by the number of requests served during the interval yields the energy consumed per request, $E_{service}$. We calibrated two cases: one where all the files were served from memory, and one where the files were served entirely from disk. In both cases, we measured the energy expended by the CPU (5V motherboard) while serving the requests. When serving files from disk, the CPU expends more energy because it needs to dispatch requests to the disk and service the ensuing interrupts. The $E_{service}$ values for a variety of response sizes are shown in Table 14.5.

Next, we use $E_{service}$ to calculate the number of CPU cycles it takes to serve a response. We measured the power consumption of the CPU to be $P_{min} = 4.97W$ when idle (i.e., halted) and $P_{max} = 26.9W$ when the CPU is fully busy with no halted cycles. When fully busy, we know the CPU executes F cycles (600 million cycles per second in our case) and consumes 26.9 Joules ($P_{max} \times 1$ second) when doing so. When halted for one second, we know the CPU executes 0 cycles and consumes 4.97 Joules ($P_{min} \times 1$ second). The CPU cycles corresponding to an energy usage of x joules can thus be computed using the following formula, which is used to compute the CPU cycle estimates in Table 14.5.

$$Cycles = x \times \frac{F}{P_{max} - P_{min}}$$

Our simulator combines the model for requests served from memory and the model for requests served from disk by simulating an LRU cache of files from the disk. We use a cache size of 472MB, which leaves 40MB to be used by the operating system and Apache webserver. Each file in the cache consumes an integral number of 4K byte pages. When processing a request, if the target file is in the RAM file cache, the simulator uses the energy model for requests served from memory. Otherwise, it uses the energy model for requests served from disk. The simulator is extremely fast, allowing us to simulate over 75,000 requests/second on a 866MHz system with a memory footprint of less than 10MBytes.

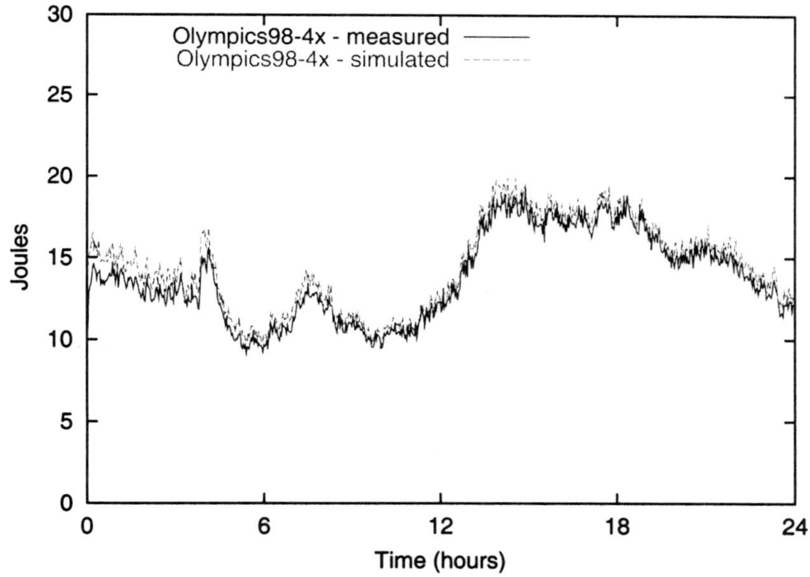

Figure 14.8. Measured vs Simulated Energy Consumption for Olympics98-4× Workload

Table 14.6. Comparison of Measured to Simulated CPU Energy for Three Workloads (Correlation coefficients were computed based on the energy used in 30 second intervals over the length of the run.)

Workload	Olympics98-4×	Finance-20×	Proxy-2×
Measured CPU Energy (Joules)	1,232,710	925,540	627,977
Simulator CPU Energy (Joules)	1,253,652	946,440	663,648
Error in Total Energy	1.70%	2.26%	5.68%
Correlation Coefficient	0.9846	0.9716	0.8485

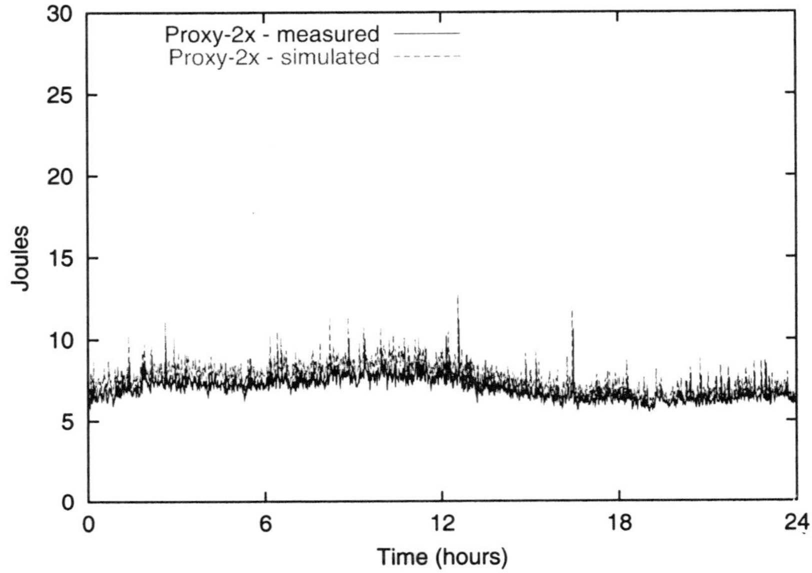

Figure 14.9. Measured vs Simulated Energy Consumption for Proxy-2× Workload

After calibration, we simulated the CPU energy consumption of the Olympics98 workload at a scalefactor of 4×. Figure 14.8 shows the measured CPU energy consumed by the 5V motherboard in the 600MHz system during the execution, overlaid with the simulator output. The simulator over–predicts energy consumption by 1.7%. Furthermore, the correlation coefficient between the measured and simulated power is 0.9846. We also simulated the Finance workload at a scalefactor of 20× and the Proxy workload at a scalefactor of 2×. Simulated and measured data for the Proxy workload is graphically shown in Figure 14.9. All the simulation results are summarized in Table 14.6. In the interests of brevity we have omitted the graphs of the results for the Finance workload.

6. Dynamic Voltage and Frequency Scaling

One way to reduce the power consumed by the CPU is to lower its operating voltage. A reduction in operating voltage generally also requires a proportional reduction in frequency [22]. This approach of varying the processor voltage in proportion with its frequency is known as voltage scaling. Voltage scaling is advantageous because the energy consumed by a processor is directly proportional to V^2, where V is the operating voltage. By varying the processor voltage and frequency, it is possible to obtain a quadratic reduction in power consumption.

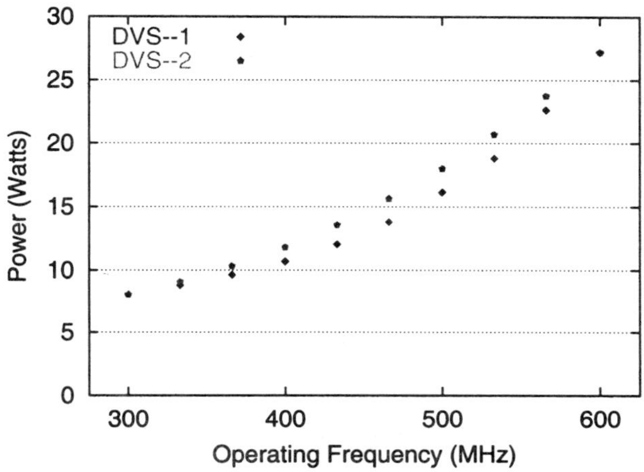

Figure 14.10. Power Consumption of Two Simulated Voltage-Scaled Pentium 600MHz III Processors

Table 14.7. Parameters of the Hypothetical Voltage-Scaled Pentium III Processor Used in the Simulation

Time quantum	20 ms
Frequency Range	300MHz - 600MHz, in steps of 33MHz
Core Operating Voltage Range	1.5V to 2V, linearly with frequency
Busy threshold	Min = 0.80, Max = 0.95

We modified our simulator to support two additional models for CPU power consumption based on data from actual processors that support dynamic voltage scaling. One model is based on publicly available data for the Transmeta TM5400 processor [9], and the other is based on data for an alternate processor [21]. We scaled the data from both processors to fit the 600MHz Pentium III's maximum frequency and core operating voltage. Our system parameters are shown in Table 14.7. Figure 14.10 shows the power consumption of each simulated processor at the different operating points. We assume that the time and energy required to change frequency and voltage is negligible. In the rest of this paper, we refer to the Transmeta based model as DVS–1 (for Dynamic Voltage Scaling 1), and the other as DVS–2.

The simulator uses a very simple policy based on the recent CPU utilization to determine the processor frequency and voltage. At the beginning of each time *quantum*, we examine the system load during the previous time quantum. If the system utilization is between the threshold values, we do nothing. If the high threshold is exceeded, we step up the frequency of the processor by 33MHz and the voltage appropriately. If the low threshold is exceeded, we step down the frequency and voltage. This type of dynamic voltage scaling algorithm was originally proposed and studied for desktop application workloads in [27]. The minimum and maximum attainable frequencies are 300MHz and 600MHz respectively.

We assume that the number of processor cycles needed to service a request is independent of the processor frequency and is always equal to the number of 600Mhz cycles needed to service the request. This is a conservative assumption since it does not consider energy savings due to a reduction in the CPU stalls that could occur as a result of other system components (particularly memory) appearing relatively faster to the processor as it is slowed down. The CPU time required to service a request is then calculated as the number of cycles to be executed multiplied by the current operating frequency of the processor. The model properly accounts for requests whose service time is spread across several time quantums with different operating frequencies.

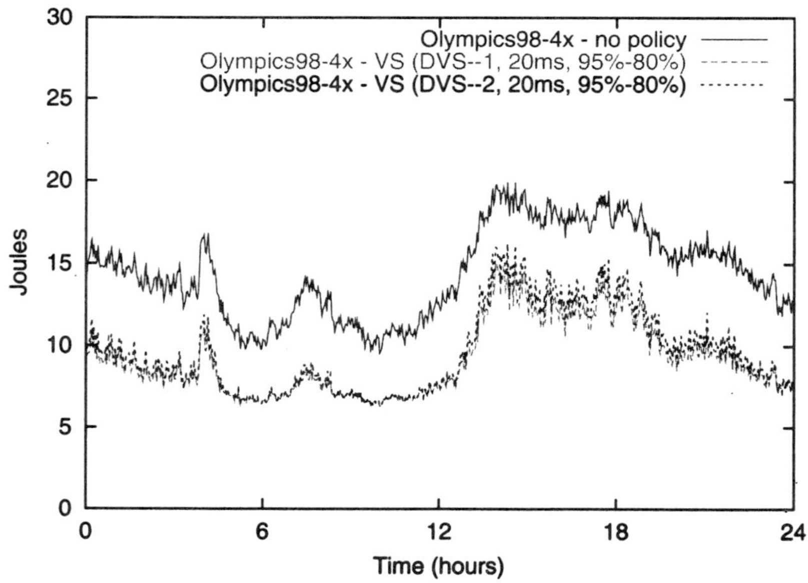

Figure 14.11. Energy Savings from Voltage and Frequency Scaling for the Olympics98 Workload

Figure 14.11 shows the results of our simulation with the Olympics98 web server workload. The top curve shows the simulated 5V energy consumption of this workload on the 600MHz system without voltage scaling. The two lower curves show the simulated 5V energy consumption of the DVS–1 and DVS–2 voltage scaled Pentium III processors.

Table 14.8. Energy Savings from Dynamic Voltage Scaling

Workload		Olympics98-4×	Finance-20×	Proxy-2×
Base energy (J)		1,253,652	946,440	663,348
DVS–1 design	Energy (J)	798,684	636,681	510,788
	Savings	36.3%	32.7%	23.0%
DVS–2 design	Energy (J)	838,436	655,426	512,710
	Savings	33.1%	30.7%	22.7%

Table 14.8 lists the energy savings we obtained for the three workloads. There is a noticeable difference in the energy savings between the two processor designs. This is because the power vs. frequency curve for the DVS–1 design is more convex than that for the DVS–2 design. Voltage scaling provides the most energy savings for Olympics98-4× and Finance-20× because those workloads exercise the CPU more than the Proxy workload.

Over the course of the execution, the average operating frequency of the voltage scaled processor correlates to the incoming request rate, which is a good indicator of the system load for these workloads. In the interests of brevity, we have omitted the graph depicting this behavior. Instead, Figure 14.12 shows the frequency of the DVS–1 voltage scaled processor over the course of 60 seconds starting at 8AM as it executes the Olympics98 workload. The frequency varies all the way between 300MHz to 600MHz, demonstrating the processor changing its frequency as it adapts to the bursty workload. The DVS–2 processor exhibits similar behavior.

7. Implications for Web Server Performance Metrics

The benefits of reduced energy consumption must be balanced against the performance impact of the power management mechanisms employed. To allow this trade off to be analyzed in a systematic manner, we need to take a fresh look at the performance metrics that are in use to characterize traditional servers. These metrics focus on absolute performance as the only relevant differentiator, and they ignore issues such as power consumption and cooling requirements. A more relevant evaluation metric for the emerging new applications would be performance relative to power consumption. To illustrate, we have conducted a comparison between two machines running the Apache

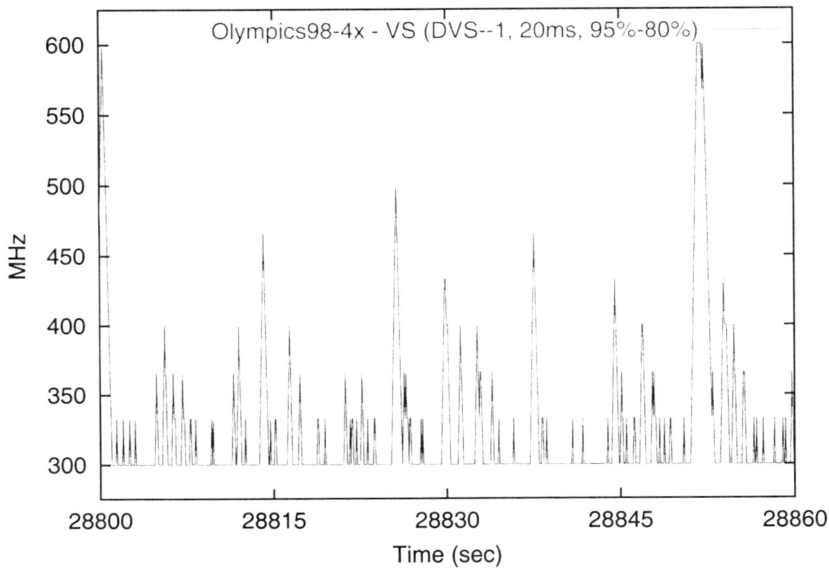

Figure 14.12. Operating Frequency of the Simulated Voltage Scaled Processor as it Executes 60 Seconds of the Olympics98 Workload. Each Data Point Represents the Processor Frequency during a 20ms Interval

Web server with the Linux operating system. The first machine is a traditional server, using a 750MHz Intel Pentium^{TM}III processor with a 16 Kbyte L1 I-cache, a 16 Kbyte L1 D-cache, a 1 Mbyte unified L2 cache, and 256MB of main memory. The second machine is power-managed server, which uses a Transmeta Crusoe^{TM}TM3200 processor with a 24 Kbyte L1 cache and 256MB of memory. The TM32000 processor is rated at 400MHz, but because of the code-morphing technology it uses, it is difficult to make comparisons based on operating frequency. The remaining components of the two systems (e.g., disk, network) are identical or comparable. (The machines used in these measurements differ from those used in the previous sections because this work was done separately, using other machines that were available to us. Our goal in this section is simply to motivate the need for new metrics of web server performance, and not to draw comparisons with the results of previous sections.) Figure 14.13 shows the results of the comparison.

Looking at absolute performance, the traditional server outperforms the power-managed server by a ratio that is roughly proportional to their relative speeds. However, when one considers the performance per unit of power consumption, the power-managed processor gives 3.6 connections/sec/Watt,

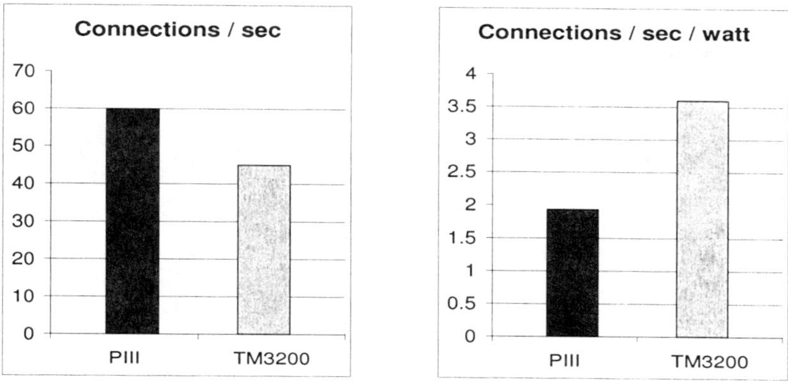

Figure 14.13. Comparison of Two Servers using a Traditional Web Server Performance Metric and Our Proposed Energy/Performance Metric

substantially better than the 2.0 connections/sec/Watt of the traditional server. Since most web applications are inherently highly parallel, and web sites are commonly implemented using clusters of web servers, simply switching to a power-managed system as the building block for these clusters could result in substantial energy savings. This reduced power consumption will also help realize further benefits in reduced cooling, and could allow more efficient packaging of server components, thus reducing the required volume of raised floor space.

8. Related Work

Both the distribution of power consumption and methods for managing it have been studied extensively in the area of portable, battery-powered computers such as laptops and personal digital assistants (PDAs). For example, one power consumption study is a detailed analysis of the energy consumed by the various components of Apple Macintosh laptop computers [19]. While similar in approach to our study, their work focuses on portable computers and workloads typical for such machines, whereas our study focuses on web servers and their typical workloads. This distinction is significant for two reasons. First, the basic set of components is different between these two platforms, and components that are present in both are frequently designed to very different specifications. Second, the workloads and the expectations about their behavior are radically different.

There have been a number of studies of specific power management mechanisms and policies, and a set of standards have been developed for the mechanisms, specifying the interfaces between power-management software and the hardware. Examples of such architectures include the industry standard Advanced Configuration and Power Interface or ACPI [6] and Microsoft's On-Now initiative [2]. Many of these mechanisms could be directly applied to a server system in a web-serving environment although there is no guarantee that the management policies designed for using them in portable, battery powered systems are equally applicable.

Among the common power management techniques is spinning down a hard disk after some period of inactivity [8]. Microprocessors have also received a considerable amount of attention, and many microprocessor architectures and microarchitectures incorporate power-saving features: examples include the mobile processors available from Intel with its SpeedStep(TM) technology and the Transmeta Crusoe processor with LongRun [9]. More recently developed and less widely deployed today are new memory chip architectures that are incorporating similar "spin down" states so that the system can actively manage the power used by main memory [3]. In addition, a number of current research efforts are focusing on new power management mechanisms employed at the operating system [25] and application layers [10] of the system. Techniques for dynamically controlling processor temperature [23] can also be applied to web servers. This results in power savings because CPU activity is decreased to lower processor temperature. Our work complements these studies by demonstrating the value of these performance management mechanisms in a web-serving environment. Furthermore, by measuring the power usage of the specific components of the machine, our work gives a valuable insight into which techniques are likely to provide the greatest benefits.

Our approach of reducing the power consumed by the server by reducing the clock frequency of the processor has been proposed and studied in a variety of contexts [27], but to our knowledge our study is the first to show the benefits of this technique in a web server environment. A variety of approaches have been proposed for determining when to change clock frequency and how to select a new frequency, and a good comparison of these policies is provided in [13]. The policy we use is based on the same basic principle as Transmeta(TM)'s LongRun(TM) power management technique [9].

Studies of web server performance commonly use synthetically generated workloads and workloads based on the web logs of actual servers. Our use of web logs to generate a web server workload differs from many prior studies in that we attempt to recreate the timing of requests exactly as they occurred on the original server. Typically, requests from the log are issued as quickly as possible to determine the maximum throughput of the server. However, there are exceptions to this practice such as the DBench tool [17], which was used in the measurements made of the Harvard Array of Clustered Computers (HACC) [7] locality-based request routing system.

Our proposed SPECWeb/Watt metric was inspired by several other efforts at extended performance metrics to include some measure of energy efficiency. Most notably, [12] considered a number of CPU-centric measures such as SPECint/W and an energy-delay metric, $SPEC^2/W$. For web servers, the SPECweb99 benchmark already includes a notion of delay in that it specifies a maximum response time for all requests in order for the connection to be considered "conforming", and thus our SPECWeb/Watt metric captures performance, energy, and delay.

Several researchers have developed tools for simulating the power consumption of computer systems. Brooks et. al. have developed Wattch, a microprocessor power analysis tool based on a microarchitecture simulator [1]. Flinn and Satyanarayanan describe PowerScope, a tool for profiling the energy usage of applications [11]. The power simulator we have developed is substantially faster, because it is targeted specifically for web serving workloads.

9. Conclusions

Over the last several years, the dominant focus of power management research has been on portable and handheld devices. Here, we have presented the case for power management in web serving systems. Typical web servers are designed to operate at a fraction of their peak operating capacity. Such over-engineering is necessary in order to handle the variable and inherently bursty nature of web server workloads. This creates opportunities for energy conservation in environments such as modern Internet data centers where computers are densely packed, and where cooling and electricity delivery are severely constrained.

Using workloads derived from the logs of three production web sites, our direct power measurements show that the CPU consumes the largest fraction of the system power. We use additional measurements to calibrate and validate a power simulator that we created. Our simulator predicts CPU energy consumption for the three workloads with errors ranging from 1.7% to 5.7%. The simulated energy also correlates highly with the measured energy.

We have used the validated simulator to measure the projected effectiveness of a power management policy, typically used only in handheld devices: dynamic voltage and frequency scaling. We find that this technique is effective for saving energy, reducing the CPU energy consumption by up to 36%. Not surprisingly, the energy savings are higher for workloads that exercise the CPU more.

We predict that technology trends will intensify the emerging need for energy-efficient servers, and we suggest that a fundamental change is necessary in the way we design and configure web servers today. Our results provide evidence that in addition to devising server–centric power management strategies, features commonly found in processors intended for mobile computing should be adapted and incorporated as standard features for server processors. The results also suggest that the performance-centric view of designing servers today must give way to a more balanced view in which energy consumption is as important as other goals of the system.

References

[1] David Brooks, Vivek Tiwari, and Margaret Martonosi. Wattch: a framework for architectural-level power analysis and optimizations. In *27th International Symposium on Computer Architecture*, pages 83–94, 2000.

[2] Microsoft Corp. *PC99 System Design Guide*. Microsoft Press, 1999.

[3] Rambus Corporation. Rambus Technology Overview, Feb 1999.

[4] M. Crovella and A. Bestavros. Self-similiarity in World Wide Web Traffic: Evidence and Possible Causes. In *1996 SIGMETRICS Conference on Measurment and Modeling of Computer Systems*, May 1996.

[5] A. Iyengar et. al. High-Performance Web Site Design Techniques. *IEEE Internet Computing*, March/April 2000.

[6] Compaq et al. ACPI Specification, version 2.0, 2000.

[7] X. Zhang et. al. HACC: An Architecture for Cluster-Based Web Servers. In *3rd USENIX Windows NT Symposium*, July 1999.

[8] P. Krishnan F. Douglis and B. Bershad. Adaptive Spin-down Policies for Mobile Computers. In *2nd USENIX Symposium on Mobile and Location-Independent Computing*, April 1995.

[9] M. Fleischmann. Crusoe Power Management: Cutting x86 Operating Power Through LongRun. Embedded Processor Forum, June 2000.

[10] J. Flinn and M. Satyanarayanan. Energy-aware adaptation for mobile applications. In *17th ACM Symposium on Operating Systems Principles (SOSP'99)*, 1999.

[11] J. Flinn and M. Satyanarayanan. PowerScope: A tool for profiling the energy usage of mobile applications. In *Workshop on Mobile Computing Systems and Applications (WMCSA)*, pages 2–10, 1999.

[12] R. Gonzalez and M. Horowitz. Energy dissipation in general purpose microprocessors. *IEEE Journal of Solid-State Circuits*, September 1996.

[13] K. Govil, E. Chan, and H. Wasserman. Comparing Algorithm for Dynamic Speed-Setting of a Low-Power CPU. In *Mobile Computing and Networking*, 1995.

[14] Akamba Inc. Velobahn product description., 2000.

[15] Mesquite Software Inc. *CSIM18 Simulation Engine*, 1994.

[16] The IRCache Project. http://www.ircache.net/. This project is supported by the National Science Foundation (grants NCR-9616602 and NCR-9521745), and the National Laboratory for Applied Network Research.

[17] A. Wharton J. B. Chen and M. Day. Benchmarking the Next Generation of Internet Servers. Can be obtained by a full text search on 'DBench' on the archives of Iris Today at http://www.notes.net/today.nsf., 1997.

[18] J. R. Lorch and A. J. Smith. Software Strategies for Portable Computer Energy Management. *IEEE Personal Communications Magazine*, June 1998.

[19] J. R. Lorch and A. Jay Smith. Energy Consumption of Apple Macintosh Computers. *IEEE Micro*, 18(6), November/December 1998.

[20] David Mosberger and Tai Jin. httperf: A Tool for Measuring Web Server Performance. In *SIGMETRICS First Workshop on Internet Server Performance*, pages 59—67. ACM, June 1998.

[21] Kevin Nowka. Private communication.

[22] T. Pering, T. Burd, and R. Brodersen. Dynamic Voltage Scaling and the Design of a Low-Power Microprocessor System. In *Power Driven Microarchitecture Workshop, attached to ISCA98*, June 1998.

[23] E. Rohou and M. D. Smith. Dynamically Managing Processor Temperature and Power. In *2nd Workshop on Feedback-Directed Optimization*, Nov 1999.

[24] Deo Singh and Vivek Tiwari. Power Challenges in the Internet World. In *Cool Chips Tutorial, held in conjunction with the 32nd Annual International Symposium on Microarchitecture*, November 1999.

[25] A. Vahdat, A. Lebeck, and C. Ellis. Every Joule is Precious: The Case for Revisiting Operating System Design for Energy Efficiency. In *9th ACM SIGOPS European Workshop*, September 2000.

[26] The World Wide Web Consortium (W3C). RFC 2068: Hypertext Transfer Protocol – HTTP/1.1, January 1997.

[27] M. Weiser, B. Welch, A. Demers, and S. Shenker. Scheduling for reduced CPU energy. In *First Symposium on Operating Systems Design and Implementation*, pages 13–23, Monterey, California, U.S., 1994.

VI

MEASUREMENTS AND EVALUATION

Chapter 15

ET^2: A METRIC FOR TIME AND ENERGY EFFICIENCY OF COMPUTATION

Alain J. Martin, Mika Nyström, Paul I. Pénzes
Department of Computer Science
California Institute of Technology
{alain, mika, penzes}@cs.caltech.edu

Abstract We investigate an efficiency metric for VLSI computation that includes energy, E, and time, t, in the form Et^2. We apply the metric to CMOS circuits operating outside velocity saturation when energy and delay can be exchanged by adjusting the supply voltage; we prove that under these assumptions, optimal Et^2 implies optimal energy and delay. We give experimental and simulation evidences of the range and limits of the assumptions. We derive several results about sequential, parallel, and pipelined computations optimized for Et^2, including a result about the optimal length of a pipeline.

We discuss transistor sizing for optimal Et^2 and show that, for fixed, nonzero execution rates, the optimum is achieved when the sum of the transistor-gate capacitances is twice the sum of the parasitic capacitances—not for minimum transistor sizes. We derive an approximation for Et^n (for arbitrary n) of an optimally sized system that can be computed without actually sizing the transistors; we show that this approximation is accurate. We prove that when multiple, adjustable supply voltages are allowed, the optimal Et^2 for the sequential composition of components is achieved when the supply voltages are adjusted so that the components consume equal power. Finally, we give rules for computing the Et^2 of the sequential and parallel compositions of systems, when the Et^2 of the components are known.

1. Introduction

With energy becoming as important as time as a criterion for computational efficiency, analytical tools are needed to evaluate computations according to both criteria simultaneously. How are energy and time traded against each other in the design process? How are algorithms compared when they have different energy and time figures? A useful metric must separate the algorithmic tradeoffs of energy against time from the physical (usually electrical) tradeoffs.

We propose an efficiency metric for VLSI computation that combines energy, E, and time, t, in the form Et^2. The choice of this metric is based on CMOS VLSI technology: in CMOS, Et^2 is independent of the voltage in first approximation. Instead of attempting to optimize a circuit for both E and t, the designer can now optimize the design for the single metric and adjust the voltage to obtain the chosen tradeoff between E and t.

We prove that the Et^2 metric is optimal for CMOS circuits under assumptions that hold approximately in the normal range of operation. Under those assumptions, energy and delay can be freely exchanged through supply-voltage adjustment. Although the metric is not adequate over the entire range of operation for CMOS transistors, we have experimental evidence that a large class of circuits exhibit a collective behavior that is more regular than that of individual transistors. We also investigate when and why the Et^2 metric is inadequate; in some cases, we can use the metric Et^n with $n \neq 2$. We shall see that most results we prove for Et^2 generalize for Et^n with $n \neq 2$.

The objection that a metric grounded in a specific technology (CMOS) cannot be general enough for the study of algorithms can be answered by observing that the CMOS model of computation is certainly as general as the "random-access machine" model, which has been used successfully in the traditional analysis of algorithms.

The Et^2 metric was originally introduced for the design of an asynchronous MIPS R3000 microprocessor [4]. The arguments about the validity of the metric and the analysis of the pipeline were published by Martin [3]. The results about transistor sizing for minimal Et^n have been described previously by the authors [7, 11].

2. Energy and Delay in VLSI Computations

Our study of energy and delay in computations is based on CMOS implementations. We consider a digital VLSI computation to be a partially ordered sequence of transitions. Each transition changes the value of a boolean variable of the computation. We consider irreversible computations only, i.e., computations in which assigning a value to a variable destroys the previous value of the variable.

In digital CMOS, a boolean variable is implemented as an electrical node whose voltage represents the present value of the variable. Each transition charges or discharges the capacitor attached to the node, bringing its voltage either to the supply voltage *Vdd* or to the ground voltage *GND*. We are interested in the *computational*, or *dynamic*, energy spent in charging and discharging the capacitances of all nodes involved in a computation. We ignore leakage energy and short-circuit energy. Since we ignore leakage, we also assume that no energy is spent maintaining the values of variables stored in registers.

Any algorithm can be implemented as a set of *production rules* [5]. A production rule is of the form: $B \to t$, where B is a boolean expression, and t is a single assignment of the value **true** or **false** to a boolean variable. Such an assignment is called a *transition;* in this example, the transition t is performed after B becomes **true**. A *logic gate* or *operator* is the physical implementation of the pair of production rules that set and reset a given variable, of the following form:

$$\begin{array}{lcl} Bu & \to & z\uparrow \\ Bd & \to & z\downarrow \end{array} \tag{15.1}$$

Let $E_{z\uparrow}$ and $E_{z\downarrow}$ be the energy spent firing the first and second production rules, respectively. (A firing is an execution of a production rule that changes the value of a variable.) The energy spent firing production rule $Bu \to z\uparrow$ is the energy dissipated charging the capacitor C_z associated with the node of z. The total energy required for charging the capacitor up to voltage V is C_zV^2; half of this energy is stored in the capacitor, and half is dissipated as heat in the pull-up network connecting the capacitor to the power supply. We have that

$$E_{z\uparrow} = \frac{C_zV^2}{2}, \tag{15.2}$$

where V is the power-supply voltage. When the capacitor is discharged to ground, the energy stored in the capacitor is dissipated in the pull-down network. Hence, $E_{z\downarrow} = E_{z\uparrow}$.

We shall not elaborate on the calculation of the capacitance C_z beyond noting that C_z depends mostly on the "load" of z, i.e., the topology of the logical gates of which z is an input, and hardly depends on the structure of the logical gate of which z is an output. In other words, the energy consumed in computing the value of z does not depend on what is computed but rather on where the result of the computation is needed.

The delay $t_{z\uparrow}$ for firing $z\uparrow$ is the ratio of the final electrical charge Q_z on C_z to the current $i_{z\uparrow}$ available for charging C_z:

$$t_{z\uparrow} = \frac{Q_z}{i_{z\uparrow}}, \tag{15.3}$$

with $Q_z = C_zV$. The current $i_{z\uparrow}$ is the current flowing in the transistor network connecting the constant power-supply to z when and only when Bu holds; similarly for the delay $t_{z\downarrow}$ and current $i_{z\downarrow}$.

In general, the transistor current is difficult to analyze. Let us look first at one single nMOS-transistor as pull-down network. (The analysis for a pMOS transistor as pull-up network is similar.) We assume that the transistor is above threshold ($V_{gs} > V_t$), and not in velocity saturation. Then, the current is either the saturation current, I_s, when $V_{ds} \geq V_{gs} - V_t$; or it is the linear current, I_l, when $V_{ds} < V_{gs} - V_t$, where V_{gs} and V_{ds} are the gate-to-source and drain-to-source voltages of the transistor, respectively, and V_t is the threshold voltage.

The formulas for I_s and I_l are well-known:

$$I_l = k\left(2(V_{gs} - V_t)V_{ds} - V_{ds}^2\right) \tag{15.4}$$

$$I_s = k(V_{gs} - V_t)^2 . \tag{15.5}$$

If we assume that the voltages V_{gs}, V_{ds}, V_t vary proportionally to the supply voltage V, then both I_l and I_s depend quadratically[1] on V, and therefore the current $i_{z\downarrow}$ discharging Q_z is of the form $i_{z\downarrow} = K_{z\downarrow}V^2$. Similarly, for a pull-up network, we have $i_{z\uparrow} = K_{z\uparrow}V^2$. Hence, we have for the delay $t_{z\uparrow}$ that

$$t_{z\uparrow} = \frac{C_z}{K_{z\uparrow}V} . \tag{15.6}$$

Combining the expressions for delay and for energy, we see that the expression $E_z t_z^2$ is independent of V.

Under certain restrictions, it is possible to extend the result that the current is quadratic in V to cover the arbitrary composition of pullup and pulldown networks. Papadantonakis has proved this result for a class of circuits called "smooth circuits" [8]. A smooth circuit is a network of transistors in which each node has a capacitance to ground, the power supplies are modeled as large capacitors, and again the threshold voltage is assumed to scale with the supply voltage.

If we assume that a CMOS circuit is a reasonable approximation of a smooth circuit, we can assume that the quadratic relation between currents and supply voltage V holds, and therefore that the delays are inversely proportional to V. For those circuits, Et^2 is independent of V, where E is the dynamic energy dissipated by a computation, and t represents either the latency or the cycle time of the computation. We shall return to the limitations of this assumption.

3. Comparing Algorithms for Energy and Delay

Given two algorithms A and B, with energy and delay (E_A, t_A) and (E_B, t_B), how do we compare them for energy and delay? In evaluating the time ef-

[1] Of course, V_{gs} and V_{ds} must vary by quite a different mechanism from the one governing V_t: V_{gs} and V_{ds} can vary "automatically" as a result of changing *Vdd*, whereas V_t must be set at the time of fabrication. The main reason that the proportional variation breaks down is that it is in practice impossible to scale V_t with *Vdd* because that would lead to unacceptably large leakage currents at the low end of the scale.

ficiency of a computation, we may be interested in either one of two delay parameters: the latency and the cycle time. For an algorithm computing the function F, the latency is the delay between the input of parameter x_i and the output of $F(x_i)$, averaged over all values of i. The cycle time is the delay between the input of parameter x_i and the input of the next parameter x_{i+1}, again averaged over all values of i.

3.1 Why Et Is Not the Right Metric

The energy-delay product Et is often used for comparing designs, but it is not usually an acceptable metric, as we shall presently demonstrate.

Let us assume that we have two circuits, A and B, that compute the same thing in two different ways. Assume $E_A = 2E_B$ and $t_A = \frac{t_B}{2}$. Then, according to the Et metric, A and B are equally good. But let us reduce the supply voltage of A by half. Let (E'_A, t'_A) be the new values of energy and delay for A. Given the dependence of energy and delay on voltage, Equations 15.2 and 15.6, we have that

$$E'_A = \frac{E_A}{4}, \tag{15.7}$$

$$t'_A = 2t_A, \tag{15.8}$$

which gives

$$E'_A = \frac{E_B}{2}, \tag{15.9}$$

$$t'_A = t_B. \tag{15.10}$$

Hence, A now has the same delay as B but at only half the energy, and therefore A is a better implementation than B, contrary to what the Et metric indicates.

These results are borne out in practice [10]. In Table 15.1, we see the results of simulating two different implementations of an eight-bit comparator with the simulator HSPICE. In each case, eight single-bit comparators perform the comparison: in the "linear" comparator, the results of the single-bit comparators are merged in a linear chain; in the "log" comparator, in a binary tree. Comparing the performance of the comparators at 3.3-V *Vdd,* we see that the linear comparator is slower than the log comparator, but using the Et metric, we find that it more than makes up for its sluggishness with its lower energy consumption. On the other hand, using the Et^2 metric, we find that the log comparator is better. Which is it?

If we adjust the supply voltage on the log comparator down to 2.15 V, we see that we can match the delay of the linear comparator while using less en-

Table 15.1. Comparison of E, t, Et, and Et^2 of Two Kinds of 8-bit Comparators. Simulations with HSPICE using Parameters for HP's 0.6-μm CMOS Process (via MOSIS)

8-bit comparator	E $[J \cdot 10^{-11}]$	t $[s \cdot 10^{-9}]$	Et $[Js \cdot 10^{-20}]$	Et^2 $[Js^2 \cdot 10^{-29}]$
Linear (3.3 V)	25.24	3.93	99.21	389.97
Log (3.3 V)	44.97	2.35	105.50	247.57
Log (2.15 V)	16.52	3.93	64.97	255.59

ergy; thus, the log comparator outperforms the linear one in both speed and energy if we are allowed to adjust the supply voltage. Even over this relatively wide range of supply voltages, Et^2 changes only by an insignificant 3.2 percent. This example illustrates that the Et^2 metric is more trustworthy for circuit comparisons when we are allowed to adjust the supply voltage.

3.2 The Θ Metric

Let us now ignore the lower and upper bounds imposed on the voltage by the technology and assume that we can always trade E and t against each other through voltage adjustment. Suppose that under these conditions there exists a function $\Theta(E,t)$ with the properties:

Property Θ1. Θ is monotonically increasing in E and t,

Property Θ2. Θ is independent of V.

Theorem 1 *Given two computations A and B with corresponding Θ_A and Θ_B:*

- *If $\Theta_A < \Theta_B$ then A is more delay-efficient than B when A and B use equal energy and A is more energy-efficient than B when A and B have the same delay.*
- *If $\Theta_A = \Theta_B$ then A is equivalent to B with respect to energy when their delays are the same.*

Proof: Through supply-voltage adjustment, we can equalize either the energy or the delay of the two computations. Let us arbitrarily choose to equalize the delays: $t_A = t_B$. Because of Property Θ2, the Θs have not changed.

We can now compare the two computations, thanks to Property Θ1:

- $(\Theta_A < \Theta_B) \Rightarrow (E_A < E_B)$, i.e., A is better than B,
- $(\Theta_A = \Theta_B) \Rightarrow (E_A = E_B)$, i.e., A and B are equally good.

Hence, for any chosen delay t, A is more energy-efficient than B. Likewise, for any chosen energy E, A is more time-efficient than B. □

3.3 The Et^2 Metric

Any expression in E and t that is monotonically increasing in E and t and that is independent of V can be used as complexity metric Θ. We have shown in Section 2 that, in CMOS technology, the following definition for Θ is valid:

$$\Theta \stackrel{\text{def}}{=} Et^2 \, . \tag{15.11}$$

Henceforth, Θ will always mean Et^2 . If we now return to the example at the beginning of Section 3.1, we compare the two computations A and B by comparing their Θs:

$$\Theta_A = E_A t_A^2 \tag{15.12}$$

$$\Theta_A = (2E_B)\left(\frac{t_B}{2}\right)^2 \tag{15.13}$$

$$\Theta_A = \frac{\Theta_B}{2} \, . \tag{15.14}$$

Hence, we can conclude that A is twice as Θ-efficient as B. For equal delays, $E_A = \frac{E_B}{2}$. For equal energies, $t_A = \frac{t_B}{\sqrt{2}}$.

3.4 Et^2 Measurements

How constant is Et^2 in reality? There are several operating modes for the CMOS transistor, each with a very different relation between current and voltage. In particular, at high electric field, the carrier velocity saturates and becomes constant; the delay becomes independent of the voltage, and Et^2 becomes quadratic in the voltage. Figure 15.1 shows the measured Et^2 for the two-million–transistor asynchronous MIPS R3000 microprocessor designed at Caltech between 1996 and 1998. It was fabricated in 0.6-μm CMOS and was entirely functional on first silicon [4]. (Measurements on other fabricated chips give similar results.)

The behavior below 1.3 V shows the effect of approaching the threshold voltage; in our calculations we have assumed that the threshold voltage would scale with *Vdd*, but we obviously cannot enforce this for HP's 0.6-μm process whose threshold voltage is fixed at 0.8 V. The positive slope from 3 V and up shows the onset of velocity saturation. The nominal voltage of this process is 3.3 V; the graph shows that Et^2 varies only about 20% around its average in the range 1.5–4.9 volts *Vdd*.

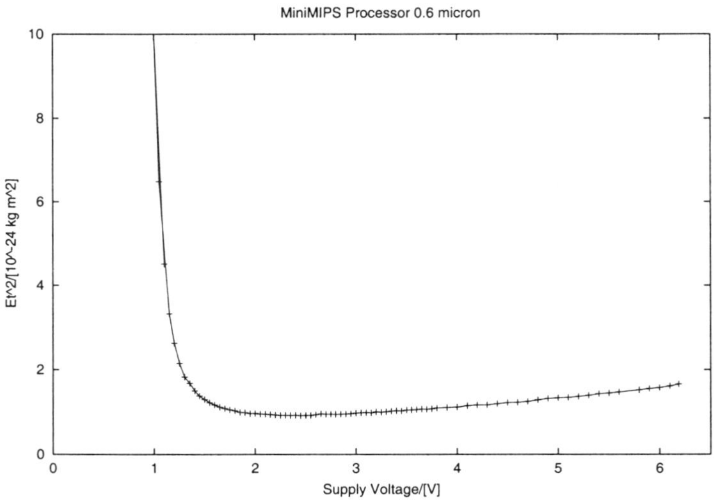

Figure 15.1. Measured Et^2 for a Two-Million-Transistor Asynchronous Microprocessor

4. The Θ-Efficiency of Designs

In this section, we use the Θ metric to determine when two standard design transformations—parallel composition and pipelining—improve the efficiency of a design compared with sequential execution.

4.1 The Θ-Efficiency of Parallelism

Given a collection of independent tasks, when does the parallel execution of the tasks improve (i.e., reduce) the Θ of the computation? For simplicity, consider m identical tasks each consuming energy E/m and using delay t/m to complete.

If the m tasks are executed sequentially, the total energy is E and the total delay (execution time) is t, giving a Θ_0 of Et^2.

Now, let us consider what happens when the m tasks run in parallel. First, we ignore the cost of the *split* circuitry that may be needed to distribute the control (and possibly data) to all tasks, and the cost of the *merge* circuitry that may be needed to gather the completion signal (and possibly some results) from all tasks. In that simple case, the total energy is still E, but the total delay is now reduced to t/m, giving a Θ_{par} of $(Et^2)/m^2$. Hence, the improvement $\Theta_{par}/\Theta_0 = 1/m^2$: parallelism reduces the Θ of the computation by a factor m^2.

Assuming we can vary the voltage of the new design so as to make the delay equal to the delay of the sequential design, then under the invariance of Et^2, the parallel transformation decreases the energy consumption by a factor m^2. (For large m, it may in practice be impossible to scale down the voltage by a factor m, and therefore it may be impossible to exploit all the potential energy improvement of parallelism.)

Theorem 2 *The parallel composition of m identical tasks without overhead gives an Et^2 reduction of m^2 compared with sequential execution, and a potential energy reduction of m^2 if the voltage can be reduced by a factor m.*

The situation is more complicated if the tasks are different. Let us assume that each task A_i now uses energy E_i and delay t_i, such that we still have $\sum E_i = E$ and $\sum t_i = t$. The parallel composition still uses energy E, but the delay is now $\max(t_i)$; let us call the task with maximal delay A_{max} and its delay $t_{max} = \max(t_i)$. Remember that we are assuming that the voltage of each task can be adjusted freely. Then it is clear that, for each task except A_{max}, the early termination of the task amounts to a waste of energy since every task but A_{max} can be slowed down to t_{max} without affecting the delay of the whole computation, but with an energy improvement corresponding to the voltage decrease.

According to the above analysis, parallelism always improves Et^2 if we ignore any overhead it introduces. Let us now examine the case when the cost of the *split* and *merge* circuitries cannot be ignored. We are looking at the simple case where we split the original task into two parallel tasks with the help of just one binary *split* and one binary *merge*. We assume that both the split and merge have an energy and delay that are a fraction k of the energy and delay of the original task:

$$E_{split} = E_{merge} = kE \ , \tag{15.15}$$

$$t_{split} = t_{merge} = kt \ . \tag{15.16}$$

With the added overhead, the energy E_{par} and delay t_{par} of the parallel execution become:

$$E_{par} = E + 2kE \ , \tag{15.17}$$

$$t_{par} = t/2 + 2kt \ , \tag{15.18}$$

which means that

$$\Theta_{par} = \Theta_0(1+2k)(1/2+2k)^2 \ . \tag{15.19}$$

The ratio Θ_{par}/Θ_0 is less than 1 only for $k < 0.18$. In other words, binary parallel composition using a split and a merge with the above characteristics decreases Θ only when the task to be parallelized is at least 5.6 times as expensive as a split or merge.

As a concrete example, the authors have investigated the possibility of improving the performance of a 32-bit four-stage carry-lookahead adder by interleaving two identical adders. For this type of circuit, k is empirically found to be approximately 0.25 (the split and merge networks are about as expensive as one adder stage). Therefore splitting the adder in this particular way does not help.

4.2 The Θ-Efficiency of Pipelining

Now let us consider a task S that repeatedly evaluates the function F for a sequence of parameter values: S receives a parameter value x from the environment, evaluates $F(x)$, sends the result to the environment, and repeats the cycle for a next parameter value.

Pipelining is the transformation that replaces the single task S with a chain P of m tasks P_j computing functions f_j, called the "stages" of the pipeline. Each stage P_j behaves exactly like the original task, except that it computes f_j instead of F, and its environment is different: P_j receives its parameters from P_{j-1} if $j > 0$, or from the environment of S if $j = 0$; it sends its results to P_{j+1} if $j < m-1$, or to the environment of S if $j = m-1$.

The fs are chosen such that $F = f_{m-1} \circ \ldots \circ f_1 \circ f_0$, with $0 < m$.

In this example, we are interested in the cycle time and the energy consumed by one cycle. We first determine the energy E_1 and cycle time t_1 for the computation of one $F(x)$ by task S.

There are two parts to the activity of a cycle: the computation of F and the communication overhead (receiving parameters and sending results). Let E be the energy and t the time to compute F. We assume that the energy of the communication overhead is kE and that the delay is kt. Putting the pieces together, we get:

$$E_1 = E + kE, \qquad (15.20)$$

$$t_1 = t + kt, \qquad (15.21)$$

$$\Theta_1 = Et^2(1+k)^3 \, . \qquad (15.22)$$

Now, want to choose the length of the pipeline, m, and the functions f so as to minimize the Θ of the pipeline in terms of the energy and cycle time for computing one $F(x)$.

The cycle time of the pipeline is the cycle time of the slowest stage. Hence, we should choose the functions f such that all stages have the same cycle time. For all j:

$$t_j = \frac{t}{m} + kt \, . \qquad (15.23)$$

But the stages do not need to have the same energy; let us say that each stage consumes G_j, with $(\sum_j G_j) = E$. For simplicity, let us assume that the communication overhead for the entire non-pipelined implementation is paid by *each stage* of the pipelined implementation. Under these assumptions, the total energy for stage j is

$$E_j = G_j + kE \; . \tag{15.24}$$

We can now compute the energy E_m and the cycle time t_m for computing one $F(x)$:

$$E_m = \sum_{j=0}^{m-1} E_j \tag{15.25}$$

$$t_m = t_j \qquad \text{for all } j, \tag{15.26}$$

which gives:

$$E_m = E + kmE \; , \tag{15.27}$$

$$t_m = \frac{t}{m} + kt \; . \tag{15.28}$$

Let Θ_m be the Θ of the pipeline. By definition, $\Theta_m = E_m t_m^2$, i.e.,

$$\Theta_m = Et^2 \frac{(1+km)^3}{m^2} \; . \tag{15.29}$$

We can express the improvement in Θ compared with the non-pipelined case as the ratio of the two Θs:

$$\frac{\Theta_m}{\Theta_1} = \frac{1}{m^2} \frac{(1+km)^3}{(1+k)^3} \; . \tag{15.30}$$

The ideal case $k = 0$ (no overhead) gives an improvement

$$\frac{\Theta_m}{\Theta_1} = \frac{1}{m^2} \; , \tag{15.31}$$

with $E_m = E$ and $t_m = t/m$. Although it looks like we have gained nothing in energy, in fact we can save up to a factor m^2 in energy if we equalize the cycle time to that of the non-pipelined case by adjusting the supply voltage.

For $k > 0$, the optimal improvement is achieved for

$$\frac{d}{dm}\left(\frac{\Theta_m}{\Theta_1}\right) = 0 \; , \tag{15.32}$$

i.e., for $mk = 2$.

In the optimal case

$$E_m = 3E \text{ , and} \quad (15.33)$$

$$t_m = \frac{3}{2}kt \text{ ,} \quad (15.34)$$

whence we derive the following theorem on the optimal length of a pipeline.

Theorem 3 *The Θ-optimal pipeline requires an energy per computation step that is 3 times the energy required for computing F. It has a cycle time that is $3/2$ the overhead's cycle time.*

Let us compute the optimal pipeline improvement as a function of the overhead ratio k, ($m = \frac{2}{k}$). We get the following result:

$$\frac{\Theta_m}{\Theta_1} = \frac{27}{4}\frac{k^2}{(1+k)^3} \text{ .} \quad (15.35)$$

The result shows that the pipeline is very sensitive to the communication overhead. For an overhead ratio of one (which obtains when the pipelining communication is as costly as the operation itself), the pipeline offers practically no gain in Et^2.

* *

In the second part of this chapter, we examine the relationship between E, t, and the two physical parameters that a designer (usually) can adjust: the supply voltage and the transistor widths. First, we address the issue of optimizing Et^2 as a function of the transistor widths. Secondly, we introduce the notion of *minimum-energy functions* $E(t)$ to express the dependence of E and t on each of the two physical parameters. We use those functions for deriving a number of important results about the sequential and parallel compositions of systems.

5. Transistor Sizing for Optimal Θ

The task of adjusting the transistor widths of a circuit is called "transistor sizing," or "circuit sizing." We are interested in sizing transistors so as to minimize Θ. Both capacitance C_z and current $i_{z\uparrow}$ in Equations 15.2 and 15.3 for E and t depend on the size of the transistors. The capacitance contributed by transistors increases linearly with the transistor widths, but the current also increases linearly. Hence, it is not immediately clear how transistors should be sized to optimize the Θ of a circuit.

We shall find that, in a Θ-optimal circuit, the transistors are sized such that the total transistor capacitance is approximately twice the total parasitic capacitance. As we shall see, the result is exact only for a restricted class of circuits; nevertheless, it is a good approximation for most circuits.

Let E be the total energy of the computation: it is the sum of all energy spent exercising the nodes of the computation. Assume, without loss of generality, that there are exactly two transitions corresponding to each node. (This amounts to "unrolling" into several nodes the nodes of the circuit that see more than two transitions and ignoring the nodes that never transition.) Let t be the cycle time of a critical cycle. We assume that the circuit is designed so that all cycles are critical; this is true in many well designed circuits, and it is true for any optimally sized circuit in the absence of additional constraints (e.g., minimum-size or slew-rate constraints) on transistor sizes.

We distinguish between two types of capacitances attached to a node j: first, the "gate" (or transistor) capacitance c_j contributed by the transistors of the operators to which node j is connected, and secondly, the "parasitic" capacitance p_j contributed by the wires connecting node j to other operators. We assume that p_j is fixed and that we can change c_j as we please by adjusting the transistors' widths.

We first consider scaling all the transistors in the circuit by the same factor: we want to determine the global scaling factor w to be applied to all transistors' widths that achieves the lowest Et^2. We have

$$E = \sum_{j \in N} (c_j + p_j) V^2, \tag{15.36}$$

where N is the set of all nodes on the chosen critical cycle, i.e.,

$$E = (C + P) V^2 \ , \tag{15.37}$$

with $C \stackrel{\text{def}}{=} \sum_{j \in N} c_j$ and $P \stackrel{\text{def}}{=} \sum_{j \in N} p_j$.

For the sake of simplicity, we assume that upgoing and downgoing transitions on a given node have the same delay. For the cycle time t of a critical cycle, $t = 2(\sum_{j \in N} t_j)$ where t_j is a transition of node j. We use Equation 15.6 to compute all t_js

$$t = 2 \sum_{j \in N} \frac{(c_j + p_j)}{k_j V} \ . \tag{15.38}$$

We first simplify the problem and assume that the circuit is "homogeneous," i.e., that all gates are identical, and hence that $\forall j : j \in N : k_j = Kw$. We get that

$$t = 2\frac{\sum_{j\in N}(c_j + p_j)}{KwV}, \tag{15.39}$$

i.e.,

$$t = 2\frac{C+P}{KwV}. \tag{15.40}$$

By definition of the global sizing factor w, we have $C = wC_{min}$, and therefore we can eliminate w from the expression of t:

$$t = \mathcal{K}\frac{C+P}{CV}, \tag{15.41}$$

$$Et^2 = \mathcal{K}^2\frac{(C+P)^3}{C^2}; \tag{15.42}$$

where $\mathcal{K} \stackrel{\text{def}}{=} 2C_{min}/K$. It is easy to check that $\frac{\partial}{\partial C}(Et^2) = 0$ for $C = 2P$. Hence the theorem:

Theorem 4 *For a homogeneous circuit, the minimal Et^2 is achieved when the total gate capacitance is twice the total parasitic capacitance.*

5.1 Using Et^n with $n \neq 2$

Sometimes, we may want to optimize Et^n for $n \neq 2$ when using a Θ-optimal circuit would not be possible because the required delay or energy would result in a supply voltage outside the practically possible range. Roughly speaking, when we perform Et^n optimization we mean that we consider a 1% improvement in speed is worth an n% increase in energy. For example, for a circuit operating in velocity saturation, we might have expend twice the energy for a 10% speed improvement. In that case, we should optimize for Et^n with $n \approx 5$.

There is another reason to examine Et^n, besides extreme supply voltages (and mathematical insight). Even though a large system may be optimized for Et^2, components of that system may not individually be optimized for Et^2. For example, speeding up critical paths while lowering the Et^2 of these paths may make the entire design run faster and actually improve Et^2 for the entire design. If multiple supply voltages are allowed, then Theorem 1 applies to each component of the system, so each component is optimally characterized by Et^2. But multiple supplies are impractical; instead we can use Et^n optimization for the different paths, with a larger n for the more critical paths and a smaller n for the less critical paths.

As an example of Et^n optimization, we generalize Theorem 4 for all n:

$$Et^n = \mathcal{K}^n\frac{(C+P)^{n+1}}{C^nV^{n-2}}, \tag{15.43}$$

and we find that $\frac{\partial}{\partial C}(Et^n) = 0$ for $C = nP$. Hence the theorem:

Theorem 5 *For a homogeneous circuit, the minimal Et^n is achieved when the total gate capacitance is n times the total parasitic capacitance.*

5.2 Optimal Energy and Cycle Time

We have seen that, for a ring of identical operators, E and t are of the following form:

$$E = (C+P)V^2 \tag{15.44}$$
$$t = \mathcal{K}\frac{C+P}{CV} . \tag{15.45}$$

When optimizing for Et^n by transistor sizing, we have established that the minimum is achieved for $C = nP$, to which correspond an energy E_n and a cycle time t_n, with

$$E_n = (n+1)PV^2 , \tag{15.46}$$

and

$$t_n = \mathcal{K}\frac{n+1}{n}\frac{1}{V} . \tag{15.47}$$

Two interesting quantities are E_0 and t_∞: $E_0 = PV^2$, and $t_\infty = \mathcal{K}/V$. By definition of n , E_0 is the theoretical minimal energy, corresponding to minimizing E without regard for t; it corresponds to the situation when the transistors are all zero-sized and the fixed parasitic capacitances constitute the entire E. Conversely, t_∞ is the theoretical minimal cycle time corresponding to minimizing t without regard for E. It is obtained when C goes to infinity, i.e., when only gate capacitances contribute to E and t. We may eliminate V from Equations 15.46 and 15.47 and restate our results in terms of E_0 and t_∞; thus,

$$E_n = (n+1)E_0 , \tag{15.48}$$

and

$$t_n = \frac{n+1}{n}t_\infty . \tag{15.49}$$

In particular

$$E_2 = 3E_0 , \tag{15.50}$$
$$t_2 = \frac{3}{2}t_\infty . \tag{15.51}$$

Theorem 6 *The cycle time for optimal Et^2 is 3/2 the theoretical minimal cycle time at that supply voltage. The energy for optimal Et^2 is three times the theoretical minimal energy at that supply voltage.*

If we eliminate n from Equations 15.48 and 15.49, we arrive at the following relation between the minimum energy and the minimum delay of a single-cycle circuit at a fixed voltage:

$$E_n = \frac{E_0 t_n}{t_n - t_\infty} . \tag{15.52}$$

5.3 A Minimum-Energy Function

We can define an antimonotonic *minimum–energy-consumption function* or *minimum-energy function* $E(t)$ that describes the effect of transistor sizing on the minimum energy required for a system to run at a given t at a fixed voltage. (Tierno has previously used a similar energy function [12].) If we rewrite Equation 15.52 with E a function of t, we get the following function:

$$E(t) = \frac{E_0 t}{t - t_\infty} . \tag{15.53}$$

It is easy to prove that Equation 15.53 satisfies the above definition of the minimum-energy function.

5.4 Experimental Evidence

Even though Equations 15.48 and 15.49 have been derived for only a very restricted class of circuits, they are in fact good approximations for a much wider class. The authors have checked the equations against the minimal Et^n obtained by applying an optimization algorithm (gradient descent) to a class of circuits. The circuits, each consisting of a ring of operators, were chosen at random with a uniform-squared distribution of parasitic capacitances; the number of transistors in series was also chosen according to such a distribution. The authors used real numbers for both parameters; they optimized the expression for Et^n using Equations 15.36 and 15.38 for E and t. The range of parasitics was [1,100] in normalized units; the range of transistors in series was [1,6]. The results show that Equations 15.48 and 15.49 hold, with very good accuracy, over a wide range of parasitics, logic-gate types, and circuit sizes.

The results of the simulations for circuits consisting of a ring of 100 operators are summarized in Figure 15.2. (Simulations for rings of 10 and 1000 operators show similar results.) The figure shows the mean and standard deviation of the error in the estimates of Equations 15.48 and 15.49 for a range of different optimization indices ($n \in 1..10$ in Et^n). The estimates get more

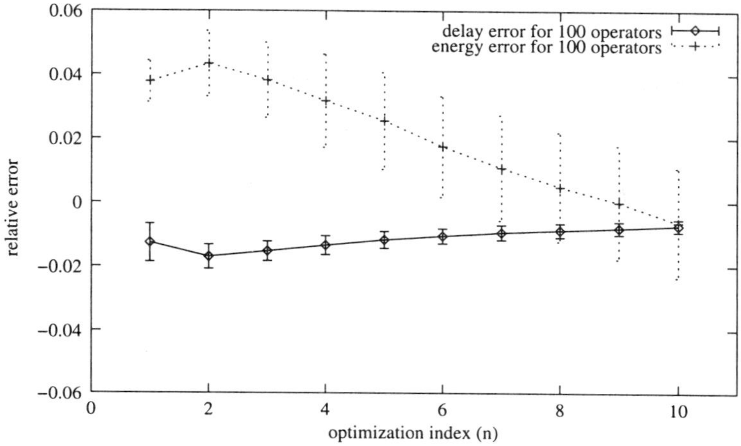

Figure 15.2. Results of Simulating Random Circuits

dependable for larger circuits, where the random variation in operators tends to average out over the cycle. Overall, the estimates are usually good to within five percent of the energy and delay values for the actual optimum Et^n.

5.5 Multi-Cycle Systems

Let us now consider a system composed of m subsystems S_i (E_{0i}, $t_{\infty i}$) executing in parallel; each subsystem S_i has minimum-energy function

$$\frac{E_{0i}t}{t - t_{\infty i}} \; .$$

These subsystems can be chains or rings of arbitrary logic gates, since our experiment shows that Equations 15.48 and 15.49 adequately describe the minimum energy and delay of a large class of circuits. Let us assume that the subsystems are synchronized so that all ts are equal. As a consequence, the minimum-energy function for the composed system is

$$E(t) = t \sum_{i=1}^{m} \frac{E_{0i}}{t - t_{\infty i}} \; . \tag{15.54}$$

Theorem 7 *For a system composed of m subsystems S_i (E_{0i}, $t_{\infty i}$) as specified above, if the system is optimally sized for Et^n then*

$$E_n \leq (n+1) \sum_{i=1}^{m} E_{0i} \tag{15.55}$$

with equality if and only if all $t_{\infty i}$s are equal.

(Note that Equation 15.48 is simply the special case of Theorem 7 that holds when all $t_{\infty i}$s are equal.)

Proof. The optimal Et^n of this composed system is reached for E and t that satisfy

$$\frac{d(E(t)t^n)}{dt} = 0\,, \tag{15.56}$$

which is achieved when

$$(n+1)\sum_{i=1}^{m}\frac{E_{0i}}{t-t_{\infty i}} = t\sum_{i=1}^{m}\frac{E_{0i}}{(t-t_{\infty i})^2}\,. \tag{15.57}$$

We may now invoke the Cauchy-Schwarz inequality $(\sum l_i r_i)^2 \leq \sum l_i^2 \sum r_i^2$, where equality holds if and only if l_i/r_i has the same value for all i. If we substitute $l_i \leftarrow \frac{\sqrt{E_{0i}}}{t-t_{\infty i}}$ and $r_i \leftarrow \sqrt{E_{0i}}$, we get that

$$\left(\sum_{i=1}^{m}\frac{E_{0i}}{t-t_{\infty i}}\right)^2 \leq \sum_{i=1}^{m}\frac{E_{0i}}{(t-t_{\infty i})^2}\sum_{i=1}^{m}E_{0i} \tag{15.58}$$

with equality if and only if all $t_{\infty i}$s are equal. Using Equation 15.57, we replace $\sum\frac{E_{0i}}{(t-t_{\infty i})^2}$ with $\frac{(n+1)}{t}\sum\frac{E_{0i}}{t-t_{\infty i}}$ in Equation 15.58, and we get the following result:

$$\left(\sum_{i=1}^{m}\frac{E_{0i}}{t-t_{\infty i}}\right)^2 \leq \frac{(n+1)}{t}\sum_{i=1}^{m}\frac{E_{0i}}{t-t_{\infty i}}\sum_{i=1}^{m}E_{0i}\,. \tag{15.59}$$

By Equation 15.54, then,

$$E(t) = t\sum_{i=1}^{m}\frac{E_{0i}}{t-t_{\infty i}} \leq (n+1)\sum_{i=1}^{m}E_{0i}\,. \tag{15.60}$$

And therefore

$$E_n \leq (n+1)\sum_{i=1}^{m}E_{0i}\,. \quad \square \tag{15.61}$$

In Theorem 7, equality holds if and only if all $t_{\infty i}$s are equal; in this situation, we also have that $t = \frac{n+1}{n}t_\infty$. This is a generalization of Equations 15.48 and 15.49 to multi-cycle systems. As we have already pointed out, the $t_{\infty i}$s are likely to be close to each other in most well designed circuits, so we should expect that usually $E(t) \approx (n+1)\sum E_{0i}$. In other words, in a multi-cycle system optimally sized for Et^2 , the gate capacitance is (close to) twice the parasitic capacitance.

We have shown that the $C = nP$ result is correct for a ring of operators. We previously observed that if a dominant term exists then $C = nP$ is approximately correct for general circuits. We have experimental evidence that the relation is true for a large class of multi-cycle systems. Such evidence is also provided by SPICE simulations of an adder published by Chandrakasan and Brodersen and summarized in Figure 4.7 of their book [2]. Their figure shows that, for the five different parasitic contributions they study, the minimum energy for a given speed (allowing supply-voltage adjustment) is achieved when the gate capacitance is very close to twice the parasitics. (They did not, however, draw the conclusion that we have reached here.)

6. Power Rule for Sequential Composition

Let us now consider the sequential composition of two systems A and B. Let us assume a sequential computation that runs A to completion and then B to completion; we assume the delay between the end of A and the start of B to be negligible. We want to know at what t_A, t_B to run A and B so as to optimize the Et^n of the sequential composition.

We now recall the concept of a minimum-energy function introduced by Equation 15.53. Equation 15.53 applies to the specific transformation of changing transistor sizes; i.e., it describes what the minimum energy of a circuit will be when that circuit is sized to achieve a certain performance. We are no longer limiting our discussion to the effects of transistor sizing, so we can allow other transformations to be used as a basis for this $E(t)$.

Theorem 8 *For the sequential composition of two systems A and B, if the composed system is optimized for Et^n, then*

$$\frac{dE_A(t_A)}{dt_A} = \frac{dE_B(t_B)}{dt_B} . \tag{15.62}$$

Proof. The latency of the composed system is $t = t_A + t_B$, while its energy is $E = E_A(t_A) + E_B(t_A)$. Hence we are minimizing

$$f(t_A, t_B) \stackrel{\text{def}}{=} (E_A(t_A) + E_B(t_B))(t_A + t_B)^n. \tag{15.63}$$

If we set the partial derivatives of f with respect to t_A and t_B equal to zero, we obtain

$$\frac{dE_A(t_A)}{dt_A} = -\frac{n(E_A(t_A) + E_B(t_B))}{t_A + t_B}$$

and

$$\frac{dE_B(t_B)}{dt_B} = -\frac{n(E_A(t_A) + E_B(t_B))}{t_A + t_B} ,$$

from which it is clear that Equation 15.62 holds. □

Theorem 8 holds for any minimum-energy function $E(t)$ and any value of the optimization index n.

If we now vary the supply voltages of the components A and B of the sequential composition so as to optimize Et^2 , we have the following theorem:

Theorem 9 *For the sequential composition of two systems A and B with power consumptions P_A and P_B, respectively, if the composed system is optimized for Et^2 by adjusting the supply voltages of the components, then*

$$P_A = P_B \ . \tag{15.64}$$

Proof. Let us define $\Theta_A = E_A t_A^2$, $\Theta_B = E_B t_B^2$; as we have established, Θ_A and Θ_B are voltage independent. Using Theorem 8 with $E_A(t_A) = \Theta_A / t_A^2$ and $E_B(t_B) = \Theta_B / t_B^2$, we get:

$$\frac{\Theta_A}{t_A^3} = \frac{\Theta_B}{t_B^3} \ . \tag{15.65}$$

Hence,

$$\frac{E_A}{t_A} = \frac{E_B}{t_B} \ . \quad \square \tag{15.66}$$

In other words, circuits composed sequentially in a Θ-optimal way should have their supply voltages adjusted so as to equalize their power use. (If the circuits are themselves Θ-optimal, then equalizing their power is a necessary and sufficient condition for making the composition Θ-optimal.)

7. Θ-Rules for Parallel and Sequential Compositions

Finally, let us consider the parallel and sequential composition rules for Et^2 , assuming that we have the freedom of independently adjusting the supply voltages of the composed systems. Given are two systems A and B with latencies t_A and t_B, energies E_A and E_B; we have $\Theta_A = E_A t_A^2$ and $\Theta_B = E_B t_B^2$.

First, consider the parallel composition of the two systems. We want to compute the minimum $\Theta_{A\|B}$ as a function of Θ_A and Θ_B .

The minimum $\Theta_{A\|B}$ is achieved when $t_A = t_B$ (see Section 4.1). With $E_{A\|B} = E_A + E_B$ and $t_{A\|B} = t_A = t_B$, we get $\Theta_{A\|B} = E_A t_A^2 + E_B t_B^2$. Hence the theorem:

Theorem 10 *For two systems A and B composed in parallel,*

$$\Theta_{A\|B} = \Theta_A + \Theta_B \ . \tag{15.67}$$

Now consider the two systems composed in sequence as in Section 6. As in the previous example, we are given $\Theta_A = E_A t_A^2$ and $\Theta_B = E_B t_B^2$, and we wish to determine the optimal $\Theta_{A;B}$ as a function of Θ_A and Θ_B . We have:

$$\Theta_{A;B} = (E_A + E_B)(t_A + t_B)^2 \ , \tag{15.68}$$

i.e.,

$$\Theta_{A;B} = \left(\frac{\Theta_A}{t_A^2} + \frac{\Theta_B}{t_B^2}\right)(t_A + t_B)^2 \ . \tag{15.69}$$

Since the optimal total Et^2 is independent of global scaling of t, it is sufficient to determine a single parameter defined as follows: $\alpha = t_A/t_B$. From Equation 15.65, we have $\alpha = \sqrt[3]{\frac{\Theta_A}{\Theta_B}}$. For this α , we compute the optimal $\Theta = (\sqrt[3]{\Theta_B} + \sqrt[3]{\Theta_A})^3$:

Theorem 11 *For two systems A and B composed in sequence,*

$$\sqrt[3]{\Theta_{A;B}} = \sqrt[3]{\Theta_A} + \sqrt[3]{\Theta_B} \ . \tag{15.70}$$

The Θ-rules are very useful for computing the Θ of a computation as we did for parallelism and pipelining in Section 4. The industrious reader that goes through the exercise of using the Θ-rules to compute the same Θ_{par} that was computed in Section 4 is in for a surprise: the result is different!

Using the Θ-rules, we get $\Theta_{par} = \Theta_0(2k + \frac{1}{2^{2/3}})^3$, which is smaller than the result of Equation 15.19. The reason is subtle but important. In the first computation, we postulate that $E_{split} = E_{merge} = kE$ and $t_{split} = t_{merge} = kt$. By fixing the E and t ratios, we implicitly assume that the voltages of the different components are identical. The computation using the Θ-rules just assumes that $\Theta_{split} = \Theta_{merge} = k^3\Theta_0$. Hence the voltages of the split and merge can be adjusted independently, which leads to a lower Θ_{par} .

(An analysis of the Θ-efficiency of parallelism can be found in a companion paper [6].)

8. Summary and Conclusion

In this chapter, we have seen that Et^2 constitutes an excellent metric for comparing computations for energy and delay efficiency when the physical behavior is that of CMOS VLSI circuits. We started by observing that the Et^2 metric for a CMOS circuit is independent of the supply voltage, as long as we can scale the threshold voltage linearly with the supply voltage and as long as we stay away from velocity saturation.

We showed that when supply-voltage adjustment is allowed, the popular Et metric is inferior. Following along these lines, we established that any metric with certain properties (we called such a metric Θ) could be used to compare designs independently of the voltage. As long as the required speed or energy lies within the threshold-voltage to velocity-saturation range of the implementations, we saw that the implementation with the better Θ is better for *any* desired speed or energy consumption. We showed that Et^2 is to first order a metric with the required properties. We then applied the metric to various circuit transformations, namely pipelining and parallelism. We also applied the metric to transistor sizing and were able to show that the optimal sizing for energy efficiency is not what is commonly used (minimal sizes). Finally, we established rules for computing the Θ of the sequential and parallel compositions of systems.

Overall, Et^2 is a very useful efficiency metric for designing CMOS VLSI circuits. Time and experience will show how applicable it is to other computations.

Acknowledgments

Acknowledgment is due to Karl Papadantonakis, Martin Rem, and Catherine Wong for their comments and criticisms. The research described in this paper was sponsored by the Defense Advanced Research Projects Agency and monitored by the Air Force.

References

[1] Carver A. Mead and Lynn Conway. *Introduction to VLSI Systems*, Addison-Wesley, Reading MA, 1980.

[2] Anantha P. Chandrakasan and Robert W. Brodersen. *Low Power Digital CMOS Design*, Kluwer Academic Publishers, Dordrecht,1995.

[3] Alain J. Martin. Towards an Energy Complexity of Computation. *Information Processing Letters*, 77, 2001.

[4] Alain J. Martin, Andrew Lines, Rajit Manohar, Mika Nyström, Paul Penzes, Robert Southworth, Uri Cummings, and Tak Kwan Lee. The Design of an Asynchronous MIPS R3000 Microprocessor. *Proceedings of the 17th Conference on Advanced Research in VLSI*, IEEE Computer Society Press, 164-181, 1997.

[5] Alain J. Martin. Synthesis of Asynchronous VLSI Circuits. In *Formal Methods for VLSI Design*, ed. J. Staunstrup, North-Holland, 1990.

[6] Alain J. Martin and Mika Nyström. The Et^2-Efficiency of Parallelism, Caltech Technical Report, October 2001.

[7] Mika Nyström. Et^2 and Multi-voltage Logic, Caltech Technical Report, April 1995.

[8] Karl Papadantonakis. A Theory of Constant Et^2 CMOS Circuits, Caltech Computer Science Technical Report 2001.004, July 2001.

[9] Karl Papadantonakis. Hierarchical Voltage Scaling for Et^2 Optimization of CMOS Circuits, Caltech Computer Science Technical Report 2001.005, July 2001.

[10] Paul I. Pénzes. *Energy-delay Efficiency of Asynchronous Circuits*, Ph.D. Thesis (in preparation), California Institute of Technology, 2002.

[11] Paul I. Pénzes and Alain Martin. Global and Local Properties of Asynchronous Circuits Optimized for Energy Efficiency. *IEEE Workshop on Power Management for Real-time and Embedded Systems*, May 2001.

[12] José A. Tierno. *An Energy-Complexity Model for VLSI Computations*, Ph.D. Thesis, California Institute of Technology, 1995.

[13] José A. Tierno and Alain J. Martin. Low-Energy Asynchronous Memory Design. *Proceedings of International Symposium on Advanced Research in Asynchronous Circuits and Systems*, IEEE Computer Society Press, pp. 176–185, 1994.

Chapter 16

CHALLENGES FOR ARCHITECTURAL LEVEL POWER MODELING

Nam Sung Kim*, Todd Austin*, Trevor Mudge*, Dirk Grunwald*

* *The University of Michigan, Ann Arbor*

{kimns,austin,tnm}@eecs.umich.edu

* *The University of Colorado, Boulder*

grunwald@cs.colorado.edu

Abstract The power aware design of microprocessors is becoming increasingly important. Power aware design can best be achieved by considering the impact of architectural choices on power early in the design process. A natural solution is to build a power estimator into the cycle simulators that are used to gauge the effect of architectural choices on performance. Cycle simulators intentionally omit considerable implementation detail in order to be efficient. The challenge is to select the details that must be put back in if the simulator is required to also produce meaningful power figures. In this paper we propose how to augment a cycle simulator to produce these power figures.

Keywords: power estimation, cycle simulator, microarchitecture

1. Introduction

Power dissipation has become a significant constraint in modern microprocessor design. In many mobile and embedded environments power is already the leading design constraint. Although it may not be so apparent, it is almost as important in the design of general purpose high-performance computers [1]. It has become one of the primary design constraints along with performance, clock frequency, and die size. In addition to extra heat removal costs, high power consumption in embedded processors also reduces the battery lifetime. Hence, a mobile computing system's quality and reliability could be affected by its high power dissipation. In case of high performance microprocessors, high power dissipation leads to thermal issues like device degradation, higher packaging cost, and reduced chip lifetime.

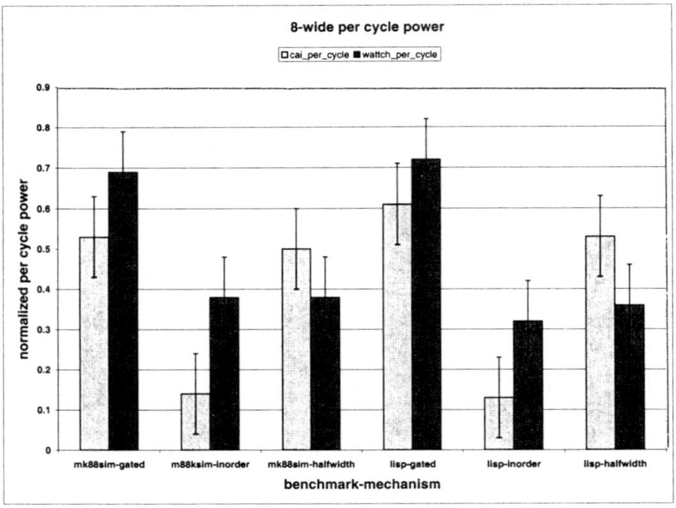

Figure 16.1. Normalized per Cycle Power Estimation Results for 8-Wide Microarchitectures. The Error Bars are Based upon the 10% Accuracy Reported by Wattch (4).

The elevation of power to a "first-class" design constraint requires that power estimation be done at the same time as performance studies in the design flow. Performance analysis for a proposed design is usually accomplished during the design exploration phase with a cycle simulator. The natural solution would be to augment cycle simulators so that they can also provide power estimates. This has been done in the case of three recent simulators [2, 3, 4]. A cycle simulator models the behavior during each clock cycle of the processor. The goal of these simulators is to assess the impact on performance of cache and memory organizations, pipelining, multi-instruction issue, branch prediction, and other microarchitectural mechanism. The more sophisticated cycle simulators can boot an operating system and run significant parts of an application program within a several hour period. This enables the user to observe the performance under a variety of workloads for billions of cycles. Such long simulations are necessary to provide the confidence in the resulting performance figures. To simulate billions of cycles within a several hour period requires a high degree of efficiency in the cycle simulator. Efficiency is achieved by abstracting away the physical behavior of the microarchitecture — the very details that are required to obtain power figure. The challenge for architectural power modeling is to add enough detail back into the simulator so that power estimates are meaningful without unduly slowing the simulator.

The difficulty of this challenge is apparent from the results that one of us obtained by calibrating two of the recent power simulators mentioned above, the Cai-Lim simulator [2] and the Wattch simulator [3]. Figure 16.1 reproduces one of the figures from [5]. It shows power estimates obtained from two SPEC benchmarks, *mk88sim* and *lisp*. Three implementations of an 8-wide issue processor are examined: 1) an out-of-order machine with gated clocks; 2) an in-order machine; and 3) a half width machine, i.e., a 4-wide issue. Clearly the two models give significantly different results. The accuracy that we can expect is probably much less that the 10% claimed in [3]. This is not unacceptable for an early design tool. However, to be useful in the design process a successful model should indicate the trends accurately. The results in Figure 16.1 show that the two models also give conflicting views on trends. The Cai-Lim model indicates the in-order machine is the least power hungry, even allowing for a high level of imprecision in the results. The Wattch model, on the other hand, does not show a clear preference. This paper will identify the causes of these inaccuracies, and propose how a cycle simulator can be augmented to correct for them.

The remainder of this paper is organized as follows. The next section reviews the power metrics required for power aware design. Section 3 discusses existing cycle simulators that measure power, and points out their weaknesses. Section 4 discusses the requirements needed by a cycle simulator to produce useful power figures. Section 5 considers the details of implementing such a power estimator. Section 6 conclude the paper and identifies some remaining open questions.

2. Power Metrics

There are three components that define the important contributions to power consumption in CMOS[1] technology:

$$P = ACV^2 f + \tau AVI_{short} + VI_{leak} \tag{16.1}$$

The first component is perhaps the most familiar. It measures the dynamic power consumption caused by the charging and discharging of the capacitive load on the output of each gate. It is proportional to the frequency of the operation of the system, f, the activity of the gates in the system, A (some gates may not switch every clock), the total capacitance seen by the gate outputs, C, and the square of the supply voltage, V. The second term captures the power expended due to the short-circuit current, I_{short}, that momentarily, τ, flows between the supply voltage and ground when the output of a CMOS

[1] We focus on CMOS because it will likely remain the dominant technology for the next 5-7 years

logic gate switches. The third term measures the power lost due to leakage current that is present regardless of the state of the gate.

The first two terms can be lumped as activity based because they are directly related to the the toggling frequency of the gates in the circuit. In contrast, the leakage term is unaffected by activity, because it is governed only by the number of gates and their threshold voltages. It is only affected by activity in the sense that leakage is reduced to zero when the gate is turned off. Unfortunately, this also results in any state in the circuit being lost. Thus to obtain power from the execution of a cycle simulator it is necessary to tie cycle behavior to activity at the gate level for the first two terms and to estimate the number of gates that the microarchitecture requires for the third term. The difficulties associated with cycle level power estimation arise directly from the difficulty of calculating these values with any level of accuracy.

If average power consumption were our main concern then the values for A and f could be calculated by sampling. Sampling would give us the time to perform more detailed power consumption calculations. This idea has been proposed in [6]. However, average power values can hide important details as can be seen in [7]. In this work the energy dissipation of individual ARM instructions is measured by running each instruction in a loop and measuring the average current drawn. The results show very little difference between the energy usage of different instructions, because the effect of averaging is to smooth out short lived effects like cache misses and pipeline stalls. In addition, there are two other power metrics that are important design constraints. The first is peak power. Typically, systems have an upper limit, which if exceeded will lead to some form of damage. The second is dynamic power. Sharp changes in power consumption can result in inductive effects that can result in circuit malfunction. The effect caused by "di/dt" noise. Equation 16.1 can be used to monitor peak power and estimate "di/dt" noise, if a running value of (16.1) is maintained. Sampling can miss these effects.

3. Previous Work

As we have noted, the abstractions necessary for efficiency make it difficult to base an accurate power estimation tool on a cycle simulator. The lack of detailed low-level physical design information such as interconnection capacitance, types of circuits for each microarchitectural block, clock trees, and I/O pads are among the principal sources of inaccuracy.

Prior work on microarchitectural level power estimation [2, 3, 4] has mainly relied on microarchitectural-activity-based calculations. A typical power dissipation figure for each block used in the target microprocessor is estimated. Then the activity of each block is recorded every cycle based on the behavior of the cycle simulator. Finally, the power dissipation of the microprocessor is estimated by combining the activity with the power figures.

In [3], Brooks et al. utilized detailed analytical power models for array structures and content addressable memories based on CACTI [8] to estimate the power consumption of memory-like microarchitectural blocks. In [2], Cai et al. introduced a power density model to estimate power dissipation of each microarchitectural block based on proprietary Intel design data. In [4], Vijaykrishnan et al. also considered bus transition-sensitivity by employing a register-transfer level power estimation technique. The power models for their blocks were implemented as look-up tables (LUTs) [9]. They also considered the power dissipation of the I/O pad and external memory bus.

In these simulators there is no accounting for changes in power due to data sensitivity. According to [10] and [11], the power consumption of a microarchitectural block is highly dependent on the input data characteristics applied to it. Figure 16.2 shows power dissipation simulation results for an 8-bit ALU

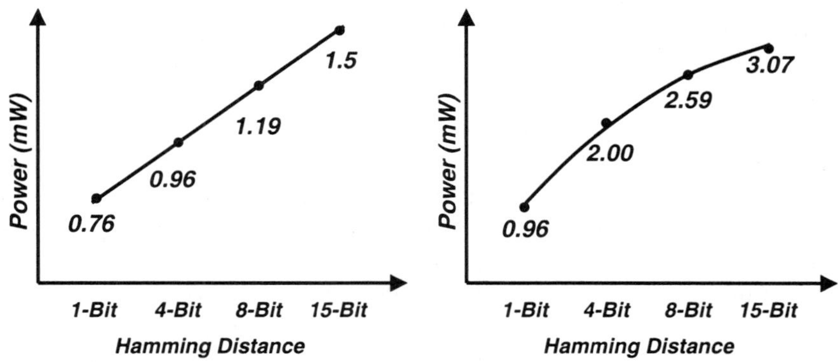

Figure 16.2. Power dissipation for a 100 mhz 8-bit alu and 8-bit multiplier. These microarchitectural blocks where designed using the TSMC 0.25μm LEDA library, the *Synopsys Design Compiler*, and *PrimePower*. Power is plotted against the Hamming distance between consecutive inputs to the blocks.

and an 8-bit multiplier at 100MHz using the TSMC 0.25μm LEDA library, Synopsys Design Compiler, and PrimePower. These measurements agree with those in [12]. It can be seen that the variation in power dissipation is quite significant for different input patterns, suggesting we should consider data-sensitivity in the power estimation of microarchitectural blocks.

All but one of the earlier work also ignored power dissipation caused by accessing the external memory, which usually consumes a significant amount of power since the I/O pads of the microprocessor typically drive very large off-chip bonding wires. Furthermore, the earlier work has ignored power dissipation caused by the clock distribution network and global interconnect — the internal buses that interconnect the blocks. This is particularly tricky to do because it requires some notion of the layout of processor. However, interconnect is becoming an ever more significant contributor to power consumption.

Clearly, there are a number of omissions in current cycle simulator-based power estimation tools that could account for their inaccuracy. In the next section we will propose a framework for augmenting cycle simulators so that the omissions can be included.

4. Augmenting a Cycle Simulator for Power Estimation

4.1 Details Omitted from Cycle Simulators

As we have noted, cycle simulators derive their speed from abstracting out many of the physical details. We touched on some of the resulting omissions in the discussion of earlier power estimators. We will illustrate the extent of this abstraction by considering *SimpleScalar* [13], a cycle-based performance simulator that is widely used and forms the basis for the earlier power estimators. SimpleScalar simulates a specified architecture running a particular benchmark and returns performance in terms of total simulated clock cycles. In order to do this it only needs to trace address streams and some architectural activity. Examples of this activity include function unit usage, and the number of cache, TLB, and branch predictor accesses. These are combined with their respective access latencies to calculate the execution times of instructions. The effect of resource and data hazards is reflected in the access latencies. Thus cycle counts of program execution are calculated without having to model the detailed structure of pipelines. The simulator executes the instructions and stores their results in the simulated register file and main memory in the issue cycle of instructions interpreted. It simply advances the simulation clock after calculating the latencies that would have resulted from the execution of a real multi-issue pipeline.

There is no modeling of the movement of instructions and data between the various pipeline stages after the issue stage, or between other microarchitectural blocks. For instance, the fetch stage fetches instructions from the simulated main memory directly, not from a simulated L1 instruction cache. The cache is modeled by keeping track of the addresses of the instructions and noting when they are no longer included in a list of cache line addresses. When this occurs, a miss penalty is accounted for, and the cache line list is appropriately updated to reflect new cache entries and evictions.

To illustrate the effects of abstraction on memory buses, consider how SimpleScalar models a memory access over a memory bus. SimpleScalar checks the current memory access status and returns the access latency and the requested data blocks. The latency is determined from the number of the memory ports, the number of the requested data blocks, and whether or not the previous memory access cycle is complete. In real microprocessors, however, the memory access transaction occurs over several cycles, and the requested data blocks are transferred from/to the memory during the pertinent cycles according to the memory transaction type and the number of the requested data blocks as shown in Figure 16.3.

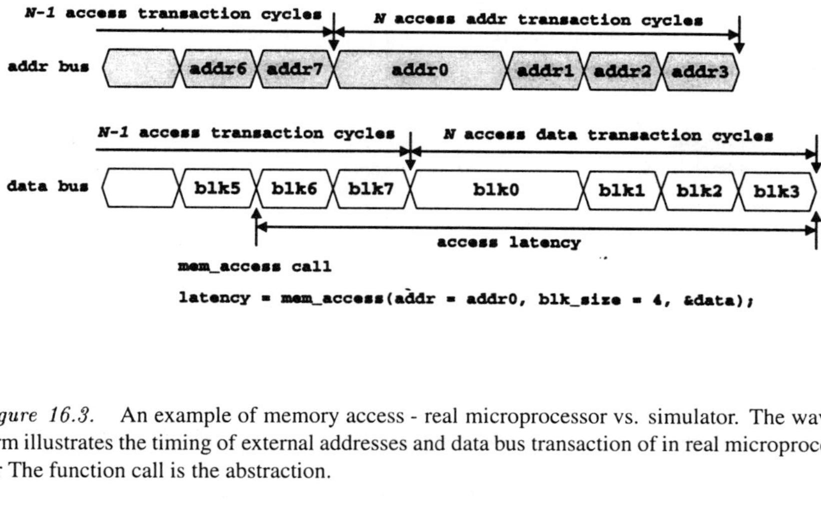

Figure 16.3. An example of memory access - real microprocessor vs. simulator. The waveform illustrates the timing of external addresses and data bus transaction of in real microprocessor The function call is the abstraction.

With only the latency information we have no idea about the details of the memory transfer cycles such as addresses and data values. This was also a shortcoming of earlier power estimators — they assumed that the transaction occurs in the currently accessed cycle regardless of the access latency, and the characteristics of the transaction cycle. To correct this, we need a mechanism for tracing data transactions on buses (internal as well as external) in a cycle accurate way. In order to provide this mechanism, it is necessary to augment the simulator to trace data bus streams. This can be accomplished with special routines at the interfaces of the microarchitectural blocks that capture the cycle accurate bus transaction cycles. In other words, we need to know what values are on the internal buses between the microarchitectural blocks and on the external I/O buses in each simulation cycle, in order to measure the switching power more accurately.

In summary, the principal abstraction in cycle simulators is to omit the modeling of data movements on the internal and external buses between microarchitectural blocks such as function units, caches, memories and pipeline stages.

4.2 Power Estimation Methodology

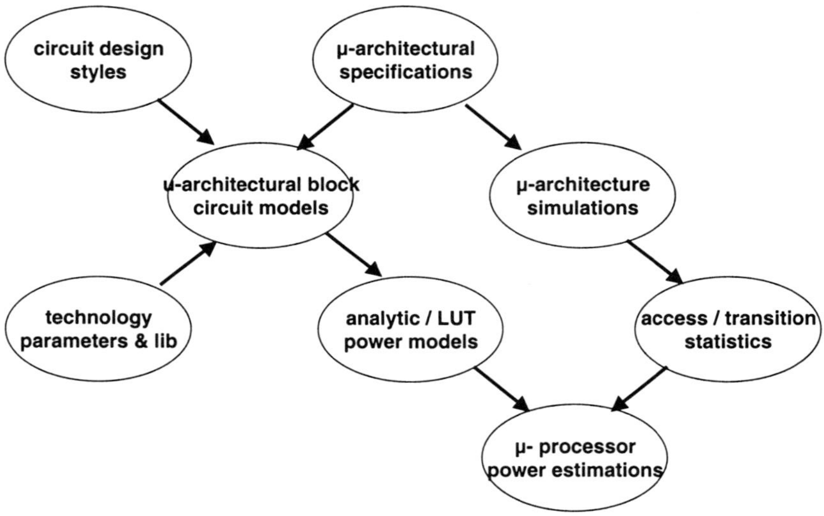

Figure 16.4. A Microarchitectural Power Estimation Methodology

Figure 16.4 shows a proposed methodology for power estimation by augmenting a cycle simulator. First, target technology parameters need to be known or estimated — the *Berkeley Predictive Technology Model* is an example estimator. Other factors that are needed include the supply voltages, threshold voltage, and capacitance per area/length and sheet resistance values of the interconnecting material. In addition, the microarchitectural specification, target operating frequency, and circuit design style for each microarchitectural block should be determined as well.

Second, we need to construct power models for microarchitectural blocks based on the circuit or sub-system design styles. For example, the datapath of the microprocessor may be designed as full-custom, through standard cell synthesis, or by using a datapath compiler. The different cases require different circuit and power models. We can generate analytical power models [8, 14], empirical models [15, 16], or by employing other power macromodeling techniques [12, 17, 18, 19, 20, 21, 22, 23]. These models need only be developed once. Modeling techniques will be discussed further in Section 5.3.

Finally, the power estimator calculates the power dissipation of each microarchitectural block by obtaining the execution statistics from runs of the cycle simulator.

5. Implementing a Cycle-Accurate Power Estimator

The power dissipation of each microarchitectural block consists of three components:

1. The switching power of the load capacitance, which is directly proportional to the number of zero-to-one transitions of outputs per cycle. The number of transitions on a bus (zero-to-one and one-to-zero) can be summarized by recording the Hamming distance between successive bit patterns on that bus. The number of zero-to-one transitions is half this number on average.

2. The power dissipation of the microarchitectural block caused by the switching of its inputs, which is approximately proportional to the number of transitions of inputs (zero-to-one and one-to-zero) per cycle. The power dissipation is actually quite specific to each type of microarchitectural block, and can be characterized in a LUT indexed by Hamming distance. The LUT could be replaced by a simple equation or a macromodel.

3. The leakage power from the reverse-biased diode current and sub-threshold leakage current that are approximately proportional to the area and the number of the transistors of the microarchitectural block. The last term is no longer negligible as supply voltages and threshold voltages decrease. We can implement a generic data structure for power estimation that can handle most of the microarchitectural block types by passing as arguments circuit parameters and LUTs indexed by the Hamming distances for the data on the input and output buses. This is combined with the switching activities from the runs of the simulator — the switching activities define the Hamming distances. As mentioned earlier, the LUT entries are calculated off-line. There are a number of well-known techniques for this, see [12, 17, 18, 19, 20, 21, 22, 23].

Thus, we need to implement an interface between the cycle simulator and the power estimator to collect the appropriate switching activities.

5.1 Implementation of the Data Structure and Microarchitectural Block Models

Figure 16.5 shows a generic data structure for the architectural power estimators that will support flexible power models, and that will interface to SimpleScalar.

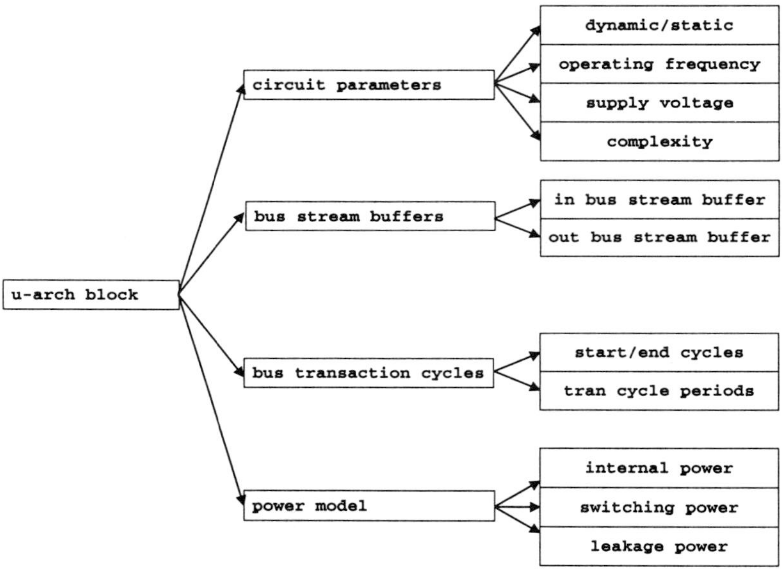

Figure 16.5. Generic Data Structure for a Microarchitectural Block

The circuit parameters in the data structure include a circuit design style — dynamic or static, a supply power voltage, an operating frequency, and complexity information such as the number of estimated transistors and area of the microarchitectural block.

The use of dynamic vs. static circuits affects the power dissipation characteristics of the block significantly. In current state-of-the-art microprocessors, multiple supply voltages are often fed to the chip. For example, 2.5V might be supplied to the internal core, and 3.3V to the I/O pads. Therefore, the data structure for each microarchitectural block should contain supply voltage information. In addition, some peripherals, which would typically be modeled as microarchitectural blocks may be integrated into the chip, and may operate at different frequencies as well. Thus, we support an independent operating frequency for each functional unit as well as a supply voltage. Furthermore, the complexity information, in particular the number of the transistors of the block determines leakage power, and is important for estimating the clock tree capacitance.

A microarchitectural block is modeled as a block with input and output buses (the internal buses between the blocks), see Figure 6. The input and output buses are implemented in the simulator as stream buffers in order to support Hamming distance calculations over complete bus transactions. In the case of blocks like ALUs, these buffers need only have one stage, but for memories and I/O buses they may need to be much deeper (recall Figure 16.4). These buffers are filled by the interface routines of the cycle simulator. In order to estimate the power of each block, we employ a method based on Hamming distance [12], as noted earlier. However, the information in the data structure does not limit us to Hamming distance methods. The availability of input and output statistics permits the use of other dynamic estimation techniques [22, 23].

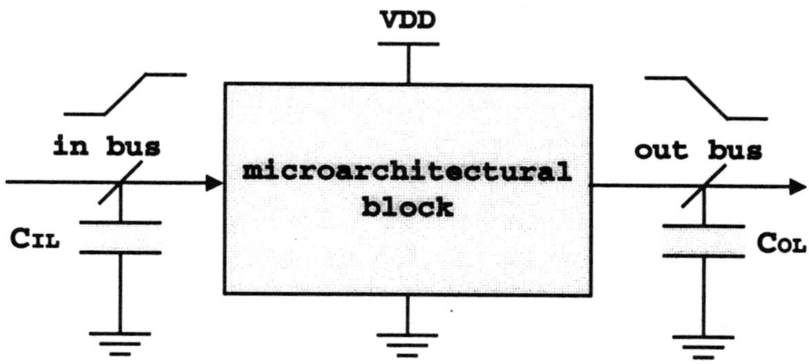

Figure 16.6. The Microarchitectural Block Model

In order to analyze Hamming distance or zero-to-one transition activity we have to keep the previous bus stream values. However, there are different types of bus states to consider — unchanged , pull-up, pull-down, and high impedance. For instance, a directional bus usually keeps its value until the output driver of the microarchitectural block drives a different value on the bus, while a bi-directional bus that is in the high impedance state when it is not driven by any block has an unpredictable value due to leakage current. This value becomes predictable if keeper logic is used to prevent excessive current flow due to leakage current of the bus. Accounting for all these cases is not difficult, but it does require that we have to specify the bus type when we initialize the data structure shown in Figure 16.7.

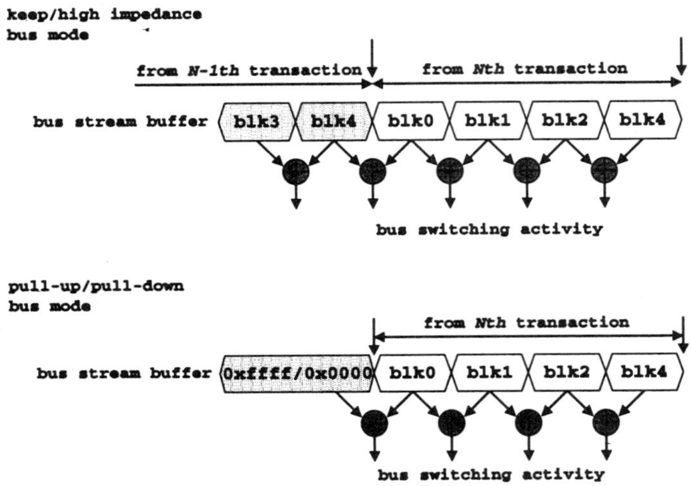

Figure 16.7. Bus Modeling with Data-Activity

Finally, microarchitectural blocks can be decomposed into more detailed blocks to increase the granularity and the accuracy of the power estimation. For instance, a cache can be viewed as one microarchitectural block, or split into decoders, tag arrays, data arrays, comparators, sense amplifiers, and output drivers, which we can model as a collection of separate blocks, assigning different power models to each block, see Figure 16.8.

5.2 Cycle Simulator and Power Estimator Interface

In Order To Track The Bus Streams For Each Block we need to implement an interface or API to collect the stream information for each microarchitectural block from the cycle simulator. The relative timing of the bus streams are dependent on the microarchitectural block and its current state. For example, reads and writes to a memory block create quite different bus switching activities, which are further modified by the occurrence of a cache miss.

Furthermore, as mentioned before, the cycle simulator often simulates the logical behavior of a microarchitectural block in a single cycle and then uses

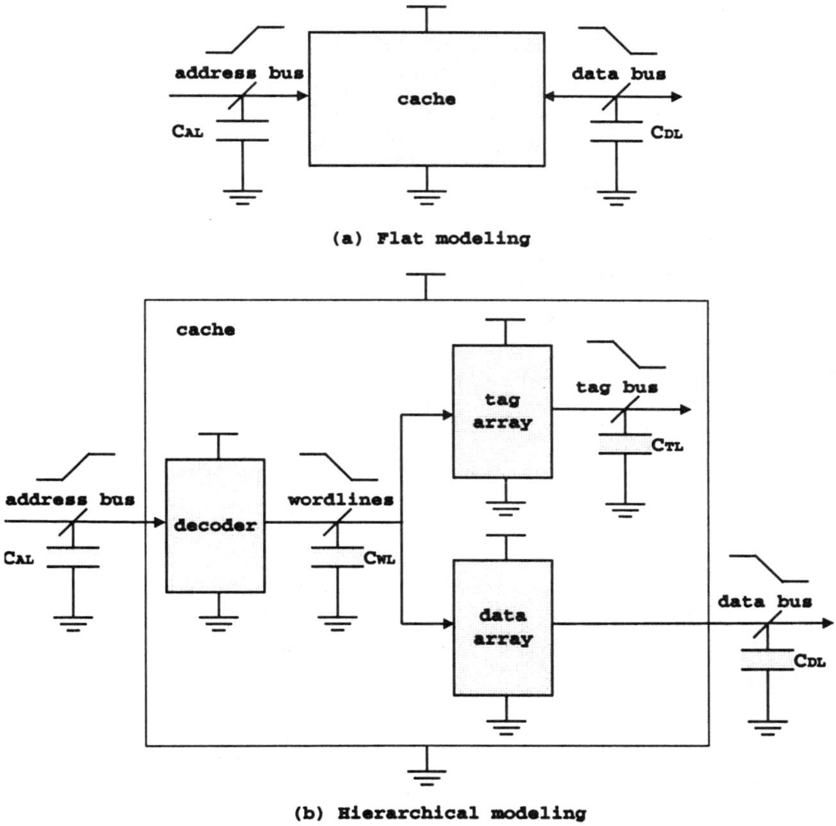

Figure 16.8. Hierarchical Modeling of Microarchitectural Blocks

latency and hazard information to model the performance impact of the microarchitectural activities that may occur over several cycles. Figure 16.9 illustrates how stream information should be retained to correctly model the real data activity occurring on the bus.

5.3 Power Modeling Techniques

The estimation of the power dissipation is quite straightforward if we know the data activity of the microarchitectural block so that we can derive the effective capacitance of the block. In this section we will examine in more detail how we can derive the effective capacitance for several important microarchi-

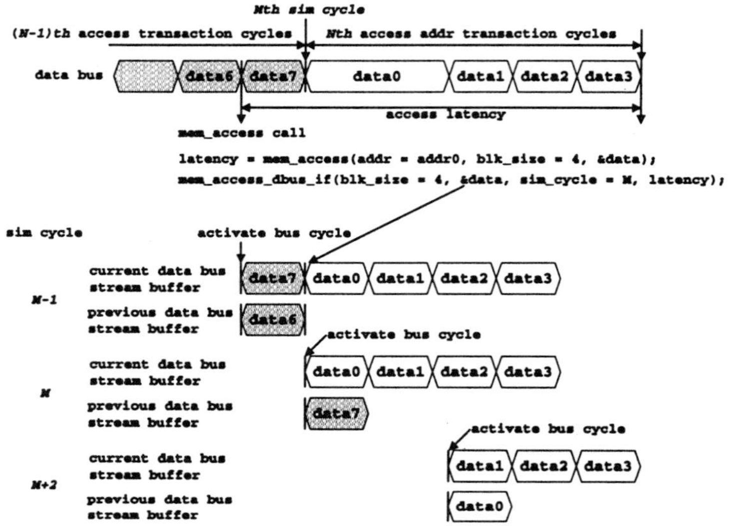

Figure 16.9. Interfaces between the Simulator and the Power Estimation Framework, and Activation of Bus Transaction Cycles

tectural blocks, the internal buses that connect them, and the clock tree that synchronizes them.

5.3.1 Memory Models. A microprocessor may contain many memory blocks including caches, TLBs, register files, reorder buffers, etc. Their fraction of the total area and their contribution to power consumption in a microprocessor can be as high as 40% [14]. Most memory structures consisting of five parts — an SRAM cell array, a row decoder/wordline driver, a column decoder, a sense amplifier, and a precharge circuit, see Figure 16.10.

The power consumption of most memory blocks can be estimated quite accurately by using a cache access time estimation model. Such models typically derive the access time from estimates of the capacitance and the resistance of the critical path during memory reads or writes. We can use these capacitances to obtain estimates of the power or energy consumption. Two examples are CACTI [8] and the analytical equations proposed in [14]. Both CACTI and the model in [14] compute the physical capacitance of each stage of the memory.

We illustrate the method using CACTI. The total switching capacitance of a bit line C_{BitTot} represented in (16.2) consists of the metal line capacitance C_{bit}

and the drain capacitance of transistors Q_1, Q_p, and Q_{pa} in Figure 16.10.

$$C_{BitTot} = N_{rows} \cdot (C_{d,Q_1} + C_{bit}) + C_{d,Q_p} + C_{d,Q_{pa}} \tag{16.2}$$

Similarly, the total switching capacitance of a word line shown in (16.3) is composed of the metal line capacitance C_{word} and the gate capacitance of the pass gates Q_1 and Q_2.

$$C_{WordTot} = N_{cols} \cdot (2 \cdot C_{g,Q_1} + C_{word}) \tag{16.3}$$

Therefore, the power dissipation of the SRAM array can be represented by (16.4). In the power dissipation of the bit line $V_{BitSwing}$ should be used instead of V_{DD} because the voltage swing of the bit line is less than the full supply voltage.

$$P_{SRAMArray} = f \cdot (N_{cols} \cdot C_{BitTot} \cdot V_{BitSwing}{}^2 + C_{WordTot} \cdot V_{DD}{}^2) \tag{16.4}$$

In addition, there are two types of the bit lines used in the memory blocks. One is a single-ended bit line structure, and the other is a double-ended bit line structure. The bit lines dissipate most of the power in the memory structure. The power dissipation of the double-ended bit line memory structure is independent of the data activity of the memory cell because the operation of each bit line is complemented regardless of the contents of the SRAM cells.

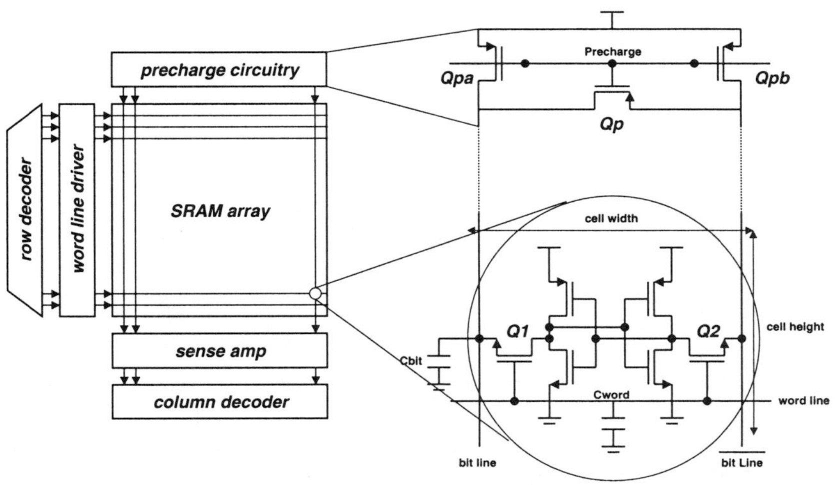

Figure 16.10. Memory Structure

In case of the single-ended bit line, its power dissipation of the bit lines depends on the values read from the cells. Thus we have to account for its data activity of the bit lines

5.3.2 Datapath Components. A datapath typically contains ALUs, shifters, multipliers, register files, etc. Although most of datapath components also have very regular structure it is not a trivial problem to estimate the power dissipation because the effective switching capacitance is a complex function of the applied input sequences, and can vary non-linearly with the bit-width of the datapath.

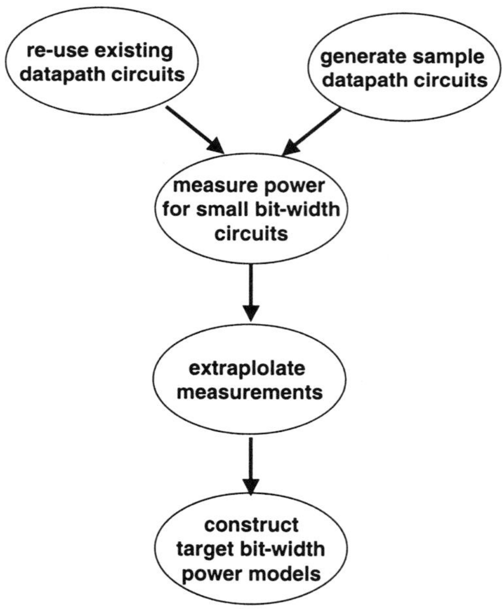

Figure 16.11. Power Model Construction Flow for Microarchitectural Power Estimation

Figure 16.11 represents the flow for the microarchitectural block power estimation. Either, we re-use small datapath component models, e.g., 4-bit, 8-bit, and 16-bit width of transistor- or gate-level existing designs or we generate combinational macro models for the datapath components using an HDL, a design library, and a synthesizer [20]. Second, we measure the average power dissipation for each possible Hamming distance by applying a large number of input sequences that have the same Hamming distance. Finally, we can extrapolate the results with regression-based techniques [17] to obtain the power model for 32-bit or 64-bit width datapath component.

This power modeling methodology has several advantages over other estimation techniques used in earlier power estimators. First, we can consider the effect of the technology parameters directly on the power dissipation because we measure the power consumption of the small block with the power simulator and the technology parameters.

Second, since the datapath components have a very regular structure we can reduce the power model construction time by using regression-based technique. The regularity makes it possible to apply regression-based estimation techniques without losing much accuracy.

Third, the Hamming distance based power estimation is simple to apply to the microarchitectural simulation if one traces the change of the input/output values presented to the blocks (the bus streams) during a simulation run.

5.3.3 Random Logic and Interconnections. There is no exact way to predict, in the early design phase, the form of the random logic used for control. However, there are some empirical models based on Rent's rule [15, 16], which rely on parameters such as transistor count, area, the number of the pins, logic depth, etc. These model parameters can be determined based on similar existing designs.

Interconnection information cannot be determined at the microarchitectural design stage, but we can estimate its length by rough floorplaning. Its capacitance can be estimated by:

$$C_{Int} = L_{Int} \cdot W_{Int} \cdot C_{Area} + 2 \cdot (L_{Int} + W_{Int}) \cdot C_{Length} \qquad (16.5)$$

In (16.5) C_{Int} is the interconnection capacitance of the internal input and output buses between the microarchitectural blocks, L_{Int} is the interconnection length, W_{Int} is the width of the interconnection line, C_{Area} is the capacitance per unit area, and C_{Length} is the capacitance per unit length of the interconnection layer. The estimated interconnect capacitance for the interconnect between each major microarchitectural block is assigned to the data structure associated with each microarchitectural block. This is combined with the simulation statistics of the internal bus transition activity derived from the cycle simulation to estimate the power dissipation of the interconnect bus, see (16.6).

$$P_{Int} = HammingDist \cdot f \cdot C_{Int} \cdot {V_{DD}}^2 \qquad (16.6)$$

5.3.4 Clock Distribution Tree. According to [24] the power dissipation of the clock distribution network can be responsible for 40% of the total power dissipation in high performance microprocessor design. As the chip size and the clock frequency increase the fraction of the power dissipation by the clock distribution network is becoming ever more significant.

There are several clock distribution styles. The most common are the H-tree and the balanced H-tree. The nodes of the clock tree includes all clocked transistors in the microprocessor core logic and memory, as well as the clock wiring and the clock driver. In addition, the clock network is distributed over the entire chip and therefore related to the overall chip area. The total load capacitance for the clock distribution tree contains three components: clock tree wiring, random logic and memory clocked nodes [15]:

$$C_{ClkTot} = C_{ClkWire} + C_{ClkLogic} + C_{ClkMemory} \tag{16.7}$$

In (16.7) $C_{ClkWire}$ is a function of the interconnection capacitance of clock wire per length $C_{Int/Length}$, the total area A_{Tot} of the estimated chip die size, and the number of the levels in the clock distribution tree, which in turn is a function of the target clock skew and chip die size, see (16.8) and Figure 16.12:

$$C_{ClkWire} = C_{Int/Length} \cdot A_{Tot} \cdot \sum_{i=1}^{N_{tree}} 2^{i-1} \cdot \frac{1}{2^{\lfloor i/2 \rfloor + 1}} \tag{16.8}$$

The effective switching capacitance of the clock distribution tree C_{ClkEff} in-

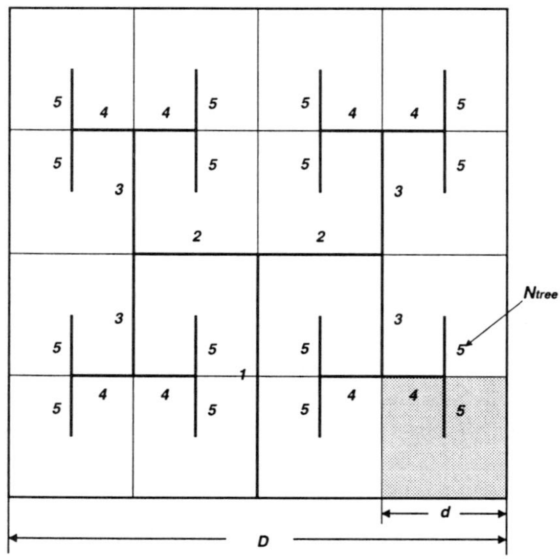

Figure 16.12. An Example of H Clock Distribution Network

cludes the switching capacitance of the chain of the inverters forming the clock driver. This can be represented by:

$$C_{ClkEff} = C_{ClkTot} \cdot \left(\frac{1}{1 - 1/a_{clkdriver}} + 1 \right) \quad (16.9)$$

where $a_{clkdriver}$ is the optimal stage ratio for the clock driver. From (16.7),(16.8), and (16.9) the total power dissipation of H-tree clock distribution network can be represented in terms of C_{ClkEff} as follows:

$$P_{Clk} = f_{Clk} \cdot C_{ClkEff} \cdot V_{DD}^2 \quad (16.10)$$

6. Conclusion and Future Work

Microarchitectural level tools are important for power estimation during the early design phase of microprocessor systems, particularly as we continue to explore new low power optimization techniques. In earlier work we found that recent power estimation tools are inaccurate enough that they may lead designers to make the wrong choices when deciding between several design trade-offs. In this paper we list the potential sources of error that could contribute to this inaccuracy. They are connected with the abstractions that are employed by cycle simulators. We propose a detailed methodology for reinstating some of the details lost in the abstraction.

Several important questions remain. The first is how much the power modeling slows simulation. We have implemented a preliminary version of this power estimation simulator on top of SimpleScalar and have observed slow downs of about a 2x. The second question is how much accuracy is regained by our proposed methodology. This requires calibration against existing chips and remains for future work.

Acknowledgments

This effort has been sponsored by the Defense Advanced Research Project Agency (DARPA) and Air Force Research Laboratory under contract No. F33615-00-C-1678. The U.S. Government is authorized to reproduce and distribute reprints for governmental purposes notwithstanding any copyright annotation thereon.

The views and conclusions contained herein are those of the authors and should not be interpreted as necessarily representing the official policies or endorsements, either expressed or implied, of the Defense Advanced Research Projects Agency (DARPA), the Air Force Research Laboratory, or the U.S. Government.

References

[1] Trevor Mudge, "Power: A First-Class Architectural Design Constraint," IEEE Computer, Vol 34. No. 4, April 2001.

[2] G. Cai, and C. H. Lim, "Architectural Level Power/Performance Optimization and Dynamic Power Estimation," Cool Chips Tutorial in conjunction with MICRO 32, November 1999.

[3] D. Brooks, V. Tiwari, and M. Martonosi, "Wattch: A Framework for Architectural-Level Power Analysis and Optimizations," Proceedings of 27th International Symposium on Computer Architecture, May 2000.

[4] N. Vijaykrishnan, M. Kandermir, M. J. Irwin, H. S. Kim, and W. Ye, "Energy-Driven Integrated Hardware-Software Optimizations Using SimplePower," Proceedings of 27th International Symposium on Computer Architecture, May 2000.

[5] S. Ghiasi, and D. Grunwald, "A Comparison of Two Architectural Power Model," Power Aware Computer Systems Workshop in conjunction with ASPLOS-IX, November 2000.

[6] T. M. Conte, M. A. Hirsh, and K. N. Menezes, "Reducing State Loss for Effective Trace Sampling of Superscalar Processors," Proceedings of International Conference on Computer Design, October 1996

[7] A. Sinha, and A. P. Chandrakassan, "JouleTrack - A Web Based Tool for Software Energy Profiling," Proceedings of 38th Design Automation Conference, June 2001

[8] S. Wilton, and N. Jouppi, "An Enhanced Access and Cycle Time Model for On-Chip Caches," Western Research Laboratory Technical Report 93/5, July 1994

[9] H. Mehta, R. M. Owens, and M. J. Irwin, "Energy Characterization based on Clustering," Proceedings of 33rd Design Automation Conference, June 1996

[10] P. E. Landman and J. M. Rabaey, "Activity-Sensitive Architectural Power Analysis," IEEE Transaction on CAD of Integrated Circuit and Systems, Vol. 15, No. 6, June 1996

[11] P. E. Landman and J. M. Rabaey, "Architectural Power Analysis: The Dual Bit Type Method," IEEE Transaction on VLSI Systems, Vol. 3, No. 2, June 1995

[12] G. Jochens, L. Kruse, E. Schmidt, and W. Nebel "A New Paramterizable Power Macro-Model for Datapath Components," Proceedings of Design Automation and Test in Europe Conference 1999

[13] D. Burger, and T. M. Austin, "The SimpleScalar Tool Set, version 2.0," ACM SIGARCH Computer Architecture News, Vol. 25, No. 3, June 1997

[14] M. B. Kamble, and K. Ghose, "Analytical Energy Dissipation Model for Low-Power Caches, Proceedings of International Symposium on Low Power Electronics and Design, August, 1997

[15] B. Geuskens, and K. Rose, "Modeling Microprocessor Performance," Kluwer Academic Publishers, 1988

[16] Bakoglu, "Circuits, Interconnections, and Packaging for VLSI," Addison Wesley, 1989

[17] A. Bogliolo, L. Benini, and G. De Micheli, "Regression-based RTL Power Modeling," ACM Transaction on Design Automation of Electronic Systems, Vol. 5, No. 3, July 2000

[18] L. Benini, A. Bogliolo, E. Macii, M. Poncino, and M. Surmei, "Regression-based RTL Power Models for Controllers," Proceedings of the 10th Greate Lakes Symposium on VLSI, March 2000

[19] S. Theoharis, G. Theodoridis, P. Merakos, and C. Goutis, "Accurate Data path models for fast RT-Level Power estimation," IEE Proceedings of Computers and Digital Techniques, Vol. 147, No. 4, July 2000

[20] A. Bogliolo, R. Corgnati, and E. Macii, and M. Poncino, "Parameterized RTL Power Models for Combinational Soft Macros," Proceedings of International Conference on Computer-Aided Design, November 1999

[21] M. M. Khellah, and M. I. Elmasry, "Effective Capacitance Macro-Modeling for Architectural-Level Power Estimation," Proceedings of the 8th Great Lakes Symposium on VLSI, 1998

[22] C. Zhanping, K. Roy, and E. K. Chong, "Estimation of Power Dissipation Using a Novel Power Macromodeling technique," IEEE Transaction on CAD of Integrated Circuits and Systems, Vol. 19, No. 11, November 2000

[23] S. Gupta, F. N. Najm, "Power Modeling for High-Level Power Estimation," IEEE Transaction on VLSI Systems, Vol. 8, No. 1, February, 2000

[24] P. E. Gronowski, W. J. Bowhill, et al., "High Performance Microprocessor Design," IEEE Journal of Solid-State Circuits, Vol. 33, No. 5, May 1998

Chapter 17

SOFTWARE ENERGY PROFILING

Amit Sinha, Anantha Chandrakasan
Massachusetts Institute of Technology
Cambridge, MA 02139
{sinha, anantha}@alum.mit.edu

Abstract A software energy estimation methodology is presented that avoids explicit instruction level analysis and predicts energy consumption to an accuracy within 3% for a set of benchmark programs evaluated on the StrongARM SA-1100 and Hitachi SH-4 microprocessors. A technique has also been proposed to separate the switching and leakage energy components of software. A web-based tool, JouleTrack, has been developed that incorporates the proposed techniques.

Keywords: software energy, low power software, leakage, JouleTrack

1. Introduction

Power aware computing is becoming fundamentally important with the proliferation of portable, battery operated systems, such as laptops, Personal Digital Assistants (PDAs), cellular phones, etc. Most of these systems consist of a substantial chunk of software running on programmable platforms such as low power microprocessors and Digital Signal Processors (DSPs). Power aware software is therefore integral to the realization of low power consumption in such systems. Software power estimation is the indispensable first step towards software power optimization. Robust models that predict the energy cost of software are non-trivial to determine. In this chapter, we propose a software energy estimation framework based on experiments done on the StrongARM SA-1100 and Hitachi SH-4 processors. In battery operated systems, energy is a preferred metric over power since battery life is directly related to energy consumption. Our proposed estimation framework has been incorporated into a web-based tool, JouleTrack, that allows a user to upload a piece of C code and outputs energy measurements on the remote web browser [1, 2].

2. Factors Affecting Software Energy

Software energy estimation is non-trivial because of the dependence of energy consumption on the following parameters.

- *Target processor and Instruction Set Architecture (ISA)* - Software energy is an obvious function of the processor, data path and instruction set used. For example, the StrongARM microprocessor does not have a dedicated floating point unit. The energy required by a floating point operation on the StrongARM will be more than that on an equivalent processor with a well designed floating point unit. This is because a floating point operation on the StrongARM uses a software emulation to implement it and requires two orders of magnitude more time to execute than an equivalent integer operation. Other architectural parameters which strongly affect the energy consumption of a processor are cache size, datapath width, number of functional units (multipliers, shifters), register file size, legacy support hardware, multimedia extension support, etc. In general, it is practically impossible to predict how much energy a piece of software will consume on another processor given the energy consumption profile on one processor, without some prior calibration and measurement on the other one.

- *Operating voltage and frequency* - Using first order CMOS transistor power and delay models, it can be shown that switching energy consumed in a digital circuit depends linearly on the operating frequency and quadratically on the supply voltage [10].

- *Leakage and duty cycle* - As the threshold voltages scale, leakage energy will become more dominant. Already, leakage accounts for 10% of energy dissipation on the StrongARM at 100% duty cycle. At 10% duty cycle the switching and leakage energy components become comparable (as we will show). Effective macro leakage models (module wise, if possible) would have to be developed that can isolate switching and leakage energy components before techniques can devised to minimize their effect without significant loss in performance.

3. Related Work

Software energy estimation through exhaustive instruction energy profiling was first proposed in [3]. The instantaneous current drawn by a CPU varies rapidly with time showing sharp spikes around clock edges. This is because at clock edges the processorś circuits switch (i.e., get charged or discharged) resulting is switching current consumption. Expensive hardware with a lots of measurement bandwidth and low noise is required to accurately track the CPU instantaneous current consumption. From a energy measurement perspective, however, we are interested in average current consumption. To a first order, battery lifetime depends on the average current and the amount of stored energy in the cell. Measuring average current is simpler and can be done using a current meter. The current meter averages the instantaneous current over an averaging window and the corresponding readings are stable average values. The average current itself varies as the program executes. The basic approach proposed in [3] can be summarized as follows:

- Measure the current drawn by the processor as it repeatedly executes a certain instruction or sequences of instructions. This is done by putting the sequence in a loop and measuring current values. The measured values are the instruction base current cost.
- The program is broken up into basic blocks and the base cost of each instance of a block is obtained by adding the base cost of instructions in the block. These costs are provided in a base cost table. Obtain a run-time execution profile (instruction trace) for the program. Using this information the number of times the basic block is executed is determined and overall base cost is computed.
- The effect of circuit state (inter-instruction effects) is incorporated by analyzing pairs of instructions. A cache simulation is also performed to determine stalls and a cache penalty is also added to the final estimate.

The principal disadvantage of this approach is that it involves elaborate instruction trace analysis. Assuming an ISA with N instructions, N^2 instruction energy profiles have to be obtained to accurately determine base and inter-instruction costs. Moreover, most instructions have a lot of variations (e.g., based on addressing modes, immediate operands, etc.) and exhaustive characterization is very time-consuming.

4. Proposed Methodology

4.1 Experimental Setup

The StrongARM setup consists of the Brutus SA-1100 Design Verification Platform which is essentially the StrongARM SA-1100 microprocessor connected to a PC using a serial link. The SA-1100 consists of a 32-bit RISC processor core, with a 16 KB instruction cache and an 8 KB write-back data cache, a minicache, a write buffer, and a Memory Management Unit (MMU) combined in a single chip. It can operate from 59 MHz to 206 MHz, with a corresponding core supply voltage of 0.8 V to 1.5 V. Power supply to the StrongARM core was provided externally through a variable voltage sourcemeter. The I/O pads run at a fixed supply voltage. The ARM Project Manager (APM) is used to debug, compile and execute software on the StrongARM. Current measurements are performed using the sourcemeter built into the variable power supply. The instruction and data caches are enabled before the programs are executed. To measure the current that is drawn by a subroutine, the subroutine is placed inside a loop with multiple iterations till a stable value of current is measured. The execution time for multiple iterations is obtained accurately using the time() utility in C and the execution time per iteration and charge consumption are subsequently computed. The core supply voltage is altered directly from the external supply while the internal clock frequency of the processor is changed via software control. Details of the StrongARM setup can be found in [4, 5]. Figure 17.1 shows the experimental setup for the StrongARM SA-1100 processor. A similar board was setup for the Hitachi SH-4 processor using the SH7750 Solution Engine [6]. The SH7750 series (SH7750, SH7750S) is a 32-bit RISC microprocessor. It includes an 8 KB instruction cache, a 16 KB operand cache with a choice of copy-back or write-through mode, and an MMU with a 64-entry fully-associative unified Translation Lookaside Buffer (TLB). The SH7750 series has an on-chip bus state controller (BSC) that allows connection to DRAM and synchronous DRAM. Its 16-bit fixed-length instruction set enables program code size to be reduced by almost 50% compared with 32-bit instructions. The SH-4 runs at 200 MHz fixed clock frequency at a core voltage of 1.95 V with 3.3 V for I/O pads. It is a 2-way superscalar CPU, i.e., it can execute 2 instructions per cycle. Unlike the SA-1100, the SH-4 features a fully integrated floating point unit with a separate bank of 16 single precision floating point registers. All this added performance comes at an increased energy cost. While the SA-1100 operates at 1 nJ/Instruction, the SH-4 requires 4.2 nJ/Instruction cycle.

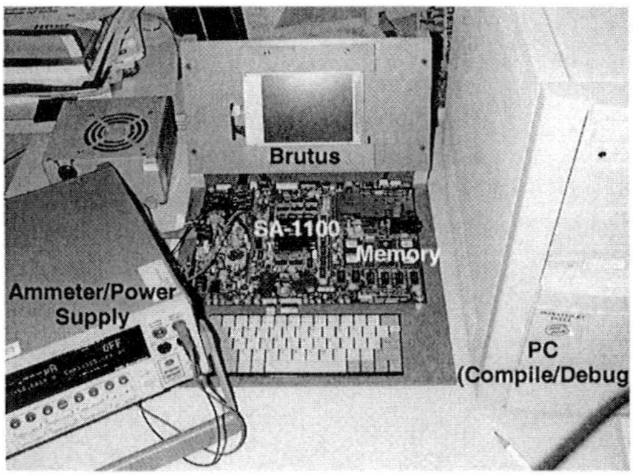

Figure 17.1. StrongARM SA-1100 Experimental Setup

4.2 Instruction Current Profiles

Our experiments on the StrongARM SA-1100 and Hitachi SH-4 microprocessors show that the variation in the current consumption across instructions is quite small. A lot of overheads are common across instructions and, as a result, the overall current consumption of a program is a weak function of the actual instruction stream and to a first order depends only on the operating frequency and voltage. Second order variations do exist but were measured to be less than 7% for a set of benchmark programs. Therefore, a complete instruction level trace analysis is unnecessary and a simple cycle accurate timing simulation can be used along with a simplified instruction/program current model to estimate overall energy consumption.

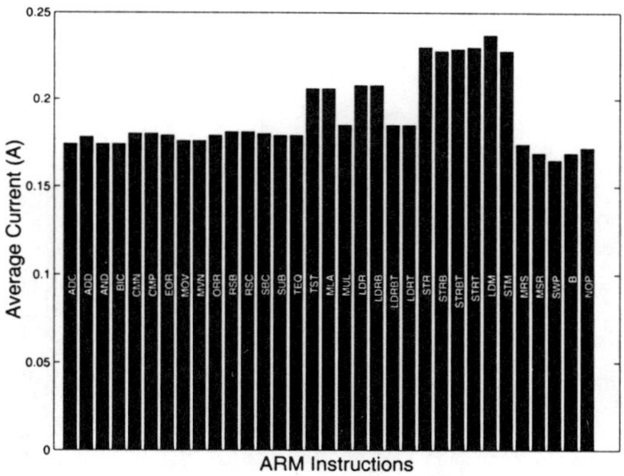

Figure 17.2. Strong SA-1100 Average Instruction Set Current Consumption

Figure 17.2 shows the current consumption of all the instructions of the ARM instruction set on SA-1100. Each of the 33 current values are themselves the averages of the various addressing modes and inputs in which the instruction can be executed, accounting for a total of about 280 data points. The important point to observe is that the current consumptions are pretty uniform. On an average, arithmetic and logical instructions consume 0.178 A, multiplies 0.196 A, loads 0.196 A, stores 0.229 A, while the other instructions consume about 0.170 A. The total variation in current consumption is 0.072 A, which is 38% of the overall average current consumption. All current measurements were done at 1.5 V core supply and 206 MHz clock frequency.

Although the small variation in instruction currents might appear surprising at first, a deeper analysis reveals that this ought to be expected. Figure 17.3 shows the distribution of power consumption across various blocks in the ARM core [7]. Between the cache, control, global clock (GCLK) and I/O cir-

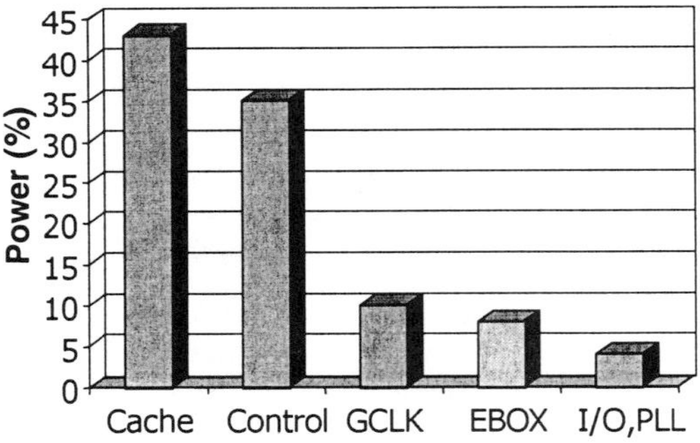

Figure 17.3. Distribution of Power Consumption in the ARM Core [7]

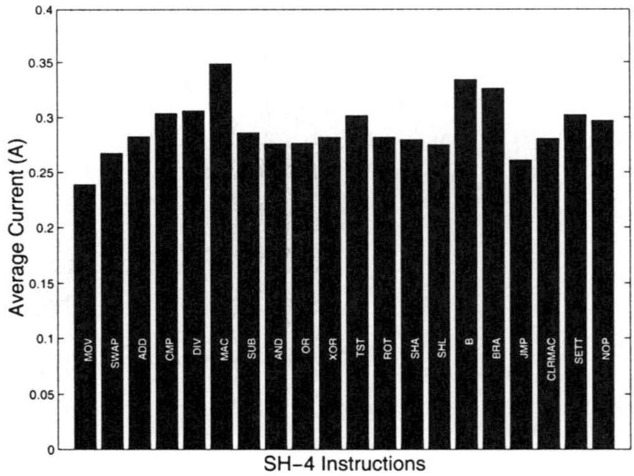

Figure 17.4. Hitachi SH-4 Average Instruction Set Current Consumption

cuits, almost 95% power is consumed. Since these overheads are common to all instructions, the variation of current consumption between instructions is small. The execution box (EBOX) is where the instruction is implemented. Figure 17.4 shows the core current consumption measured for the Hitachi SH-4 processor running at 2.0 V core power supply and 200 MHz clock frequency. The average instruction current is 0.29 A, with a variation of 0.11 A, which, once again, is 38% of the average. The small variation in instruction current can again be explained using a similar argument.

The current variation within an instruction (as a function of addressing modes and data) is even smaller. Figure 17.5 shows the current consumption of 3 different instructions as a function of various addressing modes and data on the SA-1100. In general, we observed that to a first order the instruction

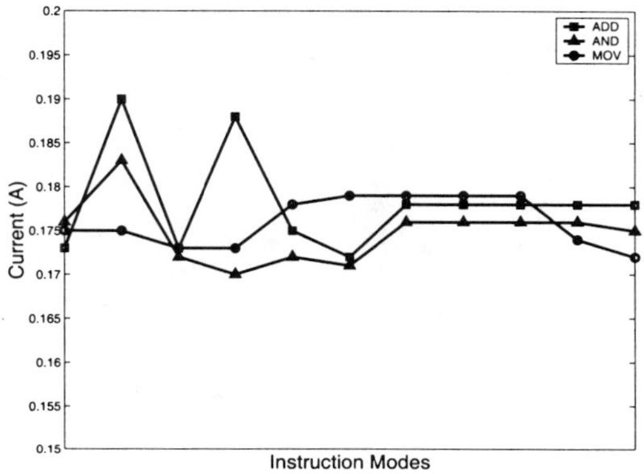

Figure 17.5. Current Variations within Instructions on the StrongARM

current consumptions are independent of the addressing modes or operands. We now propose a simple and fast technique to estimate software energy. Our experiments indicate an accuracy within 3% of actual measurements.

4.3 First Order Model

While the instruction level current consumption has a variation of about 38%, the variation of the current consumption between programs is much less. Figure 17.6 shows the current consumption of six different benchmark programs at different supply voltage and frequency levels on the StrongARM. The maximum current variation is about 8%. This implies that, to a first order, current consumption of a piece of code is independent of the code and depends only on the operating voltage and frequency of the processor. The first order software energy estimation model is then simply

$$E_{tot} = V_{dd} I_0(V_{dd}, f)\Delta t \tag{17.1}$$

where V_{dd} is the supply voltage and Δt is the program execution time.

4.4 Second Order Model

While the current variation across programs is quite small in the StrongARM, it might be significant in datapath dominated processors. For example, the current consumption of the multiply instruction in DSPs will be far greater than the current consumption of other instructions. In such cases a simple model like Equation 17.1 will have significant error. The following second order model is proposed. Let C_k ($k \in [0, K-1]$) denote K energy differentiated instruction classes in a processor. Energy differentiated instruction

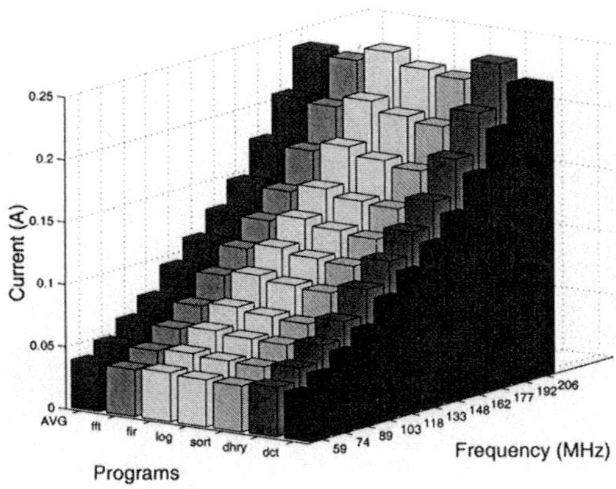

Figure 17.6. Program Current Consumption as a Function of Operating Point

classes are partitions of the instruction set such that the average current consumption of any class is significantly different from that of another class and the current variation within a class is small. Class partitions can also be done on the basis of different cycles (e.g., instruction, data access, etc.). Let c_k denote the fraction of total cycles in a given program that can be attribute to instructions/cycles in the class C_k, i.e., $\sum c_k = 1$. The second order current consumption is estimated as

$$I(V_{dd}, f) = I_0(V_{dd}, f) \sum_{k=0}^{K-1} w_k c_k \tag{17.2}$$

where w_k are a set of weights. Let $\mathbf{W}$ represent the vector $[w_0, w_1, .., w_{K-1}]$. Let P_n $(n \in [0, N-1])$ represent a set of N benchmark programs, C_n denote the cycle fractions vector for program P_n, i.e., $[c_0^n, c_1^n, .., c_{K-1}^n]$, and I_n denote its average current consumption. Based on Equation 17.2, we can express the current vector $\mathbf{I}$ in the following form

$$\mathbf{I} = I_0(V_{dd}, f)\mathbf{CW} \tag{17.3}$$

where $\mathbf{I}$ is the average current $[I_0, I_1, .., I_{N-1}]$ for the N programs, $\mathbf{C}$ is a $N \times K$ matrix with C_n as the nth row. The weights can be solved for in a least mean square sense using the pseudo-inverse

$$\mathbf{W} = \frac{1}{I_0(V_{dd}, f)}(\mathbf{C}^T\mathbf{C})^{-1}\mathbf{C}^T\mathbf{I} \tag{17.4}$$

If the instruction classes are a valid energy differentiated partition, the weighting vector $\mathbf{W}$ will reflect the energy differentiation. The maximum current prediction error will also go down considerably.

On the StrongARM SA-1100, we partitioned the cycles into 4 classes - (i) Instruction, (ii) Sequential memory access (accesses which are related to previous ones) (iii) Non sequential accesses, and, (iv) Internal cycles. Current measurements were done for six benchmark programs running at all possible frequency and voltage combinations. The weighting vector is shown in Table 17.1. The average current drawn at each operating frequency of the StrongARM is shown in Figure 17.7. The StrongARM operates at 11 discrete frequency levels. The minimum operating voltage required is also shown and is almost linear with frequency. In fact, the normalized operating voltage and frequency are related

$$\overline{V}_{dd} = 0.66\overline{f} + 0.33 \tag{17.5}$$

where $\overline{V}_{dd}$ and $\overline{f}$ are the normalized to their respective maximum values.

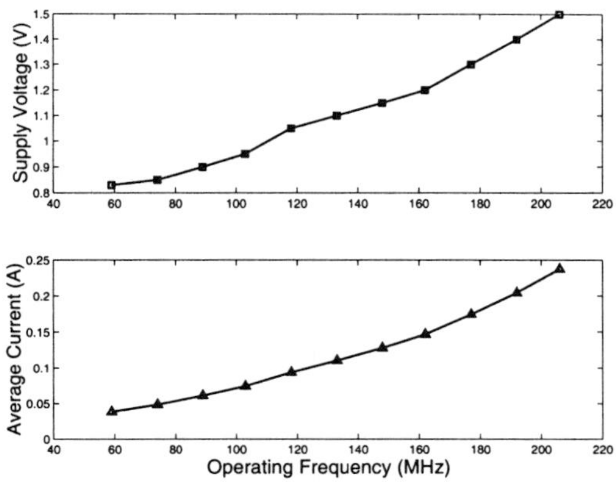

Figure 17.7. Average Current and Supply Voltage at Each Operating Frequency of the StrongARM SA-1100

The weighting factors can be interpreted as follows. For programs where instruction cycles and non-sequential memory accesses dominate, the current consumption is higher than the average current at that operating point. Internal cycles and sequential memory access dominated programs will have a lower than average current consumption. The current prediction error with a second order model can be reduced to less than 2%. The advantage is that this accuracy comes for free. No elaborate instruction level profiling is required. Such cycle counts as the ones shown in Table 17.1 are easily obtained using simulators/debuggers available with standard processors. Figure 17.8 shows the overall improvement of current prediction accuracy on 66 test points. It can be seen that the current prediction is better in every case (the maximum error of

Table 17.1. Weighting factors for K = 4 on the StrongARM SA-1100

Class	Weight	Value
Instruction	w_1	2.1739
Sequential memory access	w_2	0.0311
Non-sequential memory access	w_3	1.2366
Internal cycles	w_4	0.8289

4.7% using a first order model is reduced to 2.3%). The effectiveness of the current weighting scheme will become more pronounced in processors having a wider variation in average current consumption.

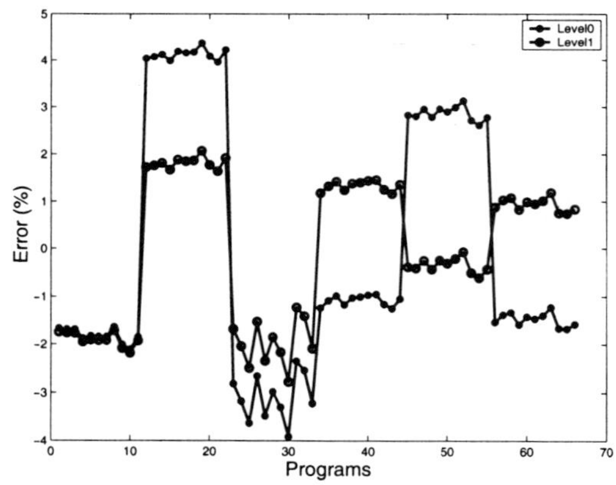

Figure 17.8. First and Second Order Model Prediction Errors

5. Leakage Energy Measurement

5.1 Principle

The power consumption of a subroutine executing on a microprocessor can be macroscopically represented as

$$P_{tot} = P_{dyn} + P_{stat} = C_L V_{dd}^2 f + V_{dd} I_{leak} \tag{17.6}$$

where P_{tot} is the total power, which is the sum of the static and dynamic components, C_L is the total average capacitance being switched by the executing program per clock cycle, and f is the operating frequency (assuming that there are no static bias currents in the microprocessor core) [10]. Let us assume that a subroutine takes Δt time to execute. This implies that the energy consumed by a single execution of the subroutine is

$$E_{tot} = P_{tot}\Delta t = C_{tot}V_{dd}^2 + V_{dd}I_{leak}\Delta t \quad (17.7)$$

where C_{tot} is the total capacitance switched by executing subroutine. Clearly, if the execution time of the subroutine is changed (by changing the clock frequency), the total switched capacitance, C_{tot}, remains the same. Essentially, the integrated circuit goes through the same set of transitions except that they occur at a slower rate. Therefore, if we execute the same subroutine at different frequencies, but at the same voltage, and measure the energy consumption we should observe a linear increase as the execution time increased, with the slope being proportional to the amount of leakage.

5.2 Observations

The subroutine chosen for execution was the decimation-in-time Fast Fourier Transform (FFT) algorithm because it is a standard, computationally intensive, DSP operation. The execution time for a $N = 1024$ point FFT on the StrongARM is a few tenths of a second and scales as $O(N \log N)$. To obtain accurate execution time and stable current readings, the FFT routine was run a few hundred times for each observation. A total of eighty different data points corresponding to different supply voltages between 0.8 V and 1.5 V and operating frequencies between 59 MHz and 206 MHz were compiled. Figure 17.9 illustrates the implications of Equation 17.7. When the operating frequency is fixed and the supply voltage is scaled, the energy scales almost quadratically. On the other hand, when the supply voltage is fixed and the frequency is varied the energy consumption decreases linearly with frequency (i.e., increases linearly with the execution time) as predicted by Equation 17.7. Not all frequency and voltage combinations are possible. For example the maximum frequency of the StrongARM is 206 MHz and it requires a minimum operating voltage of 1.5 V. The line across the surface plot demarcates the possible operating regions from the extrapolated ones (i.e., the minimum operating voltage for a given frequency).

We can measure the leakage current from the slope of the energy characteristics, for constant voltage operation. One way to look at the energy consumption is to measure the amount of charge that flows across a given potential. The charge attributed to the switched capacitance should be independent of the execution time, for a given operating voltage, while the leakage charge should increase linearly with the execution time. Figure 17.10 shows the measured charge flow as a function of the execution time for a 1024 point FFT. The amount of charge flow is simply the product of the execution time and current drawn. As expected, the total charge consumption increases almost linearly with execution time and the slope of the curve, at a given voltage, directly

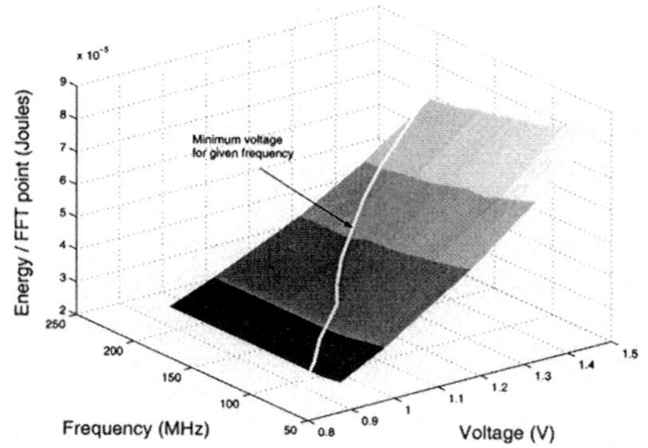

Figure 17.9. FFT Energy Consumption

gives the leakage current at that voltage. The dotted lines are the linear fits to the experimental data in the minimum mean-square error sense. At this point it is worthwhile to mention that the term leakage current has been used in an approximate sense. Truly speaking, what we are measuring is the total static current in the processor, which is the sum of leakage and bias currents. However, in the SA-1100 core, the bias currents are small and most of the static currents can be attributed to leakage. This assertion is further supported by the fact that the static current we measure has an exponential behavior as shown in the next section.

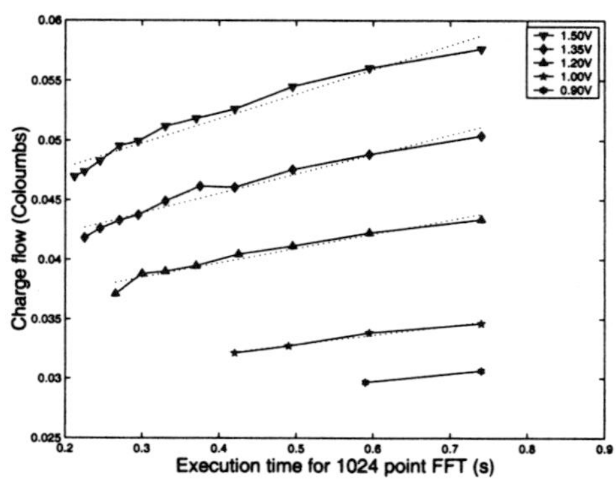

Figure 17.10. FFT Charge Consumption

From the BSIM2 MOS transistor model [8], the sub-threshold current in a MOSFET is given by

$$I_{sub} = Ae^{\frac{V_G - V_S - V_{TH0} - \acute{\gamma} V_S + \eta V_{DS}}{\acute{n} V_T}} \left(1 - e^{-\frac{V_{DS}}{V_T}}\right) \tag{17.8}$$

where V_T is the thermal voltage, V_{TH0} is the zero bias threshold voltage, $\acute{\gamma}$ is the linearized body effect coefficient, η is the Drain Induced Barrier Lowering (DIBL) coefficient and V_G, V_S and V_{DS} are the usual gate, source and drain-source voltages respectively. The important point to observe is that the subthreshold leakage current scales exponentially with the drain-source voltage.

The leakage current at different operating voltages was measured as described earlier, and is plotted in Figure 17.11. The overall microprocessor leakage current scales exponentially with the supply voltage. Based on these measurements the following model for the overall leakage current is proposed for the microprocessor core

$$I_{leak} = I_0 e^{\frac{Vdd}{nV_T}} \tag{17.9}$$

where $I_0 = 1.196$ mA and $n = 21.26$ for the StrongARM SA-1100.

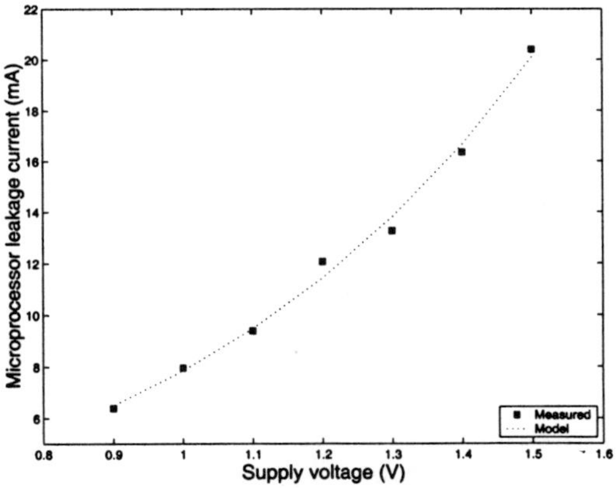

Figure 17.11. LeakagE Current Variation

5.3 Explanation of Exponential Behavior

The exponential dependence of the leakage current on the supply voltage can be attributed to the DIBL effect. Consider the stack of NMOS devices shown in Figure 17.12. Equation 17.8 suggests that for a single transistor, the leakage current should scale exponentially with $V_{DS} = V_{DD}$ because of the DIBL effect. However since the ratio V_{DS}/V_T is larger than 2, the term inside the brackets of Equation 17.8 is almost 1. It has been shown in [9] that this approximation is also true for a stack of two transistors. With three or more transistors, the ratio V_{DS}/V_T for at least the lowest transistor becomes comparable to or even less than 1. Therefore, the term inside the bracket of Equation 17.8 cannot be neglected for such cases. The leakage current progressively decreases as the number of transistors in the stack increases and for a stack of more than three transistors the leakage current is small and can be neglected. It has further been shown in [9] that the ratio of the leakage currents for the three cases shown in Figure 17.12 can be written as

$$I_1 : I_2 : I_3 = 1.8e^{\frac{\eta Vdd}{\bar{n}V_T}} : 1.8 : 1 \qquad (17.10)$$

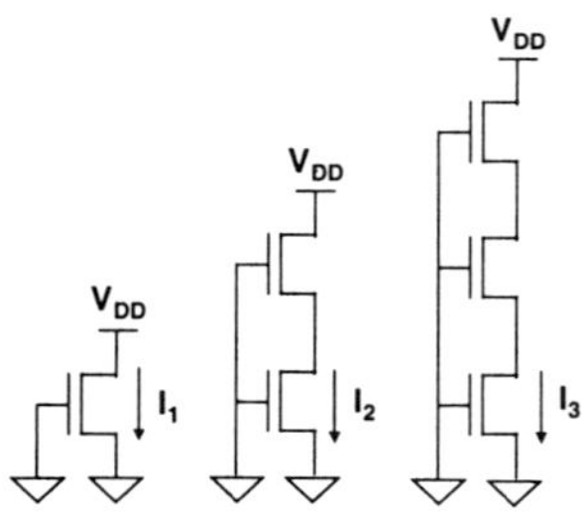

Figure 17.12. Effect of Transistor Stacking

Therefore, the leakage current of a MOS network can be expressed as a function a single MOS transistor (by accounting for the signal probabilities at various nodes and using the result of Equation 17.10). If the number of stacked devices is more than three, the leakage current contribution from that portion of the circuit is negligible. If there are three transistors stacked such that two of them are OFF and one is ON then the leakage analysis is the same as the stack of two OFF transistors. For parallel transistors, the leakage current is simply the sum of individual transistor leakages. A similar argument holds for PMOS devices. Since, the leakage current of a single MOS transistor scales exponentially with V_{DD}, using the above arguments, we can conclude that the total microprocessor leakage current also scales exponentially with the supply voltage.

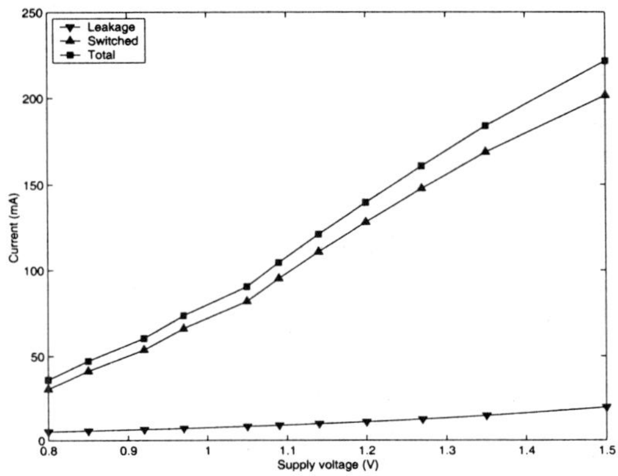

Figure 17.13. Static, Dynamic, and Total Current

Table 17.2. Leakage Current Measurements

V_{dd} (V)	Measured Leakage (mA)	Modelled Leakage (mA)	Error (%)
1.50	20.41	20.10	1.50
1.40	16.35	16.65	-1.84
1.30	13.26	13.80	-4.04
1.20	12.07	11.43	5.27
1.10	9.39	9.47	-0.87
1.00	7.96	7.85	1.40
0.90	6.39	6.53	-1.70

5.4 Separation of Current Components

Table 17.2 compares the measured leakage current with the values predicted by Equation 17.9. The maximum percentage error measured was less than 6% over the entire operating voltage range of the StrongARM, which suggests a fairly robust model. Based on the leakage model described by Equation 17.9, the static and dynamic components of the microprocessor current consumption can be separated as shown in Figure 17.13. The standby current of the StrongARM in the idle mode at 1.5 V is about 40 mA. This is not just the leakage current but also has the switching current due to the circuits that are still being clocked. On the other hand, this technique neatly separates the pure leakage component (assuming negligible static currents) from all other switching currents. For low threshold voltage microprocessors like the StrongARM, it can seen that the leakage current is quite substantial (about 10% in this case). The leakage current, as a fraction of the total current, can be expressed as

$$\beta_{leak} = \frac{I_0 e^{\frac{V_{dd}}{nV_T}}}{I_0 e^{\frac{V_{dd}}{nV_T}} + kV_{dd}^{\alpha}} \qquad (17.11)$$

where kV_{dd}^{α} models the switching current, has a very interesting profile. The leakage component, as a fraction of the total current, can be shown to have a minima at $V_{dd} = n\alpha V_T$, if we differentiate Equation 17.11 and solve for extrema. For the StrongARM, this is about 1.2 V as shown in Figure 17.14. The fact that such a minima occurs in measured current profiles further validates our model.

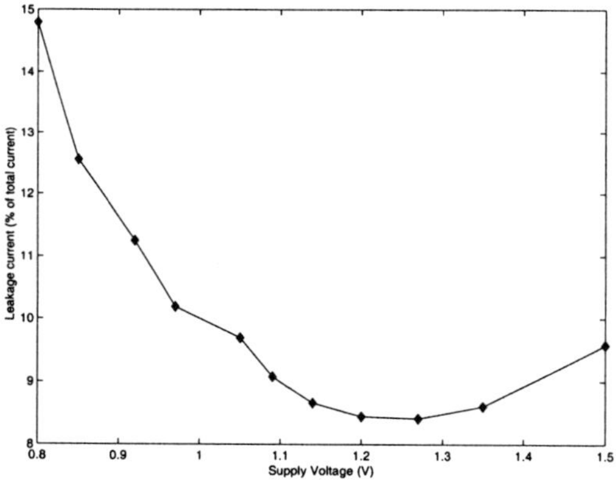

Figure 17.14. Leakage Current Fraction

5.5 Energy Trade-Offs

As the supply voltages and thresholds are reduced, system designers have to pay increasing attention to leakage currents. For the StrongARM, at maximum duty cycle and minimum voltage (for a given frequency), the leakage energy is about 10%. However, the leakage energy rises exponentially with supply voltage and decreases linearly with frequency as shown in Figure 17.15. Therefore, operating at a voltage, above the minimum possible, for a given frequency, is not advisable. In fact, if the operating voltage is sufficiently high for the particular operating frequency, leakage energy can start to dominate switching energy as shown in Figure 17.15! For low duty-cycle systems, the overall energy consumption becomes increasingly dominated by leakage effects. A fixed task consumes a certain amount of switching energy per execution while the system leaks during the idle mode between tasks. Extensive clock gating techniques,

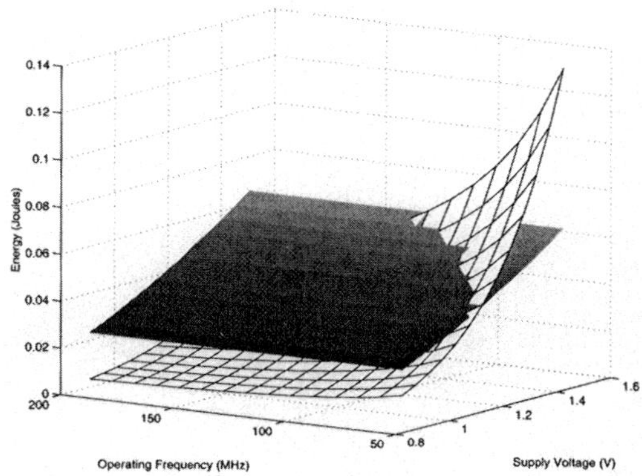

Figure 17.15. FFT Switching and Leakage Energy Components (Switching Shown as Continuous Surface)

such as those present in the StrongARM, reduce the unnecessary switching energy in the idle mode. The StrongARM also has a sleep mode where the supply voltage is reduced to zero for most circuits, and the processor state is stored. This significantly reduces the leakage problem. However, reverting to sleep mode between duty cycles may incur a lot of overhead (in terms of cycles and energy) or may not be supported by the target processor. Suppose the system

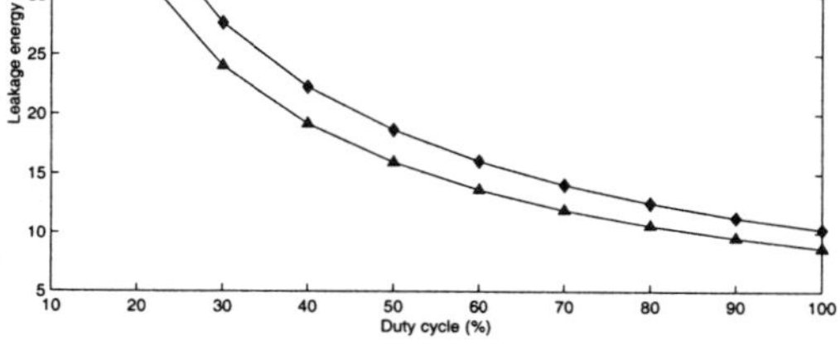

Figure 17.16. Low Duty Cycle Effects

has to do an FFT every T seconds such that the execution time for the FFT is $T_1 \leq T$. After computing the FFT, the processor enters idle mode and switching activity is reduced by clock gating techniques. Leakage on the other hand is unaffected. Figure 17.16 plots the leakage energy consumption (fraction of the total energy consumption), as a function of duty cycle, T_1/T. For a low duty cycle of 10% leakage energy is almost half the total energy consumption.

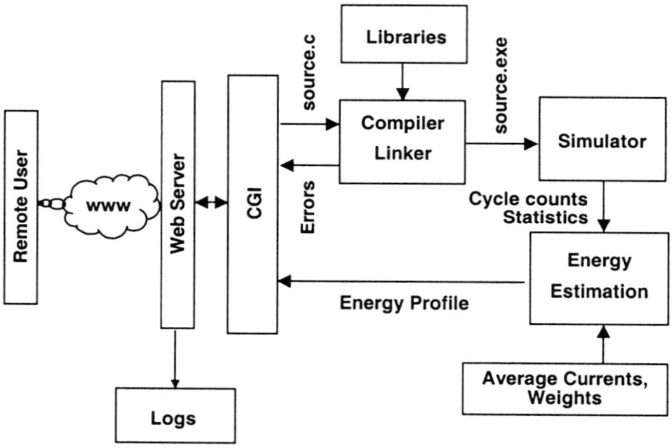

Figure 17.17. JouleTrack Block Diagram

6. JouleTrack

The estimation techniques described in the previous sections were implemented in a web-based tool called JouleTrack. The tool is available at http://dry-martini.mit.edu/JouleTrack. The broad approach in the tool is summarized in Figure 17.17. The user uploads his C source code. The webserver uses Common Gateway Interface (CGI) scripts to create a temporary workarea for the user. His programs are compiled and linked with any standard C libraries. The user also specifies any command line arguments that the program might need along with a target operating frequency. Compiler optimization options are also available. The user can choose the current prediction model. Compile/link time errors are reported back by the CGI to the user. If no errors exist the program is executed on an ARM simulator which produces the program outputs (which the user can view), assembly listing (which can also be viewed) as well as run-time statistics like execution time, cycle counts, etc. These statistics are fed into an estimation engine which computes the energy profile and charts the various energy components. Figure 17.18 shows the timing estimation engine used within JouleTrack. Energy is estimated as the product of execution time, average current and power supply voltage. The average current is estimated using the methodology described in the previous sections. To measure the execution time, a cycle-accurate simulator for the processor core is used. A memory map specifies the memory latency and the addressable memory space of the system. If no memory map is specified, a default memory map (with 32 bit addressable memory space and uniform access time is used). The timing engine also requires the operating frequency for the processor. The functionality of the simulator core can be enhanced by adding various modules (e.g., profilers, tracers, co-processor and OS models) using a Remote Debug Interface (RDI). Figure 17.19 depicts a few browser snapshots of JouleTrack.

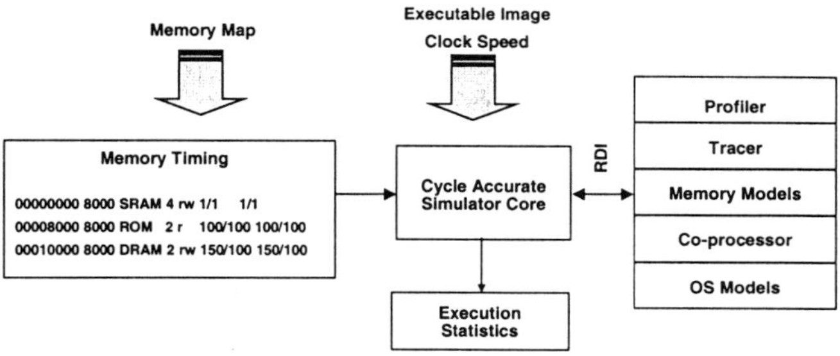

Figure 17.18. Timing Estimation Engine within JouleTrack

7. Summary

Software energy estimation is crucial for power aware computing. In this chapter, we have outlined a software energy estimation technique. Based on experiments done on the StrongARM and Hitachi processors we conclude that the common overheads present across instructions result in the variation in current consumption of different instructions being small. The variation in current consumption of programs is even smaller (less than 8% for the benchmark programs that we tried). Therefore, to a first order, we can assume that current consumption depends only on operating frequency and supply voltage and is independent of the executing program. A second order model that uses energy differentiated instruction classes/cycles is also proposed and it was shown that the resulting current prediction error was reduced to less than 2%. A methodology for separating the leakage and switching energy components

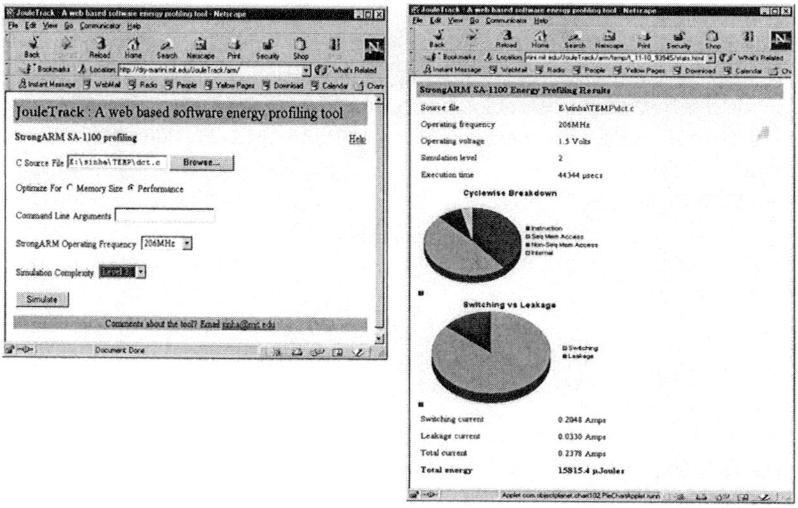

Figure 17.19. JouleTrack Screen Snapshots

is also discussed. The proposed leakage current model has less than 6% error. Based on our experiments we conclude that as supply voltages and thresholds scale, leakage (static) components will become increasingly dominant. For the StrongARM, at 100% duty cycle, the leakage energy is about 10% and increases exponentially with supply voltage and decreases linearly with operating frequency. The proposed techniques were implemented in a web-based software energy estimation tool called JouleTrack.

References

[1] *JouleTrack - A Web Based Tool for Software Energy Profiling*, http://dry-martini.mit.edu/JouleTrack/

[2] A. Sinha and A. Chandrakasan, "JouleTrack - A Web Based Tool for Software Energy Profiling", Proceedings of the 38th Design Automation Conference, Las Vegas, June 2001

[3] V. Tiwari and S. Malik, "Power Analysis of Embedded Software: A First Approach to Software Power Minimization", IEEE Transactions on VLSI Systems, vol. 2, Dec. 1994

[4] Advanced RISC Machines Ltd., *Advance RISC Machines Architectural Reference Manual*, Prentice Hall, New York, 1996

[5] Advanced RISC Machines Ltd., *ARM Software Development Toolkit Version 2.11 : User Guide*, May 1997

[6] *Solution Engine*, http://semiconductor.hitachi.com/tools/solution_engine.html

[7] J. Montanaro, et. al., "A 160MHz 32b 0.5W CMOS RISC Microprocessor", IEEE Journal of Solid State Circuits, November 1996, pp. 1703-1714

[8] J. Sheu, et. al., "BSIM: Berkeley Short-Channel IGFET Model for MOS Transistors", IEEE Journal of Solid-State Circuits, SC-22, 1987

[9] R. X. Gu and M. I. Elmasry, "Power Dissipation Analysis and Optimization of Deep Submicron CMOS Digital Circuits", IEEE Journal of Solid State Circuits, vol. 31, no. 5, May 1996, 707-713

[10] J. M. Rabaey, *Digital Integrated Circuits: A Design Perspective*, Prentice Hall, New Jersey, 1996

INDEX